RESIDUAL STRESS in Design, Process and Materials Selection

*Proceedings of ASM's Conference on Residual Stress—
In Design, Process and Materials Selection
Cincinnati, Ohio, USA
27–29 April 1987*

Edited by
William B. Young

Sponsored by the
Residual Stress Committee of the
Highway/Off-Highway Vehicles Division of
ASM INTERNATIONAL

Published by

Copyright © 1987
by
ASM INTERNATIONAL™
All Rights Reserved

No part of this book may be reproduced, stored in a retrieval system, or transmitted, in any form or by any means, electronic, mechanical, photocopying, recording, or otherwise, without the prior written permission of the publisher. No warranties, express or implied, are given in connection with the accuracy or completeness of this publication and no responsibility can be taken for any claims that may arise.

Nothing contained in this book is to be construed as a grant of any right or manufacture, sale, or use in connection with any method, process, apparatus, product, or composition, whether or not covered by letters patent or registered trademark, nor as a defense against liability for the infringement of letters patent or registered trademark.

Library of Congress Catalog Card Number: 87-071686
ISBN: 0-87170-304-1
SAN: 204-7586

Printed in the United States of America

CONFERENCE ORGANIZING COMMITTEE

William B. Young
Dana Corporation
Richmond, Indiana
Chairman

Robert W. Buenneke
Caterpillar Tractor Company
East Peoria, Illinois

Leonard Mordfin
National Bureau of Standards
Gaithersburg, Maryland

John T. Cammett, III
Metcut Research Associates
Cincinnati, Ohio

Paul Prevey
Lambda Research, Inc.
Cincinnati, Ohio

William DeHaan
BOC Flint Operations, GMC
Flint, Michigan

Fred Witt
U.S. Army Armament R&D Center
Dover, New Jersey

Charles P. Gazzara
U.S. Army Materials Laboratory
Watertown, Massachusetts

Dan Olah
Borg Warner Automotive
Sterling Heights, Michigan

Dennis W. Bloss
BOC Flint Operations, GMC
Flint, Michigan

PREFACE

The accumulation of residual stresses induced intentionally by design and manufacture or experienced in service determines a part's life. Today's technology provides the engineer with the necessary tools to identify and control such stresses. Residual stresses can be measured or estimated by one of several techniques. Finite element analysis provides a method of predicting the location of stress concentrations created during manufacturing and in its service environment.

The approach in designing a part for desired life or to solve a fracture problem, is dependent on the technology level and equipment available to each engineer. The papers presented in this conference demonstrated there are many methods to predict a part's stress condition or solve a problem. Not only did the attendees gain knowledge from the individual papers, many authors gained an understanding of other methods and/or equipment that could provide solutions. You as a reader of these papers will gain knowledge even to a greater extent because of the time limitations afforded each conference presenter.

Conventional x-ray diffraction measurement techniques are covered to a great extent in the text. Methods using focused x-ray beams, as well as techniques on preparation and procedures to determine stresses at various depths below the original surface is selectively presented. Neutron diffraction techniques, though being similar to those of x-ray diffraction, was reported as having a penetrating power a thousand times greater permitting the analysis at substantial depths, even in part assemblies, where destructive methods are undesirable.

The use of the Barkhausen effect, to pick up changes in residual and/or applied stresses on finished parts, is well described by an author. Texture and microstructural variations can help predict part quality. Sawcut displacement techniques continue to be used in several industries.

Ultrasonic techniques of measuring of residual stress using electromagnetic-accoustic transducers is presented. Data collected by non-destructive measurement techniques on plate materials are included.

The application type papers are set in two groups; thermal processes and mechanical processes.

The thermal process portion consists of studies pertaining to distortion created by an austenitizing quench, the effect of various case hardening techniques on part distortion, predicting residual stresses in grey iron castings by the use of finite element analyses; and the control of residual stresses in various types of welds. Three papers present studies of stress measurements on ceramic and refractory carbide monolith forms and coatings. One of the papers presents a modified hole test procedure for measuring stresses in a WC + Co coating.

The control of residual stresses by mechanical means is imperative in the production of parts. Techniques of grinding, grit blasting, shot peening or roll forming for extending part life is covered to a great depth. Bearing races, gears, and aerospace components are discussed in this section. The low stress grinding techniques through the use of CBN grinding wheels should be of particular interest to the reader.

A total of 125 authors, registrants and organizers attended the conference. An excellent flow of questions followed each presentation. Interest prevailed through the last paper of the third day. The great efforts put forth by the authors was demonstrated by the quality of the papers and the professionalism of the presentations.

Sincerest thanks are expressed to all the authors, session chairmen and their companies for supporting this activity. The success and quality of this conference resulted from the dedication of the organizing group, the Residual Stress Committee of the Highway/Off-Highway Vehicles Division composed of Messrs. Robert W. Buenneke, Caterpillar Tractor Company, Inc.; John T. Cammett, III, Metcut Research Associates; William DeHaan, BOC Flint Operations, GMC; Charles P. Gazzara, U.S. Army Materials Laboratory; Leonard Mordfin, National Bureau of Standards; Paul Prevey, Lambda Research, Inc.; Fred Witt, U.S. Army Armament R and D Center; Dan Olah, Borg Warner Automotive; Dennis Bloss, BOC Flint Operations, GMC and William B. Young, Chairman, Engine Products Division, Dana Corporation. A special note of gratitude goes to Mary Jean Culy, the Engine Products Division Secretary that kept up with all the changes in authors, addresses, and conference session organization with the use of the word processor. Also, Robert Uhl and Sharon Coles of ASM INTERNATIONAL for their support and consideration.

William B. Young
Engine Products Division
Dana Corporation
Conference Chairman

TABLE OF CONTENTS

Measurement Methods

Considerations of Using the Hole Drilling Method for Measuring Residual Stresses in Engineering Components 1
J.M. Boag, M.T. Flaman, J.A. Herring

Practical Aspects of Residual Stress Measurement by X-Ray Diffraction 7
C.A. Peck

The Measurement of Subsurface Residual Stress and Cold Work Distributions in Nickel Base Alloys 11
P.S. Prevey

Residual Stress Measurements in Armament-System Components by Means of Neutron Diffraction 21
H.J. Prask, C.S. Choi

Ultrasonic Measurement of Residual Stress Using Electromagnetic-Acoustic Transducers (Manuscript not available)
R.E. Schramm, A.V. Clark, P.J. Shull, J.C. Moulder, D.V. Mitrakovic

Use of Barkhausen Effect in Testing for Residual Stresses and Material Defects 27
K. Tiitto

Thermal Processes — Heat Treatment, Case Hardening

Bending Stress Relaxation of AISI 1095 Steel Strip 37
U.P. Sinha, D.W. Levinson

Selecting Quenchants to Maximize Tensile Properties and Minimize Distortion in Aluminum Parts 49
C.E. Bates

Residual Stress in Quenched Steel Cylinders 59
M.E. Todaro, M.A. Doxbeck, G.P. Capsimalis

Residual Stress in Laser Hardened Gear Material Studies 63
R.L. Frohlich

Determination of the Source of Variance in Distortion of Surface Hardened Bearing Races 77
D.L. Milam

Beneficial Residual Stresses in Gears by Incremental Contour Induction Hardening (Manuscript not available)
R. Cellitti, M. Hammond, C. Plaffmann

Thermal Processes — Casting, Welding, Metal Matrix Composites, Ceramics

Residual Stress Development in Grey Iron Castings (Manuscript not available)
J.A. Dantzig, J.W. Wiese, J.R. Howell, N.P. Lillybeck

The Influence of Residual Stresses on the Fatigue Design of Welded Steel Structures 85
J.G. Wylde

Residential Stresses at Pinch Welds in Small Stainless Steel Tubes 97
W.C. Mosley

Residual Stresses in Ceramics 103
C.O. Ruud, D.J. Snoha, C.P. Gazzara, P. Wong

Thermal Processes — Surface Films, Coatings

Stresses in Oxide Scales (Manuscript not available)
N. Jayaraman

Development of a Modified Blind Hole Test for the Measurement of Residual Stress Gradients in Thin Wear Resistant Coatings 109
L.C. Cox

Mechanical Processes: Shot Peening, Grit Blasting

Residual Stress Profiles Established by Shot Peening and Their Influence on Fatigue Properties (Manuscript not available)
B. Gillespie

Residual Stress Effects of Shot Peening of a Single Point Turn Machined High Strength Nickel Base Superalloy (Manuscript not available)
P. Allison, M. Sauby

Residual Stress Analysis and LCF Test Results for Peened Bolt Hole and Dovetail Configurations 127
M.B. Happ, D.P. Mourer, R.L. Schmidt

Fatigue Testing and Residual Stress Measurements of Grit-Blasted Aluminum Alloys 137
R. Myllymaki

Mechanical Processes: Machining, Grinding, Forming

Residual Stress Mapping of Gears and Bearings 151
C.O. Ruud, D.J. Snoha

Residual Stresses from Grinding of Some Aerospace Structural Alloys (Manuscript not available)
J.T. Cammett, W.P. Koster

Generating Compressive Residual Stress by CBN Grinding 157
G.A. Johnson

Correlation of Residual Stress to Bearing Condition 169
G.R. Kuhlman, B.S. Pardue

Residual Strains in Rolled Joints 183
S.R. MacEwen, T.M. Holden, R.R. Hosbons, A.G. Cracknell

Coldworking Fastener Holes — Theoretical Analysis, Methods of Coldworking, Experimental Results 193
W.A. McNeill, A.W. Heston

Calculation of Residual Stresses in Railroad Rails and Wheels from Sawcut Displacement 205
M.W. Joerms

CONSIDERATIONS OF USING THE HOLE DRILLING METHOD FOR MEASURING RESIDUAL STRESSES IN ENGINEERING COMPONENTS

J.M. Boag, M.T. Flaman, J.A. Herring
Ontario Hydro-Research Division,
Toronto, Ontario, Canada

ABSTRACT

The Hole Drilling Method has become an increasingly popular technique for determining surface residual stresses in components and structures. In this method, a special purpose strain gage rosette is attached to the component at the point of interest, and then a small hole is produced into the component through the rosette center. The measured relieved strains due to the hole production can then be related to the original surface residual stresses. To obtain accurate and reliable results by this method, machining stresses must not be induced by the hole production method and the hole geometry must be predictable and repeatable. The induced machining stresses and cross-section profiles resulting from four different hole producing techniques are presented for seven different metals. In addition, advanced hole drilling data analysis techniques are discussed so that stress variations with depth can be obtained. Finally, a case history including calibration using the Hole Drilling Method is presented for a typical engineering application.

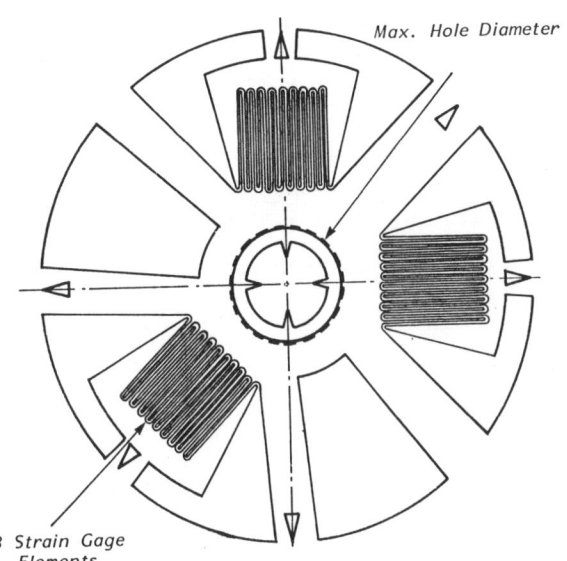

Fig. 1 - Hole Drilling Strain Gage Rosette /1/

THE HOLE DRILLING METHOD involves the application of a special three-element strain gage rosette (as shown in Figure 1) onto the surface of the component at the measurement location. A small hole (approximately 1.6 mm diameter) is then made into the component through the center of the rosette. The production of the hole in the stressed component causes a redistribution of strains to occur near the hole which can be detected and measured by the surface mounted strain gage rosette. These strain measurements can be then related to the original residual stresses in the component at the hole location.

This paper deals with some of the key considerations that should be examined prior to performing actual measurements for determining

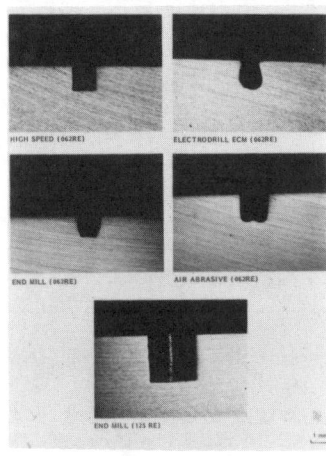

Fig. 2 - Cross-Sectional Profiles - 1020 Carbon Steel

residual stresses in components and structures. To emphasize the importance of these considerations, the results of extensive qualification tests performed on seven typical engineering metals will be discussed. Bars from each metal type were meticulously cleaned and then subjected to specialized stress-relieving procedures to obtain macroscopically "stress-free" specimens. The seven metals are: 1020 carbon steel; 304 stainless steel; 6061 aluminum; commercially pure copper; commercially pure titanium; zirconium alloy (2.5% Niobium); and stellite alloy #100.

HOLE PRODUCING TECHNIQUE

The production of the hole is critical to the usefulness and accuracy of the measurement method. The optimum hole producing procedure will produce the hole in a "stress-free" manner (i.e. with minimal induced machining stresses)/2/. In addition, the hole should be circular (in plan view) and straight sided (i.e. cylindrical) to completely satisfy the conditions in the hole drilling data analysis equations.

Four hole producing techniques were studied; ultra-high speed drilling/3/; low-speed end milling/4/; air-jet abrading/5/; and electro-chemical machining/6/.

Induced Drilling Stresses - Commercially available hole drilling strain gage rosettes/1/ were installed on each of the vacuum annealed metal specimens and a hole was produced by each technique in the center of the rosette. Stresses were calculated for each measurement technique using the analysis procedures established in the ASTM Standard E837-85. The results of the four different hole producing techniques in the seven metal specimens are listed in Table 1.

Table 1 - Measured Stresses (MPa) From Each Hole Producing Technique In Stress Relieved Specimens

	Material Yield Stress* (σ_y)	High Speed Drill	Low Speed End Mill	Air Abrasive Jet	Electro Chemical Milling
Aluminium	100	9	$>\sigma_y$	5	5
Carbon Steel	210	1	136	3	6
Stainless Steel	240	9	$>\sigma_y$	7	9
Copper	70	22	$>\sigma_y$	$>\sigma_y$	14
Stellite	850	36	-	10	22
Titanium	480	13	63	3	-
Zirconium	340	5	46	3	-

* Approximate values for annealed specimens.

In this study, a substantial effort was performed to reduce specimen residual stresses as much as possible in order to study the occurrence of induced machining stresses for each hole producing technique. The degree to which this was achieved can be readily judged in those cases where two or more of the different hole producing techniques resulted in "very low" measured stresses (as compared to the material yield stress).

Based on the high measured induced drilling stresses, it can be seen that the low-speed end milling technique is not generally suitable for use in the Hole Drilling Method relative to other hole producing techniques. The other three techniques are much more suitable (from the standpoint of induced machining stresses) with the following exceptions:

- Electro-chemical milling is limited to electrically conductive materials.
- Air-abrasive technique would not be suitable for soft materials (note copper results), and the
- Ultra-high speed drilling technique is not as suitable as other techniques for extremely hard materials (note stellite results).

Hole Geometry - All of the specimens were sectioned through the holes so that high resolution photographs could be taken. Figure 2 illustrates the typical cross-sectional profiles obtained for each hole producing technique in the 1020 carbon steel specimen. (Generally, all of the holes were observed to be circular in "plan" view.) As can be seen, the cross-sectional shape of the holes, which is required to be straight sided and flat bottomed, were variable from one hole producing technique to the other. Results are listed in Table 2.

Table 2 - Hole Geometry For Each Hole Producing Technique In Stress-Relieved Specimens

	High Speed Drill	Low Speed End Mill	Air Abrasive Jet	Electro Chemical Milling
Aluminium	✓	✓	2,3,4	1
Carbon Steel	✓	2	3	1
Stainless Steel	✓	2	2,3	1
Copper	✓	✓	3	1
Stellite	✓	4	3	1
Titanium	✓	✓	2,3,4	-
Zirconium	✓	✓	3	-

✓ Hole was circular, straight sided, and flat bottomed
1 Hole was round bottomed
2 Hole had tapered sides
3 Hole had irregular bottom profile
4 Hole was not drilled to correct depth

From the standpoint of hole geometry, only the ultra-high speed drilling and low-speed end milling techniques were judged suitable for general use in the Hole Drilling Method. However, in certain limited applications where only a series of comparative measurements are required, the air-abrasive jet and the electro-chemical milling techniques may also be acceptable.

VARIATIONS OF HOLE DRILLING EQUIPMENT

There are three commercially available versions of Hole Drilling equipment being used today with the main difference being the mechanism by which the hole is produced. Appendix 1 compares the available equipment from the useability point of view.

ADVANCED DATA ANALYSIS TECHNIQUES

Finite Element Methods (FEM)/17,18/ have been developed to extend the capability of the Hole Drilling Method to detect and determine the variation of residual stresses with depth in the drilled hole. For these types of measurements, the hole must be produced in precisely controlled flat-bottomed increments. Normally, for conventional data analysis, a full depth hole would be produced (i.e. where the hole depth is approximately equal to its diameter) and standard ASTM semi-empirical equations would equate the relaxed strains to the "average" residual stresses acting through the drilled hole depth.

Using a 3-dimensional MSC/NASTRAN FEM approach/19/, the hole in the center of the rosette was modelled in a series of precise increments while the test model was analyzed to simulate bi-axial or shear stresses. At each increment, the change in "strain" at each gage location was determined and related to the unit stresses. In this manner, a series of independent "unit" solutions were developed in the form of matrices. Thus, variable stress profiles can be solved using these unit solutions with actual strain relaxation measurements. By following this procedure, complete separate incremental stresses are made at each measurement location, rather than one "bulk" average measurement that is obtained from the conventional ASTM Standard E837-85.

The FEM unit solutions have been fully qualified and documented as a viable technique to determine subsurface stresses using incremental relaxed strains obtained from the Hole Drilling Method. In addition, an extensive sensitivity study determined:
 i) Stresses in the top depth increments can be calculated most accurately.
 ii) Generally, stresses in the bottom 40% of a full depth hole contribute an insignificant amount of strain change. Therefore, stresses in the bottom increments are usually impossible to measure by a full depth hole.
 iii) To limit "damage" to the component or structure, the hole is not required to be drilled to full depth. For example, by drilling to the fourth increment (62.5% of the full hole depth) about 97% of possible stress data can be obtained in a uniform stress field.

An example illustrating the necessity of using advanced FEM stress-with-depth analysis procedures is shown in Figure 3 for a subsurface stress reversal (obtained from a typical surface treatment process such as shot peening). The conventional ASTM Standard E837-85 would calculate an "average" compressive stress for a full depth hole of -138.6 MPa. Alternatively, the FEM analysis (which gives incremental stress information) clearly illustrates a stress gradient ranging in the first increment of -346.1 MPa (compression) to a fourth increment stress of +92.4 MPa (tensile).

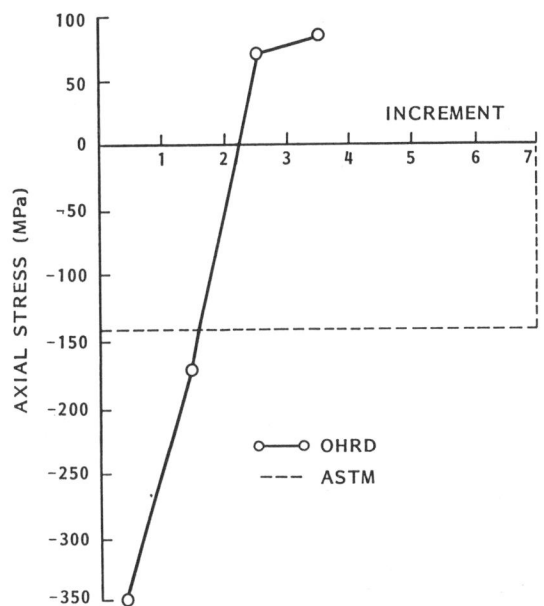

Fig. 3 - Comparison of Data Analysis Procedures For a Stress Reversal With Depth/20/

CASE STUDY

Surface residual stresses for a simulated thick-wall pressure vessel temper-bead weld repair in a 100 mm thick SA516 Grade 70 steel plate have been investigated with the Ultra High Speed Hole Drilling Technique/21/. This investigation was performed to determine the effectiveness of the temper-bead weld repair in reducing residual stresses as a result of a weld repair of a crack or flaw. A general schematic of the plate indicating the welded region is shown in Figure 4.

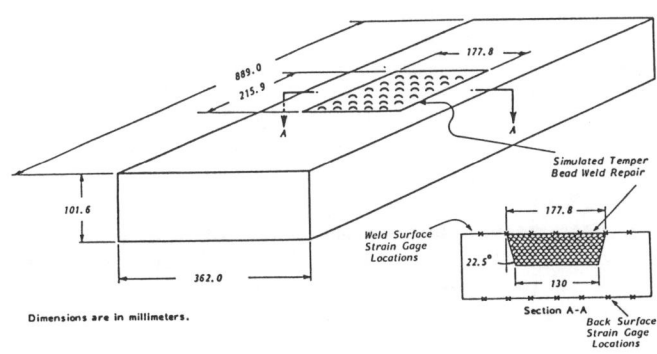

Fig. 4 - General Schematic of Welded Plate Indicating Strain Gage Locations

Prior to performing the plate measurements, qualification of the measurement technique was performed to determine the induced drilling stresses (as discussed in Section 2.1) and to evaluate the accuracy of the measurements. A bar specimen obtained from a similar SA516 Grade 70 plate (measuring 25.4 mm wide and 12.7 mm thick x 500 mm long) was carefully "stress-relieved". Two strain gage hole drilling rosettes were then installed onto the bar. The first rosette was drilled and it was determined that the induced drilling stresses by the Ultra-High Speed Hole Drilling Technique were less than 3 MPa (although this small value might relate to actual remaining stresses subsequent to the stress relieving process). These low indicated stresses were thought not to affect the accuracy of the measurements.

The bar was then placed in a tensile test machine and loaded to a known applied axial stress of 137.9 MPa. The stresses in the bar were determined by both the ASTM Standard E837-85 and FEM stress-with-depth analyses. The two data analyses were in excellent agreement with measured axial stresses of 132.4 MPa and 142.7 MPa (or approximately ±4.0 percent possible error) for the ASTM and FEM analyses, respectively.

Before the installation of the strain gages on the welded plate, the temper-bead weld crown was machined from the plate using a single point carbide tipped cutter and coolant. A qualification test on a separate stress relieved steel bar indicated that surface machining stresses of less than 30.8 MPa would be caused by this machining procedure. The magnitude (i.e. less than 10% yield stress) of these surface machining stresses and the minimal affected depth caused by the cutter was thought not to significantly alter the general temper-bead weld residual stress distributions determined by the Hole Drilling Method. In field situations, the temper-bead weld repair procedure allows for a grinding operation to remove the weld crown. However, for this study (in which the effectiveness in reducing residual stresses by this weld procedure was to be determined) grinding was thought not to be suitable because of the various uncontrolled factors associated with grinding on the surface residual stresses. The redistribution of residual stresses caused by the removal of the weld crown (possible reinforcement effects) was not accounted for.

Residual stress measurement profiles across the surface of the temper-bead weld and on the back surface of the plate are shown in Figure 5. Residual stress data are presented as a percentage of base plate material yield stress (approximately 316 MPa). It is expected that the yield stress for the actual weld material would be higher than the base plate. After performing the measurements it was determined that residual stresses did not generally vary through the measurement depth and consequently all data was analyzed using the conventional ASTM Standard E837-85 analysis.

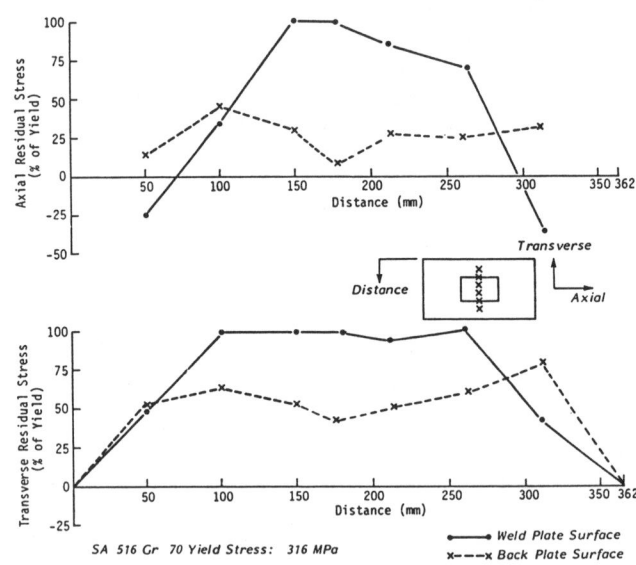

Fig. 5 - Temper Bead Weld Surface Residual Stress Results

As expected, the maximum residual stresses obtained were in the weld region on the top surface of the plate. These residual stresses were measured to be in tension close to the yield stress. All transverse residual stresses across the weld surface were also tensile. Axial residual stresses across the weld were determined to be compressive at the outer plate edges and were found to have increased to tensile yield in the weld. Both axial and transverse surface residual stresses on the back surface of the welded plate were measured to be approximately half of the tensile yield stress.

Based on the stress results obtained by the Hole Drilling Method, it was determined that no significant reduction in residual stresses were obtained in the weldment or surrounding material using a temper-bead weld repair. However, parallel studies showed that this weld repair resulted in excellent metallurgical properties and fracture toughness properties in the weld heat affected zone (HAZ).

ACKNOWLEDGEMENTS

The authors wish to thank the Canadian Electrical Association for their advice and support of this program.

REFERENCES

1. Strain rosette EA-XX-062RE-120, Micro Measurements Inc, Raleigh, NC.

2. Flaman, M.T. "Brief Investigation of Induced Drilling Stresses in the Center-hole Method of Residual Stress Measurement," M.T. Flaman, *Experimental Mechanics*, January, 1982.

3. Ultra-high speed drilling technique refers to the high speed air turbine drill rotating at approximately 350,000 rpm using a tungsten-carbide dental burr of 1.6 mm diameter developed by Ontario Hydro.

4. Low-speed end milling technique refers to Vishay RS-200 milling guide for use with a hand-held drill and special HSS end mills, Micro Measurements Inc, Raleigh, NC.

5. Air-jet abrading technique refers to the air abrasive jet machining equipment as marketed by Franklin-Stoller Ltd, Bath, England.

6. Electro-chemical machining refers to a technique to produce small, dimensionally accurate holes as developed by the Johnson International Corp. Roseville, Michigan, USA.

7. Mathar, J. "Determination of Initial Stresses by Measuring the Deformations Around Drill Holes". *Trans. ASME Vol 56*. 1934.

8. Rendler, N. and Vigness, I. "Hole Drilling Strain-Gage Method of Measuring Residual Stresses". *Experimental Mechanics*. 577. December 1966.

9. Nawwar, A.M., McLachlan, K., and Shewchuck, J. "A Modified Hole Drilling Technique for Determination of Residual Stresses in Thin Plates". *Experimental Mechanics*. 226. June 1976.

10. Flaman, M.T. and Herring, J.A. "Ultra-High Speed Center Hole Technique for Difficult Machining Materials", *Experimental Techniques*, January 1986.

11. Bush, A.J. and Kromer, F.J. "Simplification of the Hole-Drilling Method of Residual Stress Measurement". *ISA Transactions*. Vol 12, No 3. (P. 249-259). 1973.

12. Beaney, E.M. "Accurate Measurement of Residual Stress on Any Steel Using the Centre Hole Method". CEGB Report. October 1980.

13. Beaney, E.M. "The Air-Abrasive Centre-Hole Technique for the Measurement of Residual Stresses". CEGB Report. October 1980.

14. Bynum, J.E. "Modifications to the Hole-drilling Technique of Measuring Residual Stresses for Improved Accuracy and Reproducibility". *Experimental Mechanics*. January 1981.

15. Wnuk, S.P. "Residual Stress Measurement in the Field Using the Air-Abrasive Hole Drilling Method". Presented at SESA Spring Meeting. Dearborn, Michigan. June 1981.

16. Keen, W.P. and Beaney, E.M. "Instructions for Using the Air-abrasive Centre Hole Equipment to Measure Residual Stress". CEGB Report. June 1976.

17. Flaman, M.T. and Manning, B.H. "Determination of Residual Stress Variation With Depth by the Hole Drilling Method", *Experimental Mechanics*, September 1985.

18. Lu, J., Niku-Lari, A., and Flavenot, J.F. "New Developments of the Hole Drilling Method as a Means of Measuring Residual Stress Distributions in Depth", Proceedings of the V International Congress on Experimental Mechanics, Montreal, P.Q., June 10-14, 1984.

19. MSC/NASTRAN, McNeal-Schwendler Corporation, 7442 N. Figueroa St., Los Angeles, CA 90041.

20. Boag, J.M. and Flaman, M.T. "Comparison of Hole Drilling Residual Stress Analysis Techniques", Proceedings of the 1987 SEM Spring Conference on Experimental Mechanics, Houston, Texas. June 14-19, 1987.

21. Boag, J.M. "Residual Stress Measurements in a Simulated Temper-Bead Weld Repair", OHRD Report No B83-82-K, December 13, 1983.

APPENDIX 1

COMPARISON OF HOLE DRILLING EQUIPMENT

	HOLE DRILLING EQUIPMENT		
	LOW-SPEED DRILLING EQUIPMENT	ULTRA-HIGH SPEED DRILLING EQUIPMENT	AIR-ABRASIVE HOLE DRILLING EQUIPMENT
REFERENCES	7-9	2,10	11-16
Application Areas	Lab/Field	Lab/Field	Lab & Limited Field Use
Time Req'd	1 to 2 hours	1 to 2 hours	2 to 4 hours
Ease of Use	Simple	Simple	Skilled Operator
Applicability	-for most low hardness/toughness metals	-for most metals and non-metals	-for most metals and non-metals
Portability	-ideally suited for field use	-ideally suited for field use	-bulky equipment awkward for field measurements
Limitations	-induced stresses may be very high for common metals (i.e. stainless steels)	-induced stresses in extremely hard materials may be unacceptably high	-somewhat difficult and time consuming to perform -hole shape and depth not well defined -not suited to certain environmental conditions i.e. humidity
Comments	-not recommended for general use	-recommended for most lab/field applications	-greatest advantage is small induced stresses in hard materials -costly equipment

PRACTICAL ASPECTS OF RESIDUAL STRESS MEASUREMENT BY X-RAY DIFFRACTION

Charles A. Peck
Metallurgy Dept.
GM Research Labs
Warren, Michigan USA

ABSTRACT

A general discussion of the most common techniques for residual stress determination by X-ray diffraction will be presented.

VARIOUS TECHNIQUES EXIST for the determination of residual stress using X-ray diffraction. Engineers must be familiar with these techniques and their practical differences to select a method that provides both speed and accuracy while maintaining technical soundness.

Different techniques for stress determination will be examined, including the two-exposure and the $\sin^2\psi$ methods. In addition, various X-ray set-ups for measuring residual stress will be compared including parafocusing, side-inclination and parallel beam optics, and conventional, solid-state and position-sensitive detectors.

THEORETICAL BASES

When stress persists after the removal of external forces, the lattice spacings differ from a stress-free value owing to this residual stress. These internal strain gauges are the basis for residual stress measurements by X-ray diffraction in crystalline materials. A minimum of two measurements of lattice spacings, one typically being normal (d_n) and the other inclined (d_i) to the sample surface, are required to calculate surface residual strain. The strain values are converted to stress values using a proper elastic constant.

It is usually assumed in the stress determination by X-ray that the stress normal to a free surface is zero, which reduces the system to two principal stress components lying within the plane of the sample surface. It may be shown [1] using the elasticity theory that the residual stress is related to the difference in the measured d-spacings or alternatively the diffraction angles as expressed in Equations 1 and 2:

$$\sigma_\phi = \frac{E}{(1+\nu)\sin^2\psi} \frac{(d_i-d_n)}{d_n} \quad \text{Eq. (1)}$$

$$\sigma_\phi = K\Delta 2\theta \text{ (deg), where}$$

$$K = \frac{E\cot\theta}{2(1+\nu)\sin^2\psi} \frac{\pi}{180} \quad \text{Eq. (2)}$$

and ϕ is the direction of the residual stress, σ, being determined, E, the Young's modulus, and ν, the Poisson's ratio.

Determination based upon two tilt angles (commonly 0 and 45 degrees) is referred to as the two-exposure technique. The main advantage of this method is speed. Two alternative tilt angles are 30 and 60 degrees. Measurement using the 60° tilt angle provides increased sensitivity in the calculation since the stress constant in Eq. 2 decreases with increased tilt. Therefore a smaller stress produces a greater shift in peak positions with higher tilt angles.

On the debit side for the greater ψ values is the increased contribution from instrumental errors such as goniometer and sample misalignments. The two-exposure method is also limited in instances when crystallographic preferred orientation or texture confounds the selection of a value for the modulus of elasticity.

Whenever time permits and maximum accuracy is required, multiple tilt angles are employed. The technique, also called the $\sin^2\psi$ method, typically involves 5 or more tilt angles and uses linear regression to solve for the slope of $\sin^2\psi$ against $(d_i-d_n)/d_n$ from Eq.1. This method is statistically more accurate than the two-exposure technique and also yields significant additional information about the presence of texture, shear stress or instrument misalignment.

Four basic variations of the $\sin^2\psi$ curve are illustrated in Figure 1. The common $\sin^2\psi$ plot is depicted in Figure 1a. A positive slope represents tension and a negative slope indicates compressive stress. The straight line is obtained in the absence of a shear component, which is the stress component normal to the surface. The basic $\sin^2\psi$ plot is unaffected whether the angle ψ is positive or negative. Yet, if different stress values are obtained from ψ angles of equal magnitude but opposite signs, it is indicative of the presence of a shear stress. Then the $\sin^2\psi$ curve assumes the form of an ellipse as shown in Figure 1b. The degree of the ψ-splitting relates to the magnitude of the shear stress [2].

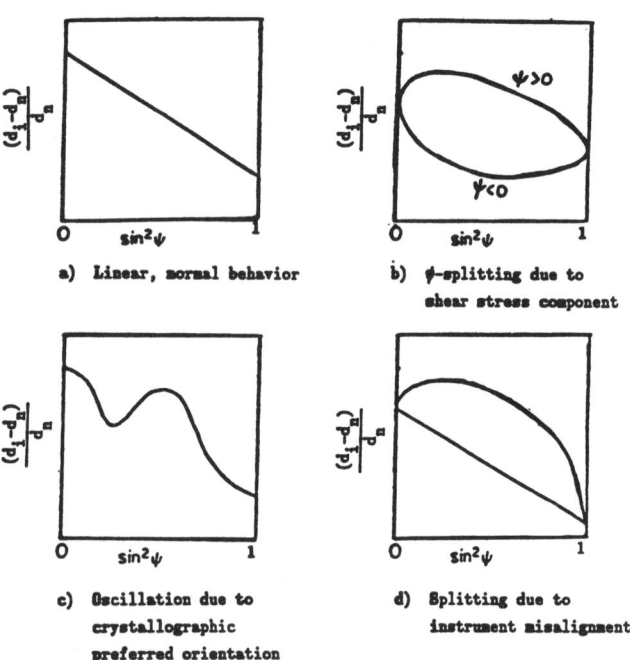

Figure 1. Four variations of the plot of lattice spacing versus $\sin^2\psi$.

A non-linear $\sin^2\psi$ plot showing oscillation (Figure 1c) is indicative of the presence of strong crystallographic texture [3]. Lastly, the kind of $\sin^2\psi$ plot in Figure 1d can arise where there is a gross instrument misalignment. In this case only one of the $\sin^2\psi$ plots, either with positive or negative ψ angles, would show curvature.

In the majority of laboratories, textbook bulk property values are used for E and ν. However, the bulk values usually do not apply to the specific crystallographic planes used in the residual stress measurement. These elastic constants are better determined experimentally with a single crystal stressed in the direction of the reflecting planes used in the stress analysis. Such experimentation is out of the realm of most laboratories. This ambiguity in the residual stress measurement often leads to comparative rather than exact stress analysis. It becomes therefore important that X-ray laboratories interact and uniformly agree on the values of elastic constants used in stress determination!

INSTRUMENTAL VARIATIONS

COUNTERS/DETECTORS: Essentially the residual stress measurement becomes one of measuring peak locations with different ψ tilt angles. Strip chart recorders yield output that requires visual determination of the peak position. The strip chart data must be corrected for the background intensity and with the modified Lorentz-polarization factor [4]. This intensity correction is usually carried out with prepared plastic overlays and is quite cumbersome. Yet skipping these correction steps can introduce substantial error when analyzing rather broad peak profiles. The continuous scan yields rather poor statistics unless slow chart speeds are used along with long ratemeter time constants. This can be significantly enhanced by step counting near the peak maximum and determining the peak position by employing a parabolic fit to the data. Care must be taken to restrict the parabolic fit to the upper fifteen percent of the peak intensity for accurate peak profiling [5].

Step counting substantially increases the measuring time with conventional proportional and scintillation counters. This time can be radically reduced with the use of a position sensitive proportional detector (PSPD) outfitted with a multi-channel analyzer. The PSPD counter remains in a fixed position while simultaneously counting over a restricted region of the scan. These detectors can usually accommodate any peak breadth used in residual stress measurement. In addition, the spatial resolution offered by modern PSPD's compares favorably with conventional detectors [6].

Solid-state detectors, Si/Li and Ge/Li types, are sometimes used to take advantage of their capabilities to reduce fluorescence and increase the peak-to-background ratio. A disadvantage of the SSD's is the necessity for long counting times due to low counting efficiency [7].

Lastly, a completely different type of detector—the position sensitive scintillation detector (PSSD)—was used in the development of a miniature X-ray diffraction instrument for stress analysis [8]. This PSSD unit is by far the most compact and portable stress measuring system yet developed.

OPTICS: When the sample surface is tilted in the stress analysis, there is an accompanying change in the diffraction circle and the focusing point of the diffracted beam. Maintaining the detector slit at the focus throughout the stress analysis is known as the parafocusing optics. Focusing of the diffracted

X-ray beam becomes a major consideration for a weak signal due either to thin sample width or to weak reflecting planes. These instances will require increased counting times and necessitate the use of conventional parafocusing optics to maximize output. The choice of the parafocusing optics requires a precise radial movement of the counter to the prescribed focus for each tilt angle. However, the required precision in the radial movement is difficult to attain, and more often the detector assembly is left stationary and measurement is made at a slight expense of accuracy.

The tilt axis for the parafocusing optics is normal to the diffraction plane. An alternative method known as the side-inclination method [9], has the tilt axis in the diffraction plane (Figure 2). The focus remains on the diffraction circle regardless of ψ, thus eliminating the need to move the counter. The method of side-inclination may be advantageous for examination of regions of a sample that are difficult to access (i.e. gears). A disadvantage is a relatively weak X-ray intensity and associated long counting time due to the limitation on the height of the incident beam.

The availabilty of a large, flat sample area allows for the use of the parallel-beam optics [10]. This non-focusing optics obviates the need to move the counter/detector assembly for each tilt angle. Consequently, the parallel-beam optics is very well suited for automated residual stress programs where the focusing distance need not be manually positioned (Figure 2). The greatest advantage of the parallel beam optics is that the peak position is little affected by the sample surface displacement.

SUMMARY

The measurement of surface residual stress involves a proper technique to meet the needs of speed and precision while minimizing assumptions and systematic error. The method of calculation, selection of set-up (i.e. counter/detector, optics etc.) and choice of material constants have been presented to help make intelligent choices for residual stress analyses in X-ray laboratories.

REFERENCES

1. F. Gisen, R. Glocker and E. Osswald, Z. Tech. Physik, 17, 1936, p. 145.
2. W. Lode and A. Peiter, Harterei-Techn. Mitt., 35 (3), 1980, p. 148.
3. R. H. Marion and J. B. Cohen, Adv. X-Ray Anal., 18, 1975, p. 466.
4. B. D. Cullity, Elements of X-Ray Diffraction, 2nd edition, Addison-Wesley, 1978, p.463.
5. D. P. Koistinen and R. E. Marburger, Trans. ASM, 51, 1959, p. 537.
6. M. R. James and J. B. Cohen, Adv. X-Ray Anal., 22, 1979, p. 241.
7. R. L. Heath, Adv. X-ray Anal., 15, 1972, p. 1.
8. D. A. Steffen and C. O. Ruud, Adv. X-Ray Anal., 21, 1978, p. 309.
9. B. D. Cullity, Elements of X-Ray Diffraction, 2nd edition, Addison-Wesley, 1978, p. 466.
10. Shuji Taira, ed., "X-Ray Studies on Mechanical Behavior of Materials," Society of Materials Science, Kyoto, Japan, 1974.

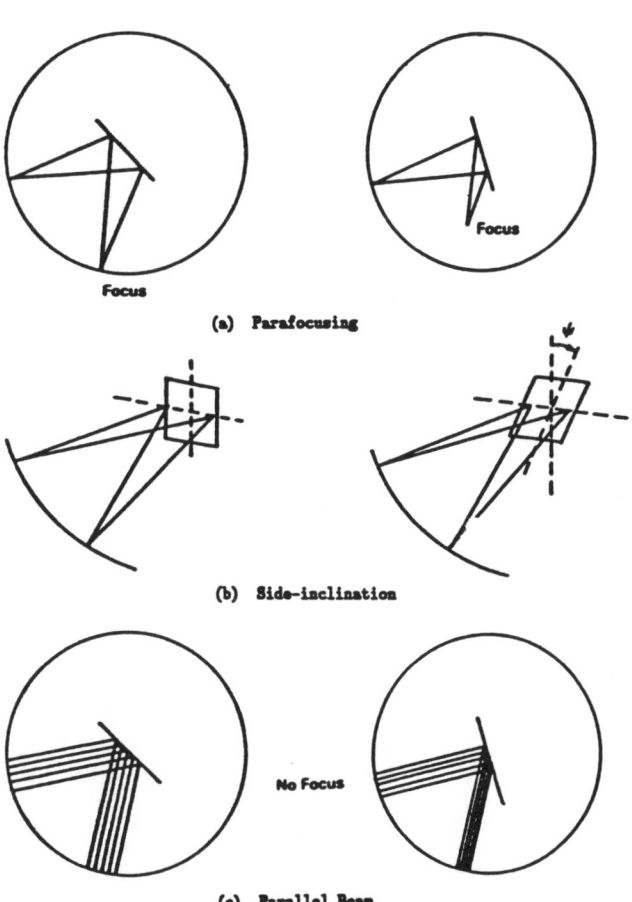

Figure 2. Parafocusing, side-inclination and parallel beam optics.

THE MEASUREMENT OF SUBSURFACE RESIDUAL STRESS AND COLD WORK DISTRIBUTIONS IN NICKEL BASE ALLOYS

Paul S. Prevey
Lambda Research, Inc.
Cincinnati, Ohio USA

Abstract

A method of determining the diffraction peak width accurately and rapidly in conjunction with x-ray diffraction residual stress measurement using Pearson VII function profile analysis is described. An empirical relationship between the (420) diffraction peak width and the degree of cold work is developed for four nickel base alloys. The peak width produced was found to be independent of the method of deformation. Examples of the concurrent determination of subsurface residual stress and cold work distributions are presented for samples of abraded and shot peened Inconel 718 and ground Inconel 600.

X-RAY DIFFRACTION TECHNIQUES can be used to determine both macroscopic and microscopic residual stresses. Macrostresses, which extend over ranges large compared to the dimensions of the crystals in the material, are determined from the shift in the position of the diffraction peak. Microstresses, which extend over distances on the order of the unit cell, cannot be directly measured individually, but can be quantified in the aggregate from the broadening of the diffraction peak which they produce.

Macrostresses are tensor properties measurable by mechanical means which are additive to applied stresses. The average magnitude of the microstress in a sample can be treated as a scalar property, and is related to the hardness or degree of cold work of the material. As a metal is cold worked, the dislocation density increases, reducing the size of the perfectly crystalline regions or crystallites (coherent diffracting domains), and increasing the average (root mean square) microstrain in the crystal lattice. The reduced crystallite size and increased microstrain both produce broadening of diffraction peaks which can be conveniently measured as a means of quantifying the degree to which the material has been cold worked.

A number of practical difficulties are encountered in determining the diffraction peak width in conjunction with macroscopic residual stress measurement. First, the $K\alpha$ radiation generally used for residual stress measurement produces two overlapping diffraction peaks. The $K\alpha$ doublet (consisting of peaks produced by the $K\alpha_1$ and $K\alpha_2$ radiations) can be separated to determine the width of the stronger $K\alpha_1$ peak, using the Rachinger correction.[1] However, this method requires collection of a large number of data points to define the entire diffraction peak profile, and may not provide sufficient accuracy for residual stress measurement.[2] Second, approximation of the diffraction peak profile by a parabola fitted to the top of the peak, a common method of peak location used for stress measurement in hardened steels,[3] may result in significant error when applied to diffraction peaks of intermediate width because the calculated peak position and width are dependent upon the portion of the diffraction peak included in the regression analysis.[4] Third, the background intensity must be known in order to calculate the diffraction peak width, but may be difficult to measure accurately for broad or overlapping diffraction peaks. Fourth, the observed diffraction peak width is actually the convolution of broadening due to the cold work present in the specimen and instrumental broadening. The two components may be separated using Stokes method,[5] but this approach requires extensive data collection to completely define the diffraction peak profile and a reference specimen which is free of cold work.

This paper describes a simple, accurate x-ray diffraction technique for determining the degree of cold work in nickel base alloys from the measured diffraction peak width. The (420) diffraction peak width is demonstrated for nickel base alloys to be independent of the mode of plastic deformation, and to be additive. An empirical relationship is described which allows the degree of cold work to be determined from the diffraction peak width measured in conjunction with the macroscopic residual stress. The method is rapid, requires minimum data collection, and is suitable for the routine analysis of surfaces deformed by machining, grinding, or shot peening where separation of the contributions to the diffraction peak width caused by crystallite size and microstrain[6] is not required.

PEARSON VII FUNCTION PEAK PROFILE FITTING

Most of the difficulties encountered in determining both the diffraction peak position and width can be overcome if the measured diffraction peak profile can be accurately described by a suitable function fitted by regression analysis. Pearson VII functions, which are bell-shaped curves ranging from Cauchy (Lorentzian) to Gaussian distributions, have been shown[2,8] to accurately describe the profiles of diffraction peaks in the back-reflection region used for residual stress measurement.

The diffracted intensity at any angle, 2θ, can be approximated by a superposition of the $K\alpha_1$ and $K\alpha_2$ diffracted intensities and a linearly sloping background:

$$I(2\theta) = f(2\theta) + \alpha f(2\theta - \delta) + C \cdot 2\theta + D \quad (1)$$

where:

$$f(2\theta) = A \left(1 + \frac{B^2}{M}(2\theta - 2\theta_0)^2\right)^{-M} \quad (2)$$

is an optimized Pearson VII function describing the $K\alpha_1$ diffraction peak profile. Eq. (1) can be fitted to the diffracted intensity measured at a relatively small number of Bragg angles, 2θ, by non-linear least squares regression. The fit is optimized by letting M be a real number ranging from M = 1 for a Cauchy to M = ∞ for a Gaussian profile.

The regression parameters are:

A   maximum net intensity of the $K\alpha_1$ peak
B   a peak width parameter
M   a decay rate parameter
$2\theta_0$ the $K\alpha_1$ peak position
C   linear background slope
D   linear background intercept

The constant α is the ratio of the $K\alpha_2$ to $K\alpha_1$ peak intensities, typically fixed at 0.5, and δ gives the separation of the $K\alpha$ doublet which is dependent upon $2\theta_0$ and the x-ray wavelength.

Examples of Eq. (1) fitted to (420) diffraction peak profiles produced with copper $K\alpha$ radiation are shown in Figures 1 and 2, where the measured diffracted intensity is plotted as a function of the diffraction angle. The linear background, $K\alpha_1$ and $K\alpha_2$ components, which are combined to model the $K\alpha$ doublet peak profile, are shown separately. Figure 1 shows the relatively narrow diffraction peak obtained in the ψ = 45 deg. orientation on an electropolished Inconel 718 surface well beneath the deformed surface layers. Figure 2 shows the broad diffraction peak produced by the cold worked surface of a shot peened Inconel 718 sample in the ψ = 0 orientation.

In Figure 2, the (331) diffraction peak centered at a lower diffraction angle to the left of the figure, produces overlap with the low angle side of the (420) peak. By eliminating the first six data points on the low angle side of the data set from the regression analysis, the effect of overlap of the adjacent diffraction peak has been effectively eliminated.

The Pearson VII function profile was fitted by linearization and successive approximation in seven iterations to the 26 and 28 data points included in the regression analysis for the diffraction peaks shown in Figures 1 and 2, respectively. The root-mean-square errors for the fitted profiles were 0.8 and 0.6 percent, respectively, which is on the order of the uncertainty in the measured intensity.

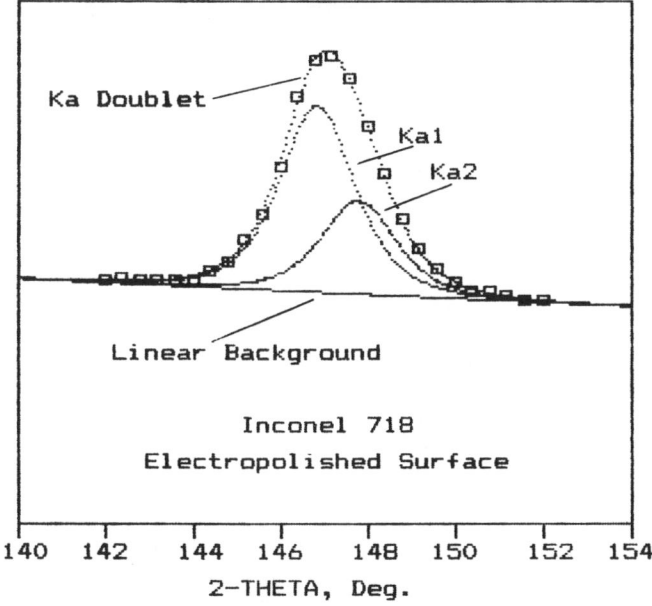

Fig. 1 - Pearson VII function profile fitted to the (420) peak from electropolished Inconel 718.

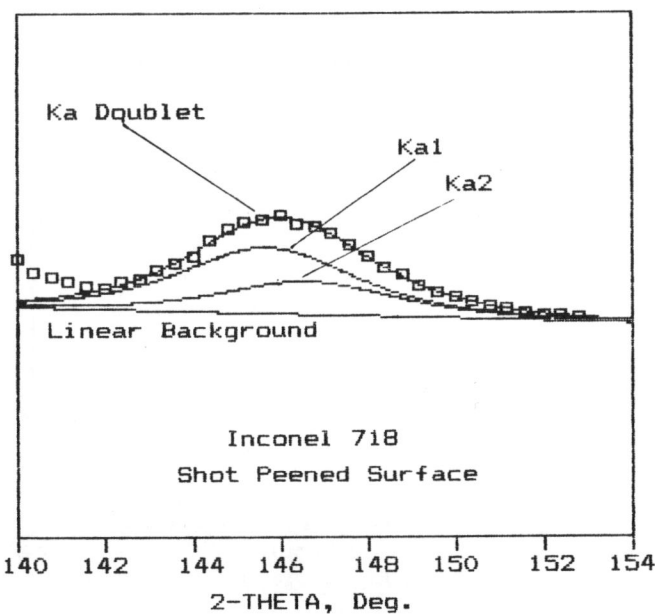

Fig. 2 - Pearson VII function profile fitted to the (420) peak from the surface of shot peened Inconel 718.

Pearson VII function regression analysis offers a number of advantages over other methods of diffraction peak location. Because the diffraction peak can be accurately described with as few as seven data points, only the data used for routine macroscopic residual stress measurement is needed for simultaneous determination of the peak width. The absolute background intensity need not be determined independently, allowing measurements to be made with the broad diffraction peaks commonly encountered on deformed metal surfaces, even when there may be partial overlap from adjacent peaks. Corrections for a linear background intensity and the superimposed $K\alpha$ doublet are achieved directly by the regression analysis. Finally, the $K\alpha_1$ diffraction peak position, width, and intensity can all be immediately determined algebraically from the fitted function.

DIFFRACTION PEAK WIDTH - The diffraction peak width is usually taken to be either the width of the peak at some fraction of its height above background, or the integral breadth (the integrated intensity divided by the peak height). For most purposes, the choice is arbitrary. For the $K\alpha_1$ diffraction peak described by an optimized Pearson VII function, the width at half height (FWHM) is given by:

$$W_{\frac{1}{2}} = \frac{2}{B}\sqrt{M\left(2^{\frac{1}{M}} - 1\right)} \quad (3)$$

and the integral breadth is given by:

$$W_{int} = \frac{\sqrt{M\pi}}{B} \frac{\Gamma(M - \frac{1}{2})}{\Gamma(M)} \quad (4)$$

where $\Gamma(x)$ is the Gamma function. For simplicity of calculation, the half width given by Eq. (3) was used in this investigation.

SEPARATION OF INSTRUMENTAL BROADENING - It is generally desirable to separate the contribution to the peak width due to instrumental effects, such as focal spot size, incident beam divergence, slit width, defocusing, etc. from the contribution caused by cold working of the specimen. Several approaches were considered. For purely Cauchy (M=1), or perfectly Gaussian (M=∞) peak profiles, the total peak breadth measured, B, can be shown to be a simple function of the instrumental broadening, b, and the specimen dependent broadening, β:

Cauchy: $B = b + \beta$ (5)

Gaussian: $B^2 = b^2 + \beta^2$ (6)

Although the diffraction peaks encountered in residual stress measurement tend toward pure Cauchy, M may commonly vary from approximately 1 to 3. The applicability of Eq. (5) and Eq. (6) was tested by assuming a simple Cauchy or Gaussian peak shape, and calculating the specimen dependent broadening using different x-ray optics. The results demonstrated that the instrumental and specimen dependent distributions are not adequately approximated by either purely Cauchy or Gaussian profiles, and that Eq. (5) and Eq. (6) are not generally adequate for the separation of instrumental broadening.

If the variance of the distribution function is used as a measure of diffraction peak breadth, the instrumental and specimen contributions to peak broadening can be easily be separated.[7] However, the variance is not defined for a Pearson VII function with $M < 1.5$ due to the extended tails of the diffraction peak. Because many of the peak profiles encountered in stress measurement of deformed metals are best described by M approximately equal to 1, the variance cannot be used to quantify the peak width.

Stokes method of deconvolution[5] to separate instrumental and specimen dependent broadening was not attempted because of the extensive data collection necessary, the dependence of the method on accurate definition of the tails of the diffraction peak, and the need for a cold work-free reference specimen.

Because no simple, accurate method of separating the instrumental and specimen contributions to peak breadth appeared feasible, a technique was developed using fixed x-ray optics to hold the instrumental contribution constant, and separation was not attempted.

EMPIRICAL RELATIONSHIP BETWEEN DIFFRACTION PEAK WIDTH AND PERCENT COLD WORK

To develop an empirical relationship between the amount of cold work present and the diffraction peak width, a series of coupons were prepared for each alloy by first heat treating or annealing, as appropriate for the alloy of interest, and then deforming the samples known amounts. The percent cold work was taken to be the absolute value of the true plastic strain calculated from changes in the dimensions of the specimens. For Inconel 718 and Rene 95, specimens were prepared by solution treating and ageing, followed by deformation in compression or tension. For Inconel 600 and Inconel 690, specimens were annealed and deformed in tension only.

For Inconel 718 and Rene 95, it was observed that the diffraction peak width produced by a given amount of cold work was independent of the mode of deformation. Whether deformed in tension or compression, the data fell in the same curve, as shown in Figure 3. The points plotted are the mean values of from four to ten repeat measurements on each sample. The standard deviation about the mean peak width ranged from approximately 0.02 deg. to 0.2 deg., increasing with peak width. Further, the peak width was accumulative for additional plastic strain induced on an already deformed specimen.

The empirical curve relating peak width to cold work could, therefore, be extended to levels of cold work beyond the range which could be produced directly in tension or compression. Specimens were prepared by grinding or shot peening the surface to induce a degree of cold work at the high end of the range previously covered by tension and compression samples. The initial surface percent cold work was first calculated from the empirical curve derived for the tension and compression samples. The ground or shot peened specimens were then deformed further in tension to extend the empirical curves for Inconel 718 and Rene 95 beyond 30 percent cold work.

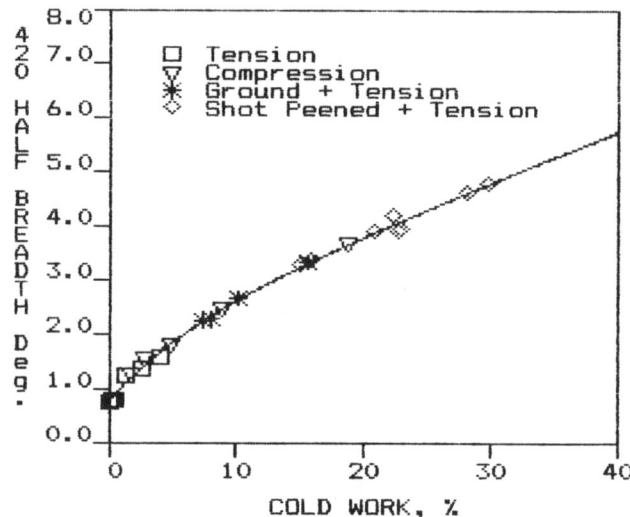

Fig. 3 - Empirical curve relating the (420) diffraction peak width to percent cold work for Rene 95 samples deformed by various means.

For all four alloys investigated, it was observed that the peak width increased monotonically, but not linearly, with the percent cold work induced. The peak width initially increased rapidly with cold work, and then approached a linear dependence for a highly cold worked state. The dependence of peak width on the percent cold work can be described for the range of cold work investigated by a function consisting of exponential and linear terms:

$$W_{\frac{1}{2}} = A\left[1 - \exp(-B \cdot \epsilon_p)\right] + C \cdot \epsilon_p + D \quad (7)$$

where the cold work is expressed as the absolute value of the percent true plastic strain. Eq. (7) was fitted by non-linear least squares regression to determine the regression parameters A through D. Although the function describes the dependence of peak width on cold work accurately throughout the entire range of cold work investigated, the function is transcendental. Reverse solution to determine the degree of cold work from measured peak width was achieved using Newton's method. The empirical functions fitted to data sets for all four alloys investigated are shown on a common scale in Figure 4. The data points are eliminated for clarity.

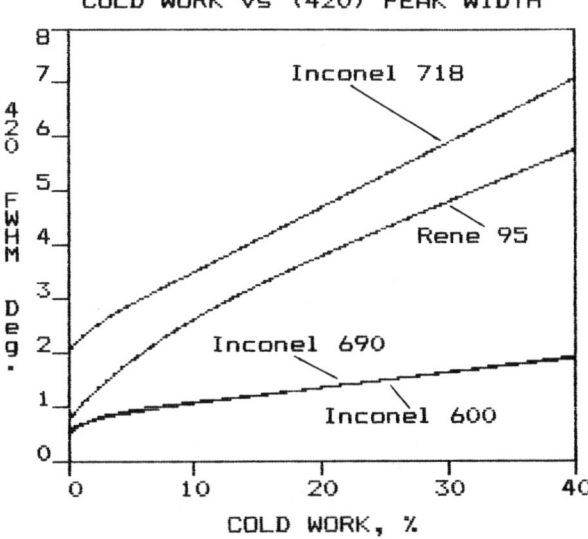

Fig. 4 - Empirical curves relating the (420) diffraction peak width to percent cold work for four nickel base alloys.

APPLICATIONS

The empirical relationship given by Eq. (7) has been applied in numerous investigations to determine simultaneously the cold work and macroscopic residual stress distributions as functions of depth induced in nickel base alloys by a variety of surface treatments.

Measurements were made on focusing horizontal diffractometers using copper Kα radiation and Si(Li) solid state detectors. The x-ray optics (focal spot size, take-off angle, incident beam divergence, receiving slit, etc.) were the same as were used for the development of the empirical curve relating cold work to peak width. Material was removed for subsurface measurement by electropolishing. The residual stress results were corrected for both the penetration of the radiation into the subsurface stress gradient[9] and for stress relaxation caused by layer removal.[10] The x-ray elastic constants in the (420) direction were determined empirically.[11]

INCONEL 718 - Figure 5 shows the residual stress and cold work distributions near the surface of Inconel 718 produced by an abrasive cut-off saw. A layer of compression reaching a maximum of approximately -600 MPa exists to a depth of approximately 50 microns, followed by a deep tensile layer peaking in excess of approximately 500 MPa, and extending to at least 300 microns. The associated cold work distribution ranges from 12 percent at the surface to approximately 2 percent at a depth of only 20 microns. Some cold work is evident to a depth of approximately 200 microns.

Figures 6 and 7 show the residual stress and cold work distributions produced in Inconel 718 by moderate and heavy shot peening intensities. Both peening intensities produced approximately 20 percent cold work at the surface. The cold work distributions differ primarily in the depth of the cold worked layer, which extends to approximately 70 microns for the 6 - 8 A Almen intensity, and to approximately 150 microns for the 5 - 7 C intensity. In both cases, the cold work distribution diminishes almost linearly near the surface of the sample, and extends to a depth approximately equal to the depth of the uniform compressive layer. The cold work appears to be insignificant beyond approximately the maximum depth of uniform high compression.

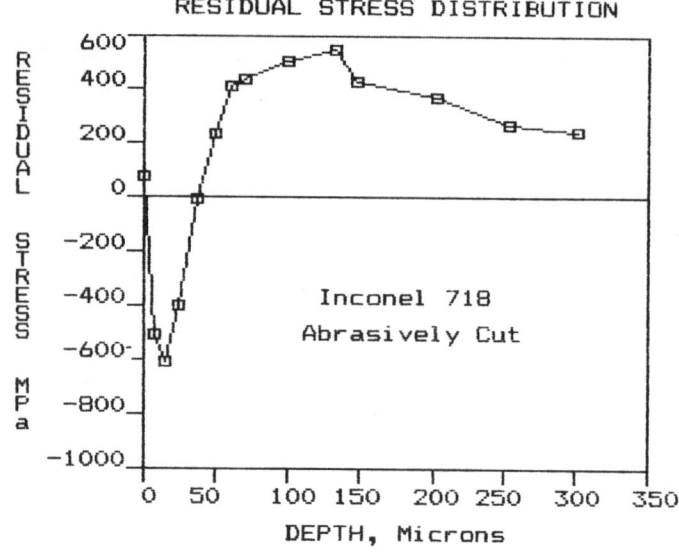

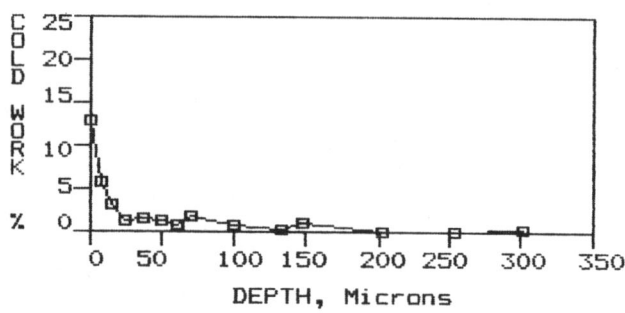

Fig. 5 – Residual stress and cold work distributions produced in Inconel 718 by an abrasive cut-off saw.

Residual stress distributions produced by shot peening of nickel base alloys typically exhibit the "hook" seen in Figures 6 and 7, with lower magnitude compression at the immediate surface after moderate to heavy shot peening. The reduced compressive stress achieved at the surface may be related to the extensive cold working of the material and a resulting increase in yield strength at the surface. The surface may be cold worked by prior machining or by the shot peening operation itself.

INCONEL 600 – Figure 8 shows the subsurface residual stress distributions in both the longitudinal and circumferential directions produced by grinding the surface of Inconel 600 tubing. The residual stress distributions are of similar form, with compression in both directions near the surface and substantial tension beneath. The circumferential stress distribution is shifted toward tension, and both distributions show substantial tensile stress beneath a depth of approximately 100 microns. The associated cold work distribution reveals a highly cold worked surface, estimated in excess of 40 percent, requiring extrapolation of the empirical relationship beyond the range which could be produced in test coupons. The material is seen to be significantly deformed to a depth of 100 microns, approximately the extent of the compressive layer.

Cold working of nickel base alloys to the extent observed in Figures 5 through 8 can result in significant work hardening and alteration of mechanical properties, such as yield strength, in the near surface layers.

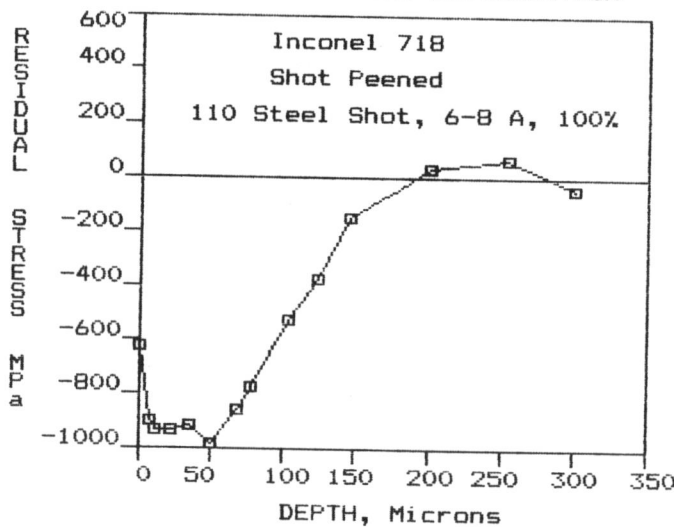

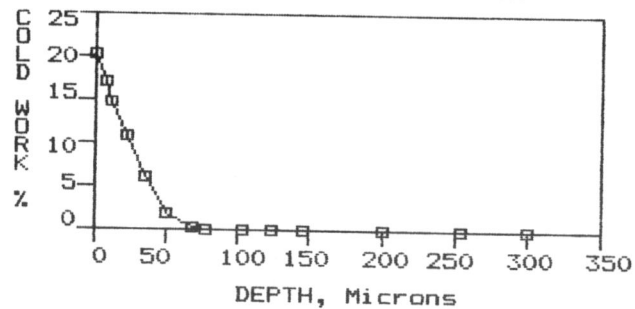

Fig. 6 – Residual stress and cold work distributions produced by moderate (6-8 A, 100%) shot peening of Inconel 718.

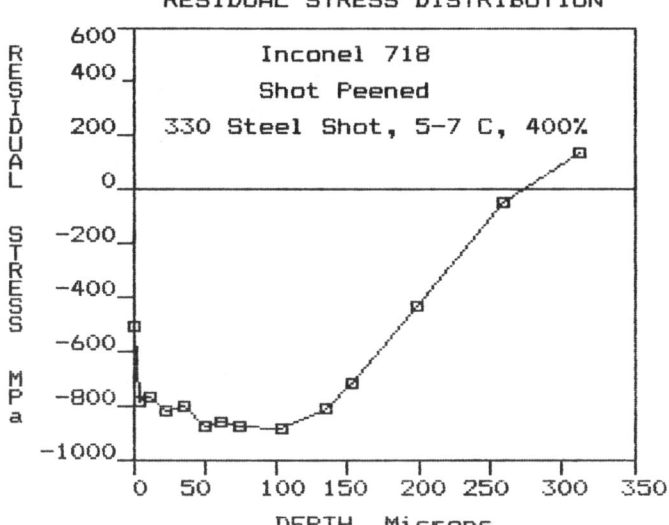

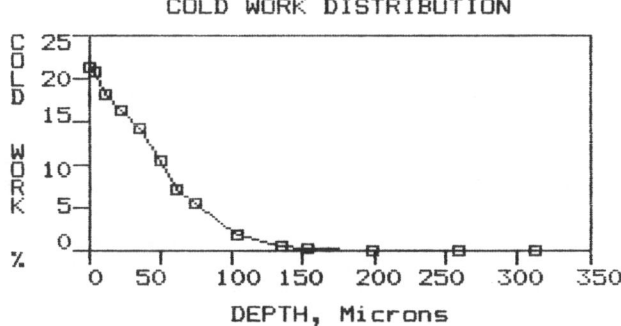

Fig. 7 - Residual stress and cold work distributions produced by high intensity (5-7 C, 400%) shot peening of Inconel 718.

From a true stress-strain curve for the alloy, it is possible to estimate the yield strength distribution from the cold work distribution. Figure 9 shows a true stress-strain curve measured at room temperature for Inconel 600 tubing in the condition supplied by the mill. Because the percent cold work was taken to be the true plastic strain, the resulting yield strength at each depth can be estimated from Figure 9. Figure 10 shows the estimated subsurface yield strength distribution produced by grinding the surface of the tubing. At the surface, the yield strength is estimated to be approximately twice the value for the core material. The distribution drops at first nearly linearly to a depth of approximately 50 microns, and then more gradually to reach the yield strength of the undeformed material at a depth of approximately 150 microns.

Figures 8 and 10 show that the ground tubing consists of a thin shell of compressively stressed, highly cold worked material on the O.D. surface with a yield strength substantially higher than the core, which is in residual tension. This inhomogeneity induced in the tubing by grinding the O.D. surface can have a significant effect upon the residual stress distributions developed in subsequent forming operations, such as expansion and bending of the tubing during fabrication.

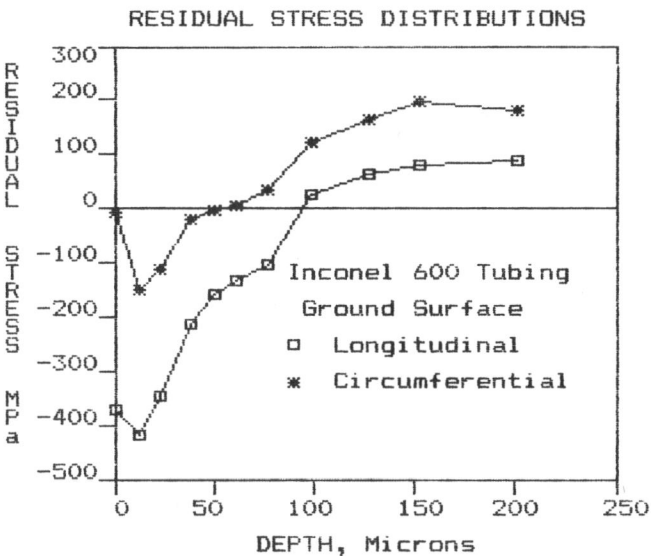

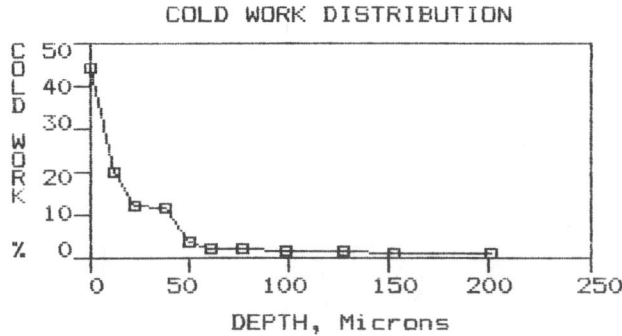

Fig. 8 - Residual stress and cold work distributions produced by grinding the O.D. surface of Inconel 600 tubing.

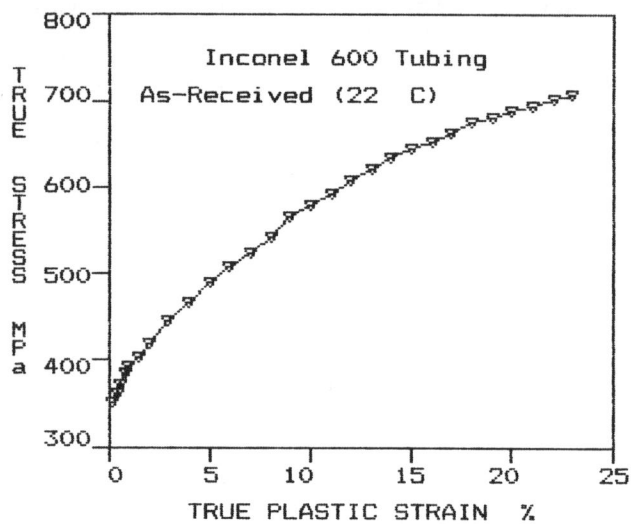

Fig. 9 - Monotonic true stress-strain curve for Inconel 600 tubing in the mill annealed condition at room temperature.

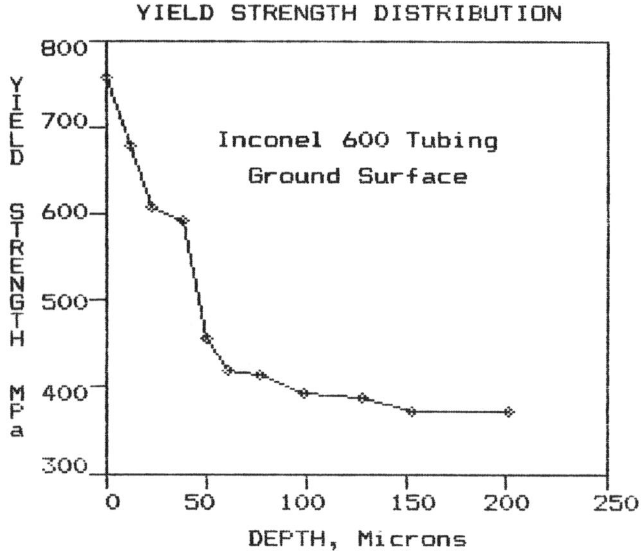

Fig. 10 - Estimated yield strength distribution for the O.D. surface of ground Inconel 600 tubing.

SUMMARY AND CONCLUSIONS

A rapid, accurate method of determining the diffraction peak width simultaneously with the peak position during residual stress measurement has been developed using Pearson VII function peak profile analysis. The method provides a number of advantages over methods previously used:

1. Separation of the $K\alpha$ doublet and background subtraction are provided directly by the regression model.

2. Independent measurement of the background intensity is not necessary, allowing the use of broadened or partially overlapping diffraction peaks.

3. Only the data routinely collected for the determination of diffraction peak position during the measurement of macroscopic residual stresses are required.

An empirical relationship has been developed relating the (420) diffraction peak width to the percent cold work present in the nickel base alloys, Inconel 718, Rene 95, Inconel 600 and Inconel 690. For all four materials, the diffraction peak width was found to approach a linear dependence upon the amount of cold work for highly cold worked material. The peak width produced for a given amount of cold work was found to be independent of the manner in which the cold work was induced, whether by simple uniaxial tension or compression, or as a result of the complex deformation produced during shot peening or grinding.

The method has been used to determine simultaneously the residual stress and cold work distributions produced on abrasively cut and shot peened surfaces of Inconel 718, and on the ground surface of Inconel 600 tubing. A method of estimating the subsurface yield strength distributions from a true stress-strain curve and the measured cold work distributions has been demonstrated for ground Inconel 600 tubing.

ACKNOWLEDGEMENT

The author wishes to acknowledge the assistance of the Metal Improvement Company in providing the shot peened and abrasively cut Inconel 718 specimens and the Babcock & Wilcox Corporation for providing the true stress-strain data for Inconel 600 tubing shown in Figure 9.

REFERENCES

1. W. A. Rachinger, J. Sci. Instr., <u>25</u>, 254 (1948)

2. K. S. Gupta and B. D. Cullity, "Advances in X-Ray Analysis," <u>23</u>, 333 (1980)

3. D. P. Koistinen and R. E. Marburger, ASM Trans., <u>51</u>, 537 (1959)

4. P. S. Prevey, "Advances in X-Ray Analysis," <u>29</u>, 103 (1986)

5. A. R. Stokes, Proc. Phys. Soc. (London), <u>A61</u>, 382 (1948)

6. B. E. Warren and B. L. Averbach, J. Appl. Phys., <u>21</u>, 595 (1950)

7. H. P. Klug and L. E. Alexander, X-RAY DIFFRACTION PROCEDURES, 2nd Ed., p. 299, John Wiley & Sons, NY (1974)

8. R. A. Young and D. B. Wiles, J. Appl. Cryst., <u>15</u>, 430 (1982)

9. M. E. Hilley, ed., SAE, J784a, "Residual Stress Measurement by X-Ray Diffraction," p. 61, Society of Automotive Engineers, Inc., Warrendale, PA (1971)

10. M. G. Moore and W. P. Evans, SAE Trans., <u>66</u>, (1958)

11. P. S. Prevey, "ADVANCES IN X-RAY ANALYSIS," <u>19</u>, 709, (1976)

RESIDUAL STRESS MEASURMENTS IN ARMAMENT-SYSTEM COMPONENTS BY MEANS OF NEUTRON DIFFRACTION

H.J. Prask, C.S. Choi
Energetics and Weapons Div., ARDEC
Picatinny Arsenal, New Jersey USA
and
Reactor Radiation Division, NBS
Gaithersburg, Maryland USA

ABSTRACT

Neutron diffraction is a measurement technique which closely parallels x-ray diffraction in methodology and analytical formalism. However, in the normal diffraction wavelength range neutrons are, in general, about a thousand times more penetrating than x-rays. In addition the coherent scattering cross-section for neutrons varies in a random manner with atomic number in contrast to the essentially monatonically increasing dependence of x-rays. We have made use of these neutron properties to nondestructively determine sub-surface residual stress in a variety of samples. More specifically, we have utilized energy (or wavelength) dispersive neutron diffraction to improve resolution and because it appears to reduce anomalous peak shifts in textured samples. We have examined aluminum and steel calibration samples to test the technique, and have applied the technique to the characterization of sub- and near-surface residual stresses in two types of armament-system components: U-0.75 wt% Ti cylindrical rods and 7075-T6 aluminum alloy ogives. Within each sample type, samples with different thermo-mechanical histories were examined. The measured residual stress distributions are presented and discussed.

SEVERAL TECHNIQUES ARE AVAILABLE for the nondestructive determination of residual stress, e.g. x-ray diffraction, eddy current, ultrasonics. However of these, the best established for quantitative characterization is x-ray diffraction which is, generally, a surface probe. Although surface stresses are of prime concern in many armament applications, certain alloys present serious difficulties for the x-ray technique because of grain size, surface contamination, or texture; among these are uranium alloys and, in some cases, aluminum alloys.

Neutron diffraction closely parallels x-ray diffraction in methodology and analytical formalism. However, because neutrons interact with nuclei and x-rays with electrons, neutrons are typically about a thousand times more penetrating than x-rays in the wavelength range for diffraction ($0.7 \text{ Å} \leq \lambda \leq 4 \text{ Å}$). In addition, different elements exhibit significantly different relative scattering powers for neutrons and x-rays. A utilization of the unique aspects of neutron diffraction for texture characterization and residual stress analysis was suggested and partially demonstrated by us previously [1]. Since 1982 a number of tests and applications of the neutron diffraction technique to sub-surface residual stress determination have been reported [2-6]. These proceedings also contain another contribution in this area [7].

In the present paper we review fundamentals of the technique and describe the application of neutron diffraction to the characterization of sub-surface tri-axial stress distributions in depleted uranium - 0.75 wt % titanium alloy samples with different thermo-mechanical histories, and to a 7075-T6 aluminum alloy component in production.

EXPERIMENTAL

GENERAL CONSIDERATIONS - In both x-ray and neutron diffraction determination of residual stress, what is measured is strain which is manifested by changes in the distance, d, between atomic planes in the sample. A unique advantage of neutron diffraction arises from the different relative scattering cross-sections and penetration relative to x-rays. This is illustrated in Table 1 in which $t_{1/2}$, the thickness at which half the beam intensity is lost through scattering and absorption processes, is listed for selected metals. The values are based on cross-sections from standard references and the difference in wavelengths used for neutrons (1.08Å) and x-rays (1.54Å) is not significant.

Table 1. X-Ray/Neutron Comparison

Element(At.No.)	$t_{1/2}$(X-rays)	$t_{1/2}$(Neutrons)
Al (13)	0.0530 mm	71.0 mm
Ti (22)	0.0076	15.9
Fe (26)	0.0027	6.1
Cd (48)	0.0035	0.057
W (74)	0.0021	6.5
U^{238} (92)	0.0015	13.6

It should be mentioned that the $t_{1/2}$ values in Table 1 do not represent the depth of penetration in a residual stress measurement. This is dependent on a number of factors such as source intensity, coherent scattering cross-section, and beam spot size. However, the $t_{1/2}$ values clearly show that neutrons in the normal diffraction wavelength range are several orders-of-magnitude more penetrating than x-rays. Also, the penetration does not decrease monotonically with atomic number as with x-rays but is, essentially, a random function of atomic number. It is clear from the Table that both depleted uranium and aluminum are especially good materials for neutron examination.

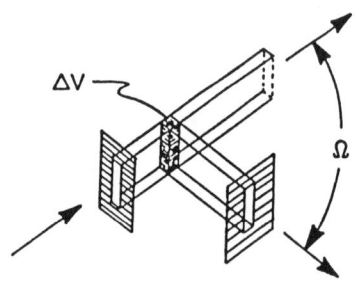

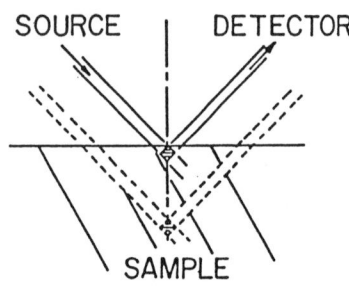

Figure 1. Schematic of how absorbing masks are used with Bragg's Law to define the examined volume, ΔV (upper); and plan view of ΔV translation through the sample (lower) in the EDND measurements.

The properties of neutrons suggested the possibility of measuring sub-surface residual stresses in metallurgical samples employing tight collimation and the scattering geometry shown schematically in Figure 1. A scattering angle, Ω, of ~90° is used to minimize the examined differential volume, ΔV.

The perspective view in Figure 1 shows ΔV defined by two rectangular apertures in an absorbing material, e.g. cadmium; however, the apertures could be any shape including circular. The plan view in Figure 1 indicates how the sample can be translated in the beam so that ΔV can be examined as a function of depth.

The strains from which residual stresses are inferred are obtained from measured d-spacings through Bragg's Law: $\lambda = 2d(hkl)\sin \Omega/2$ where λ is wavelength, $d(hkl)$ is the separation of atomic planes with Miller indices hkl, and Ω is stepped - usually with the sample orientation stepped by $\Omega/2$ - and sharp resonances in scattered intensity are observed at scattering angles where the Bragg condition is fulfilled. Precise determination of $\Omega(hkl)$, the peak position, yields $d(hkl)$ directly. Although success has been achieved with this mode of measurement, as reviewed in reference 4, some problems have been encountered with highly attenuating or highly textured samples. Here, texture means the existence of preferred crystallographic orientation of grains or crystallites relative to a coordinate system fixed in the sample.

With reference to the plan view in Figure 1, it is clear that as Ω is varied, path length to and from ΔV changes and intensity as a function of Ω is possibly distorted leading to a false shift in $d(hkl)$. Similarly, gradients in preferred grain orientation in the sample over the changing beam-in/beam-out paths can produce intensity variations which shift the apparent $\Omega(hkl)$. Since the strains, $\Delta d/d$, are on the order of 0.0001, a small anomalous shift nullifies the stress measurement.

ENERGY-DISPERSIVE NEUTRON DIFFRACTION - In our measurements, we have made use of the fact that Bragg-condition resonances can also be observed at fixed Ω with varying wavelength. With the scattering angle fixed, changing attenuation and texture gradients are less distortive. In addition, the examined volume ΔV, remains exactly the same throughout each scan.

The instrument used for energy-dispersive neutron diffraction (EDND) is called a triple-axis spectrometer. Crystals of known d-spacing are placed before (monochromator) and after (analyzer) the sample; the Bragg relation is then used to select and step the wavelength incident on the sample. In principal, the analyzer crystal - which we step at the identical wavelength as the monochromator - is not needed. However, utilization of the analyzer significantly enhances instrumental resolution and produces peak profiles which are Gaussian in shape and straightforward to analyze with least-squares curve-fitting techniques.

In our system, pyrolytic graphite crystals were used for monochromator (002 plane) and analyzer (004) with a pyrolytic graphite filter between monochromator and sample to reduce higher order wavelength contributions and background. The collimation employed was 50'-20'-27'-80' from source to analyzer, with a resultant resolution $\Delta\lambda/\lambda = 0.0073$

at $\lambda = 2.692$Å. The Cd absorbers were cut with 4 mm x 4 mm square apertures.

STRESS-STRAIN RELATIONS - The relation between stress and strain applicable to diffraction measurements has been presented, for example, by Evenschor and Hauk [8]. With reference to Figure 2, r, θ, and z are specimen-fixed axes, and the strain $\epsilon'_{\phi\Psi}$ is measured along $\vec{L}'_3$; then

$$\epsilon'_{\phi\Psi} = (d_{\phi\Psi} - d_o)/d_o$$
$$= (\epsilon_{11}\cos^2\phi + \epsilon_{12}\sin 2\phi + \epsilon_{22}\sin^2\phi)\sin^2\Psi \quad (1)$$
$$+ (\epsilon_{13}\cos\phi + \epsilon_{23}\sin\phi)\sin 2\Psi + \epsilon_{33}\cos^2\Psi$$

where $d_{\phi\Psi}$ is the lattice spacing along $\vec{L}'_3$ and d_o is the unstressed lattice spacing. The stresses are related to the measured strains through

$$\epsilon'_{\phi\Psi} = 1/2 S_2(hkl)[\sigma_{11}\cos^2\phi\sin^2\Psi + \sigma_{22}\sin^2\phi\sin^2\Psi$$
$$+ \sigma_{33}\cos^2\Psi + \sigma_{12}\sin 2\phi\sin^2\Psi + \sigma_{13}\cos\phi\sin 2\Psi \quad (2)$$
$$+ \sigma_{23}\sin\phi\sin 2\Psi] + S_1(hkl)[\sigma_{11} + \sigma_{22} + \sigma_{33}].$$

The $S_i(hkl)$ are diffraction elastic constants ("XEC") for the (hkl) reflection which, in general, depend on the material and the reflection examined. For an elastically isotropic solid the XEC are given by

$$1/2 S_2(hkl) = (1+\mu)/E \text{ and } S_1(hkl) = -\mu/E \quad (3)$$

where μ, E are Poisson's ratio and Young's modulus, respectively.

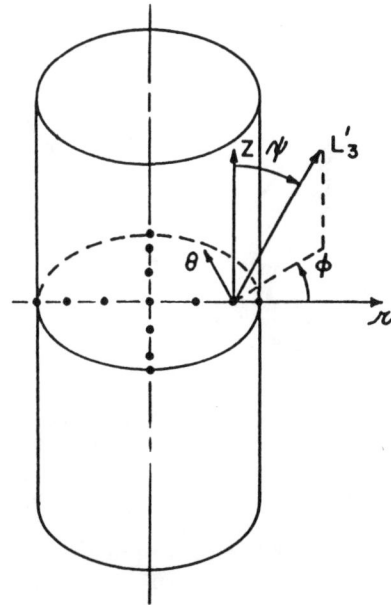

Figure 2. Coordinate system and measurement mesh for the 2.5 cm diameter by 10 cm long DU samples. The solid circles represent the centers of the 4 mm X 4 mm X 4 mm ΔVs examined.

RESULTS

U - 0.75 wt% Ti ("DU")

General - DU is an alloy of considerable importance in certain applications. In order to improve performance and decrease production costs, fabrication-process changes are dictated which can affect desired material properties such as yield strength, ultimate strength, ductility, and susceptibility to SCC. These properties depend in varying degrees on the residual stress distribution produced by the fabrication method which includes gamma-phase solutionizing, quenching, aging, and cold work. In following subsections we describe the application of EDND to the characterization of residual stress in DU specimens prepared by differing processing methods.

Samples - Two different types of DU samples were studied. For convenience, the residual stress determinations were made on 10 cm long pieces cut from the mid-point of 46 cm long, 3.3 cm (starting) diameter rods. However, with slight modification of the instrument uncut rods could also be examined. One group was γ-phase solutionized in a vacuum furnace, water-bath quenched, and rotary straightened ("mild" cold work). Of these, some were machined to the final 2.5 cm diameter, and one was moderately cold worked and machined to the final 2.5 cm diameter. In addition, one sample was solutionized in an induction furnace and water-spray quenched, aged, rotary straightened and machined to the final diameter. In all cases aging after quenching was 1-2 hours at 400°C or less.

Residual Stress - Since the determination of residual stress in technological samples by means of Eq. 2 depends directly on measurement of strain values, a precise value for the unstressed d-spacing, d_o, is essential. We have been successful in determining unstressed lattice parameter from conventional neutron measurements in most cases; however, cold working produces texture which makes lattice parameter determinations too uncertain relative to the precision required for residual stress chacterization. Alternatively, we utilize the overall equilibrium conditions required by elasticity theory to determine d_o. That is, since the body is static with no external force applied, residual stresses normal to any plane must balance such that in cylindrical geometry:

$$\int \sigma_{zz} r dr d\theta = 0 \quad (4a)$$

$$\int \sigma_{\theta\theta} dr dz = 0 \quad (4b)$$

and at any surface the stress orthogonal to that surface must vanish. Stresses inferred from measured strains can be adjusted, by adjusting d_o, to fulfill the equilibrium conditions.

One critical element in the above procedure is the determination of near-surface strains since stress gradients, especially for σ_{zz}, can be large near the cylindrical surface. In this work we have carefully measured d-spacings in a strain-free copper powder in cylindrical geometry identical to the DU cylinders. We find that measured strains, even for ΔV only half inside the curved surface (see Figure 2), agree within two standard deviations with the mean d-value. From this we conclude that no systematic instrumental errors limit the determination of d_o using the overall equilibrium approach.

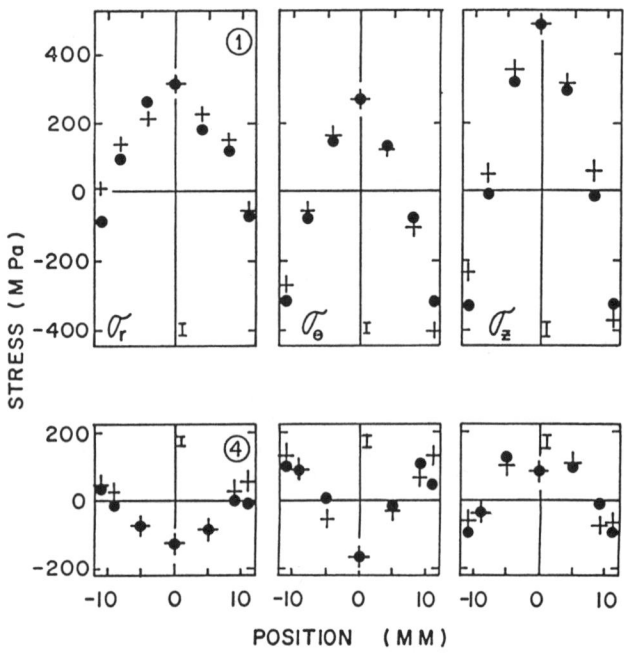

Figure 3. EDND determined residual stress distributions ("1") for mildly cold-worked DU specimens (typical), and ("4") the moderately cold-worked specimen. The crosses and solid circles correspond to measurements along the two orthogonal diameters of Figure 2. Representative standard deviations are shown near the centerline of each stress distribution.

Utilizing the (112) reflection we have determined strains in the mid-point r-θ plane of the 10 cm long DU pieces. The measurement grid for the moderately and highly cold-worked samples is shown in Figure 2. Determination of stresses was made using Eqs. 2-4 and recently determined values of Poisson's Ratio (0.267) and the elastic modulus (176600 MPa) for α'-DU [9]. We find that the results for all of the mildly cold-worked samples, including the induction-furnace solutionized/spray-water quenched sample, exhibit stress distributions of the type shown in Figure 3(1). The moderately cold-worked specimen exhibits a stress distribution which is quite different as shown in Figure 3(4). Examination of Figures 3(1) and 3(4) shows the following:

1. All of the specimens which underwent mild cold work (i.e. only rotary straightening) retain the residual stress distribution expected for a quenched-rod, with tensile stresses at the center and compressive hoop and axial stresses at the surface;
2. Moderate cold work significantly alters the residual stress distribution of quenched rods; at the center the change in stress distribution corresponds, approximately, to the superposition of a 400 MPa compressive stress on each stress component; away from the center the stress distribution changes to maintain overall equilibrium;
3. Moderate cold work can cause tensile residual stresses at the surface [cf. $\sigma_{\theta\theta}$, Fig. 3(4)].

7075-T6 OGIVES

<u>General</u> - In the past few years several ogive failures have occurred during ballistic acceptance testing of the 155 mm M483A1 projectile. The purpose of this projectile is to deliver submunitions. It is constructed of an ogive at the forward end that contains a small explosive charge, which is threaded to the projectile body and sealed from it by a pusher plate. The aft end is sealed by a shear plate. At detonation, the pressure developed in the ogive cavity creates an aft directed force that acts on the pusher plate transmitting it through the submunitions to the aft-end shear plate. For effective operation, the force must be sufficient to fracture the shear plate and accelerate the submunitions, thus scattering the submunitions behind the forward traveling projectile body. The malfunctions occurred when the ogive body failed before the aft-end pusher plate.

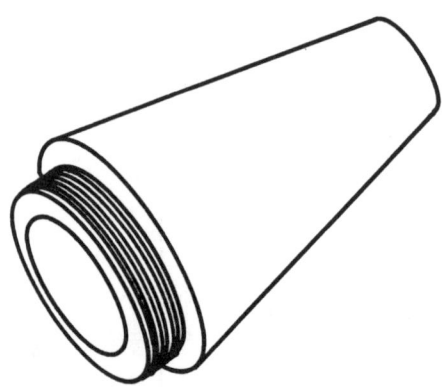

Figure 4. M483A1 ogive with forward end at upper right. Overall length is 19.3 cm, ID at aft end is 4.7 cm. The region of interest is under the most forward thread shown.

The ogive is manufactured from 7075-T6 aluminum and is pictured in Figure 4. Production ogives are manufactured by two different suppliers who use somewhat different manufacturing methods. Manufacturer B produces the ogives by cold forging the cavity to finished dimensions, heat treating, and machining the outside dimensions. Manufacturer A produces a preform by forging at 332°-382°C, machines the cavity, heat treats, then finish machines the outside dimensions. Both suppliers use ALCOA aluminum. In full-scale tests the primary failure mode, exclusive to the B-type ogive, is a circumferential fracture at the first loaded thread. A single failure of an A-type ogive, by longitudinal fracture in the conical region, has also been reported.

Several material characterization studies and simulation tests have been conducted (summarized in reference 10). The materials studies showed some difference in the microstructure of A versus B ogives, but no substantial material property differences. The results of a simulation study, in which ogives were sealed and a charge exploded within, showed substantial differences in behavior. The B ogives failed at the thread at containment pressures substantially below that which is required to fail A ogives. Furthermore, these tests showed that the A ogives did not fail at the thread. These ogives failed by ductile rupture with the crack running in the longitudinal direction.

was representative of the full circumference. The axial direction equilibrium was then determined through the thickness (8.6 mm) by two averaging procedures: one in which the three measurement positions were given equal weight (except for the circumferential area element subtended), and a second in which the center-point contribution was subtracted out of the inner and outer points. The d_0s arrived at to balance σ_{zz} stresses were in reasonable agreement for both types of averaging and for both types of ogives. The diffraction elastic constants used in the stress/strain calculations were the theoretical values of Bollenrath et al.[11]

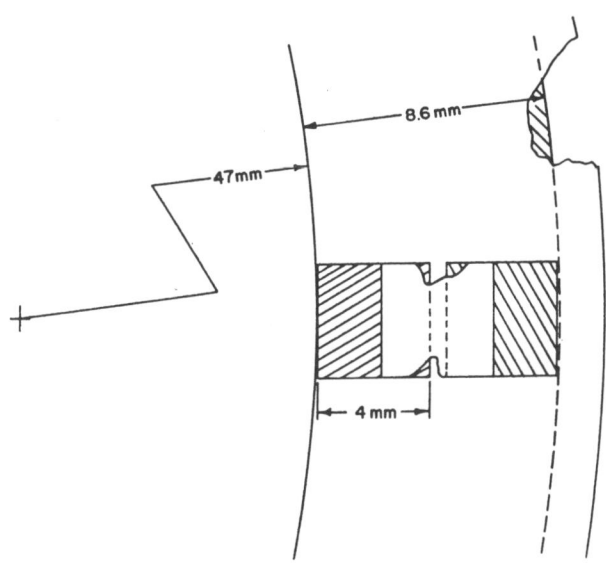

Figure 5. Partial, sectioned schematic of ogive at the failure plane as viewed parallel to the cylinder axis. Cross-sections of the 4mm x 4 mm x 4mm beam spots at the measurement positions are shown.

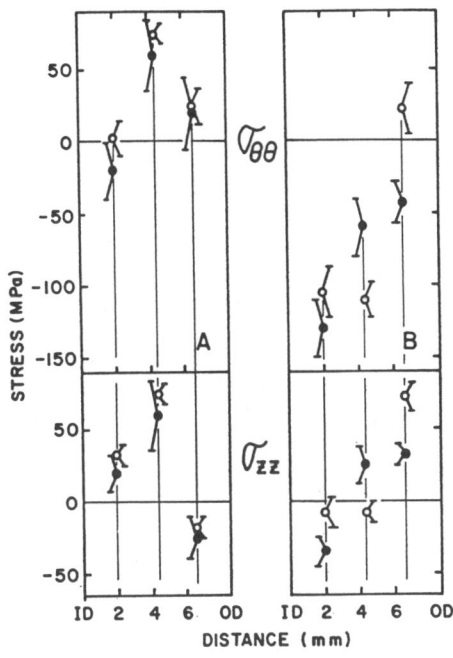

Figure 6. $\sigma_{\theta\theta}$ and σ_{zz} residual stress distributions of the A and B type ogives as determined by EDND. "OD" corresponds to the thread-root position. The open and closed circles correspond to measurements made 180° apart.

Experimental - One ogive of type B and one ogive of type A were studied by EDND. In Figure 5, a view of the three 4 mm x 4mm x 4mm beam spot positions in the plane of the potential failure site is shown. An identical set of measurements was made at 180° to the arbitrarily chosen 0° position. Partial measurements performed on a second type B ogive gave strain values very similar to the first type B ogive.

In the ogive case, overall equilibrium (Eq. 4a) was used to determine $d_0(200)$. It was assumed that the average of the 0° and 180° data

Residual Stress - In Figure 6, final absolute residual stress values are shown for both ogive types for 0° and 180° positions at the first thread position. In contrast too previous measurements, the nondestructive EDND stress determination indicates very significant differences in the two ogive types. In the A-type ogives it is found that just below the thread root ("OD") the σ_{zz} stresses are about -20 MPa (compressive) whereas the B-type show +30 to +70 MPa (tensile) stresses. This is completely consistent with the principal failure mode differences that have been observed. In contrast, σ_{zz} at the ID of A-type ogives are zero or somewhat tensile, while in the B-type σ_{zz} stresses are highly compressive at this position. Qualitatively, this is consistent with longitudinal ductile rupture being the primary failure mode for A-type ogives, however, other material anisotropies (e.g. texture) may be more important factors in controlling this.

It is also of interest that the measured stresses are very nearly symmetric [$\sigma(0°) \approx \sigma(180°)$] in A-type ogives, and noticeably asymmetric in the B-type. In terms of failure it would seem that material strength in the axial direction woud depend more on the maximum tensile stress at any point around the OD rather than the average σ_{zz}. If so, the σ_{zz} stress differential for failure is at least 90 MPa (13 ksi) for the two ogive types examined.

CONCLUSIONS

In the present work we have shown that neutron diffraction techniques, in particular energy dispersive neutron diffraction, can provide information essential to failure analysis and/or improvement of performance of armament systems. For the first time the three-dimensional stress distributions in aluminum- and DU-alloy armament-system components have, in part, been mapped. These measurements have shown a dramatic effect due to finish-machining and cold-work on near- and sub-surface residual stress distributions. Finally, because of the nondestructive nature of EDND, characterized items can be further processed and re-characterized to develop a complete understanding of fabrication steps and induced residual stresses.

ACKNOWLEDGMENT

The authors thank Mr. W. Sharpe and Mr. M. Carlini of ARDEC for many helpful discussions concerning the components under study.

REFERENCES

1. C. S. Choi, H. J. Prask, S. F. Trevino, H. A. Alperin, and C. Bechtold, NBS Tech. Note 995, 34-39 (1979); C. S. Choi, H. J. Prask, and S. F. Trevino, J. Appl. Cryst. 12, 327-331 (1979).

2. M. J. Schmank and A. D. Krawitz, Met. Trans. 13A, 1069-76 (1982).

3. L. Pintschovius, V. Jung, E. Macherauch, and O. Vohringer, Mat. Sci. Eng. 61, 43-50 (1983).

4. H. J. Prask and C. S. Choi, J. Nucl. Matls. 126, 124-131 (1984).

5. T. M. Holden, B. M. Powell, G. Dolling and S. R. MacEwen, Proc. 5th Intn'l. Symp. Mets. and Matls. Sci., Riso, 1984, pp. 291-294; R. A. Holt, G. Dolling, B. M. Powell, T. M. Holden and J. E. Winegar, ibid, pp. 295-300.

6. A. J. Allen, M. T. Hutchings, C. G. Windsor and C. Andreani, Adv. Phys. 34, 445 (1985).

7. S. R. MacEwen, T. M. Holden, R. R. Hosbons and A. G. Cracknell, "Residual Stresses in Rolled Joints", these proceedings.

8. P. D. Evenshor and V. Hauk, Z. Metallkde. 66, 167-8 (1975).

9. G. M. Ludtka, Oak Ridge Y-12 Plant, unpublished work.

10. J. A. Kapp, R. J. Fujczak, and R. T. Abbott, "An Evaluation of the Service Failure of Aluminum Nose Cones Using Four Test Techniques", ARDEC Tech. Rept. ARCCB-TR-87006, 1987.

11. F. Bollenrath, V. Hauk, and E. Muller, Z. Metallkde. 58, 76-82 (1967).

USE OF BARKHAUSEN EFFECT IN TESTING FOR RESIDUAL STRESSES AND MATERIAL DEFECTS

Kirsti Tiitto
American Stress Technologies, Inc.
Bethel Park, Pennsylvania USA

ABSTRACT

Barkhausen noise is created by the abrupt changes in the magnetization of materials under applied ac magnetizing field. These changes are known to be affected by residual and/or applied stresses. Monitoring the Barkhausen noise under controlled conditions then provides a means of evaluating the stress state of the material.
It is also known that the intensity of the Barkhausen noise can be sensitive to microstructural features such as the density and distribution of dislocations and texture. While absolute determination of residual stress may become difficult in the presence of extensive variations of the microstructure, relative information of the stress is still available and detrimental changes of stress will be observable. Furthermore, the behavior of materials under e.g. cyclic loading is determined not only by residual stress but by the generation and distribution of dislocations. Being sensitive to these features, Barkhausen noise carries information which can be used to predict the behavior under such conditions.

In this paper, Barkhausen noise theory will be reviewed and examples will be given to show how and when this technique can be used to evaluate residual stress. Further examples will show how simultaneous stress and microstructural changes can be exploited to nondestructively detect material defects such as grinding burns, or can be used to predict the behavior under cyclic loads.

FERROMAGNETIC MATERIALS consist of magnetic domains, separated from each other by domain walls. The domains are magnetized to saturation along <100>, which are the directions of easy magnetization in bcc materials. The net magnetization of the bulk material is the average of the magnetization within all domains. In demagnetized state, it will be zero.

If, for any reason, a domain wall is made to move, the magnetization within the area swept by the wall will change to other direction generating an electric pulse. In bulk sample when all individual wall movements and pulses generated are counted, a noise-like signal, called Barkhausen noise, results (1).

Domain walls can be made to move by applying magnetic field to the part. Under an AC field, the domain walls will move back and forth. Figure 1 shows how one Barkhausen noise burst is created for a magnetizing half cycle (2). At the same time the magnetic induction of the material changes along the hysteresis loop. It is noticed that the amplitude of the noise pulses is highest when the change in magnetic induction is at its maximum, i.e. when the slope of the hysteresis loop is steepest. It is thus evident that the amount of Barkhausen noise correlates with the hysteresis loop: the steeper the slope of the loop, the higher the Barkhausen noise level, and vice versa.

For practical applications, the level of Barkhausen noise can be quantified in several ways. It can be expressed, for instance, in terms of the maximum noise amplitude, the rms value of the noise, or the noise envelope.

It is known that the movement of domain walls and hence the amount of Barkhausen noise under applied ac field are controlled by the state of stress (3-7) and microstructure (2,8). In this paper, their influence on the Barkhausen noise will first be discussed. Several examples on practical applications incorporating stress or microstructural changes or both will then be presented.

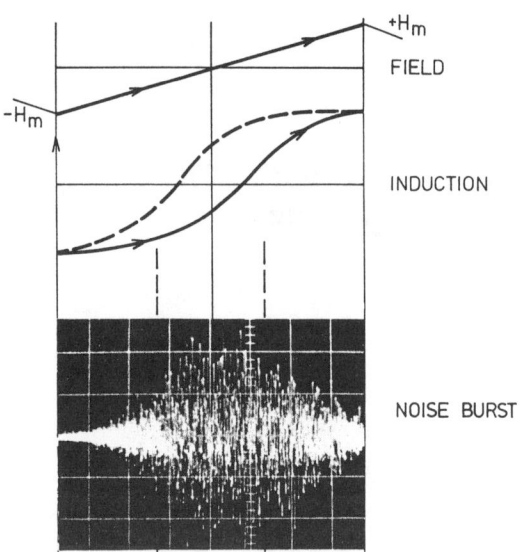

Fig. 1 - Sensor magnetizing field, sample induction and excited Barkhausen noise burst for one magnetizing half-cycle (2).

EFFECT OF STRESS - A simplified example is given in Figure 2, which represents a hypothetical piece of ferromagnetic material consisting of four equal magnetic domains. The magnetization directions of the domains are aligned so that the overall magnetization of the piece is zero. Figure 2a describes the changes in the domain configuration when the piece is exposed to tensile stress. The domains with magnetization directed parallel to the direction of the tensile stress will grow at the expense of the other domains and will finally consume the whole volume. In Figure 2c, it is shown that the domains with magnetization perpendicular to the direction of compressive stress will grow and annihilate the other domains. Under applied magnetic field, the domains with parallel magnetization to the direction of the magnetic field will grow and the domains with transverse magnetization will shrink as shown in Figure 2b.

The combined effect of stress and magnetic field on domain wall movement and the level of Barkhausen noise can be described in a simplified manner as follows. For example, the tensile stress from Figure 2a, and the magnetic field from Figure 2b, generate the same kind of change in the domain configuration. The domain walls can therefore move readily creating a high level of noise. The compressive stress from Figure 2c, and the magnetic field from Figure 2b, have opposing effects. Little wall movement will take place, and only little noise is generated.

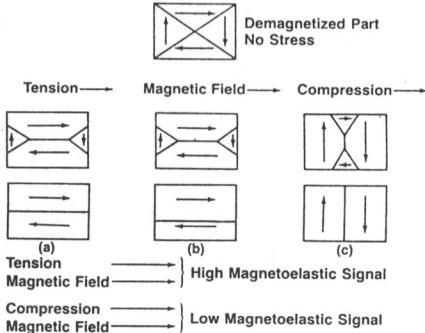

Fig. 2 - Schematic presentation of magnetic domain wall movements under stress and magnetizing field, and their effect on the Barkhausen noise or the magnetoelastic signal.

The above can also be illustrated with the aid of Figure 3, which shows the relation between magnetic induction (hysteresis loop), residual stress and Barkhausen noise. High compressive stresses generate fat hysteresis loops and low levels of Barkhausen noise. With decreasing compressive stress, the Barkhausen noise intensity is gradually increased. As the stress changes from compression to zero stress and further to tension, the noise increases simultaneously with an increase in the slope of the hysteresis loop.

Based on the above interaction of stress and magnetic field on the noise, the stress level of a ferromagnetic material can be determined by applying a known alternating magnetizing field and measuring the magnetic noise, also called magnetoelastic signal (MP = magnetoelastic parameter), created. Quantitative stress data can be obtained by calibrating the noise level to the absolute stress value e.g. with a cantilever beam test.

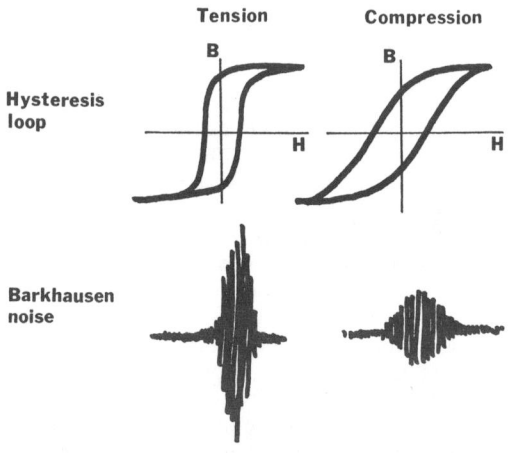

Fig. 3 - Relation between sample induction (hysteresis loop), residual stress and magnetoelastic signal (Barkhausen noise).

Some calibration curves obtained in this fashion for three different materials are shown in Figure 4 (7). It is seen that the level of magnetic signal will vary from material to material. For this reason, different materials must be calibrated separately.

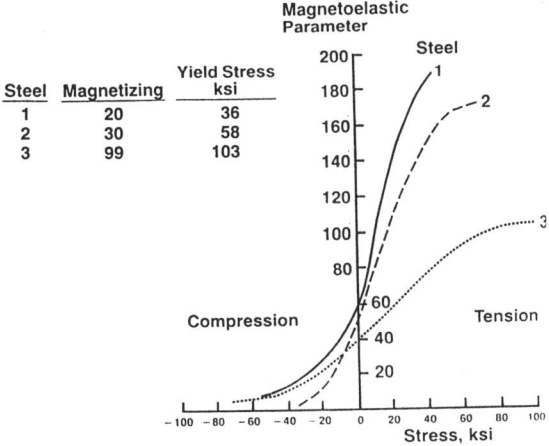

Fig. 4 - Curves for three different materials used to convert the level of Barkhausen noise (magnetoelastic parameter) to stress in ksi (7).

Once established, the calibration curves can be used to convert the Barkhausen noise data into absolute stress. This conversion takes place either manually or automatically by microprocessor controlled equipment.

The shape of the calibration curve is also affected by the measurement parameters such as the strength of the magnetizing field, as shown in Figure 5. Saturation of the calibration curve can be observed for both tensile and compressive stresses, if the magnetizing field is either too low or too high. It is therefore imperative to find the optimal value of magnetizing for the material under study. Well-established procedures are available to properly adjust the magnetizing values.

EFFECT OF MICROSTRUCTURE - The effect of microstructure on Barkhausen noise is in a simplified manner described in Figure 6.

For high hardness materials (quenched, quenched and tempered, carburized, nitrided parts) with fat hysteresis loops, low levels of Barkhausen noise are generated. When the hardness of these materials decreases because of overtempering, for instance, hysteresis loops will become slim and tall, and high levels of Barkhausen noise can be detected. There is, in fact, a continuous increase in this magnetic noise with a decrease in hardness, as can be observed in Figure 7 (9). Below HRC 57.5 high increases in Barkhausen noise were realized in this application, whereas above HRC 57.5 much less change was observed.

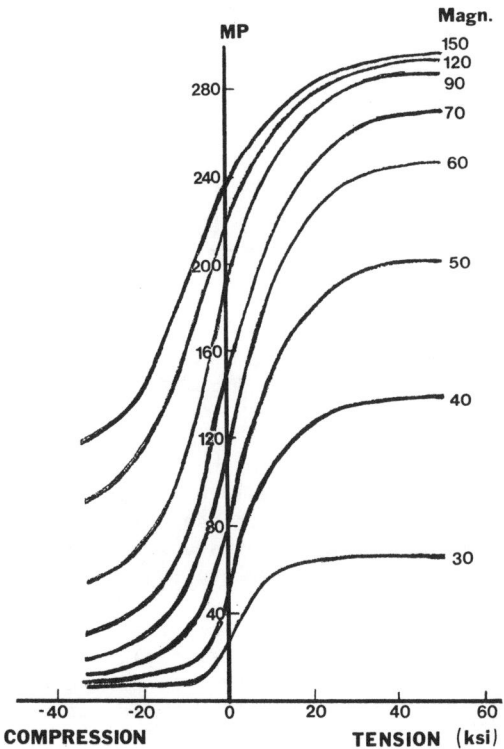

Fig. 5 - Calibration curves for steel 1022 in a stress relieved condition at magnetizing values of 30 to 150 (relative scale). Stresscan 500C.

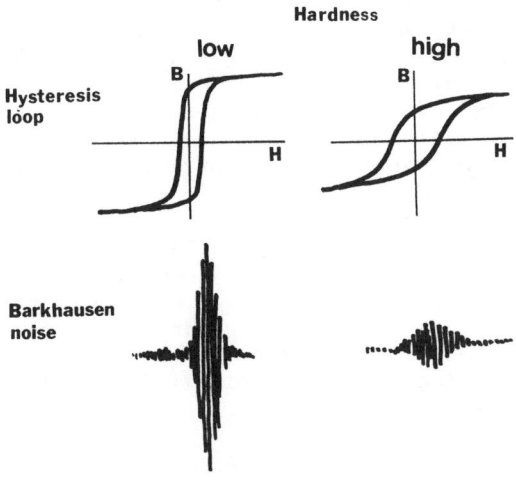

Fig. 6 - Relation between sample induction (hysteresis loop), hardness and Barkhausen noise.

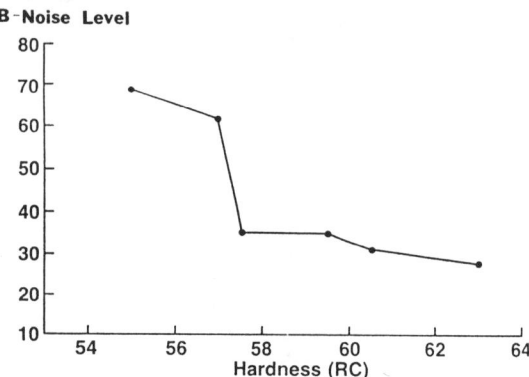

Fig. 7 - Effect of hardness on the Barkhausen noise level on gears. The change in hardness was obtained by tempering case-hardened gears at different temperature (9).

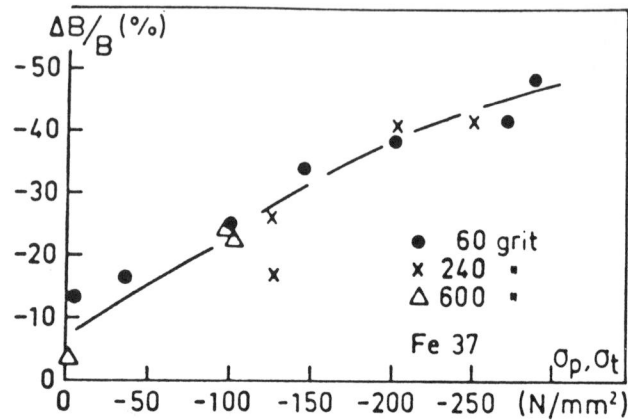

Fig. 8 - Relation between Barkhausen noise data and stress obtained by x-ray diffraction on ground surfaces of steel Fe 37. Uniaxial stress field. Karjalainen et al. (11).

It can be observed in Figures 3 and 6 that the highest Barkhausen noise amplitudes are obtained at field strengths corresponding the coercivity force, which in turn is affected differently by the stress and microstructural features. This offers a means to separate the stress and microstructural contributions by notifying the Barkhausen noise signal as a function of the field strength around the hysteresis loop (10). If there is no shift in the noise location, no microstructural changes are involved. If, on the other hand, new Barkhausen noise maxima are found at a different field value, microstructural constituents with a different coercivity are present.

On the basis of above, the following conclusions on the effect of stress and microstructure (hardness) on the Barkhausen noise level can be made: 1. A change from high compressive stress to zero stress and further to high tensile stress systematically increases Barkhausen noise. Some materials may exhibit saturation close to the yield limits. Saturation can also be found under too high or too low strengths of magnetizing field. 2. A decrease in hardness of hardened parts is accompanied by an increase in Barkhausen noise. 3. The stress and microstructural contributions to Barkhausen noise can be separated by observing the location of maximum noise as a function of the magnetic field.

COMPARISON TO CONVENTIONAL TECHNIQUES - Karjalainen et al. (11) have conducted a thorough investigation into the relation between Barkhausen noise technique and conventional x-ray diffraction method to determine stress. Two different steels were ground by various grinding papers to introduce uniaxial stress fields and evaluated by both methods, see Figure 8 (11). It is obvious that qualitatively the correlation is very good.

The situation is more complicated when biaxial or multiaxial stress fields are present, since Barkhausen noise is influenced by the transverse stress component, see Figure 9 (11). The Barkhausen noise data plotted against x-ray diffraction results in such cases fall in fact on two lines, one for parallel magnetizing (PM), the other for transverse magnetizing (TM). A correlation is then required to account for the perpendicular effects. Taking the transverse contributions into account, the authors were able to obtain a good correlation between the two techniques also in this case, Figure 9b.

Figure 10 (12) exhibits a correlation between Barkhausen noise data and data from hole drilling technique obtained on a cast iron roll.

It should be noted that good correlations between the various techniques can be obtained only if the stress fields are uniform. In the presence of high stress gradients obvious discrepancies are observable, since the area and depth to be evaluated are different for the different techniques.

DESCRIPTION OF INSTRUMENTATION

To use Barkhausen noise to stress and material defect evaluations, a controlled ac magnetic field is applied to the material. The Barkhausen noise created by the material under the ac external field is detected and calibrated to give information on stress and material defects.

The systems include a sensor to excite and detect the magnetic noise, and a central unit to control the sensor and process the Barkhausen noise signal. The sensor includes three pole pieces: see Figure 11, two outer ones are used to apply the ac field and one in the middle to detect the Barkhausen noise. Many designs of

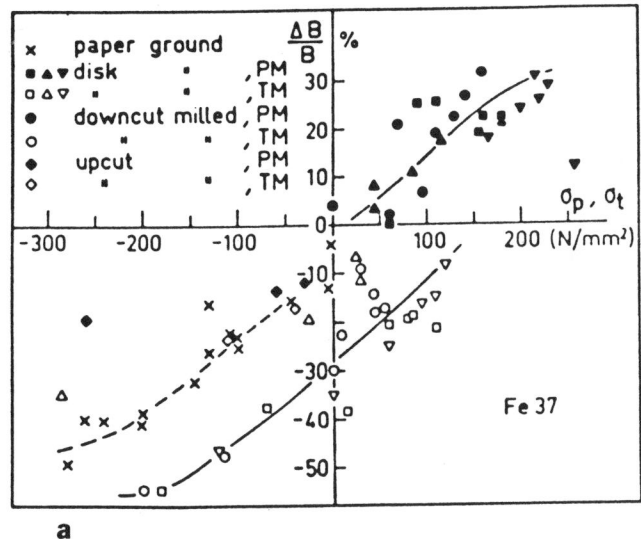
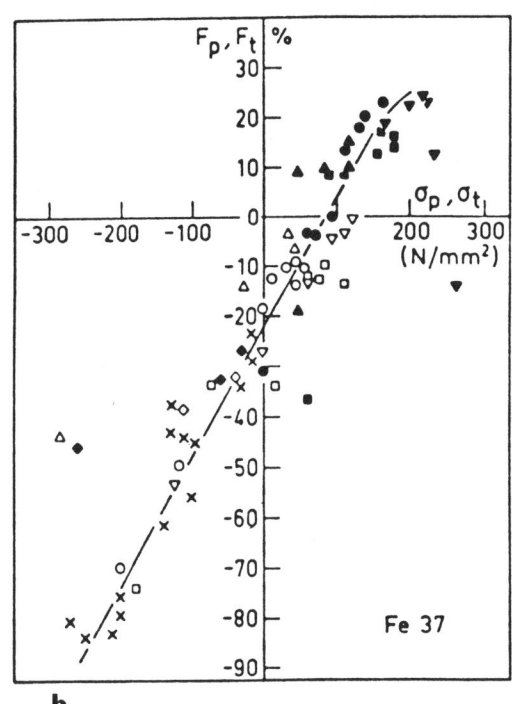

Fig. 9 - Correlation between Barkhausen noise data and stress obtained by x-ray diffraction on ground surfaces of steel Fe 37. Biaxial stress field. Karjalainen et al. (11).

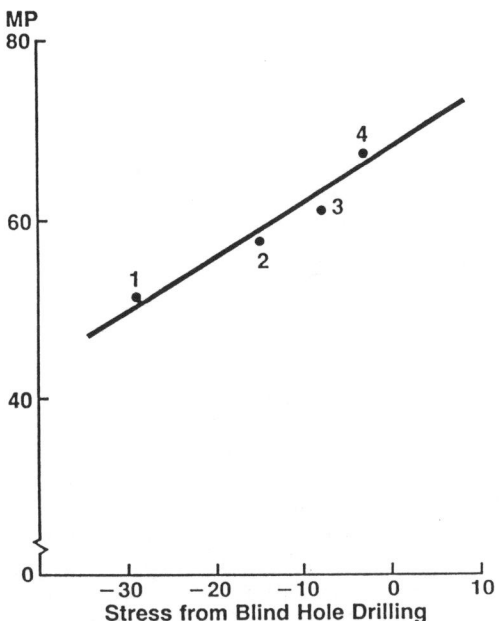

Fig. 10 - Calibration between blind hole drilling and magneto-elastic stress measurement methods. Indefinite chill DP NiHard roll.

the pole pieces are possible depending on the size and shape of the part to be tested. It should be noted that the evaluations are directional due to the alignment of the sensor pole pieces, which is very important especially in stress testing.

The central unit automatically evaluates the Barkhausen noise and displays the measurement result. In stress evaluations, the stress is expressed in terms of magnetoelastic parameter number (MP) which is a relative value of stress, or directly in MPa or Ksi.

When evaluating heat treat defects and/or grinding burns, the difference in Barkhausen noise between a good and a defective part is first assessed and rejection limits are established. The Barkhausen noise levels from unknown parts to be inspected are then compared to the reference data, and any parts showing unacceptable levels of noise are rejected. This procedure can take place either manually or automatically.

The depth of evaluation is dependent upon the frequency range of the Barkhausen noise analyzed, and can hence be varied. Due to damping effects, the feasible depth of evaluation can range up to approximately 0.2 mm.

Both static and dynamic evaluations are possible. In dynamic evaluations, speeds up to 5 m/s can be tolerated. In such tests, a controlled air gap is maintained between the part and the sensing head.

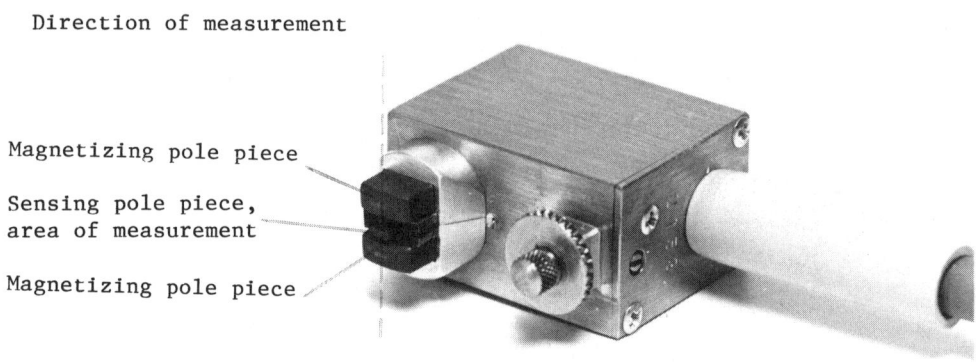

Fig. 11 - General purpose sensor.

EVALUATION EXAMPLES

Typical examples for Barkhausen noise technique are the evaluation of residual stress and microstructural defects.

HEAT TREAT DEFECTS - In carburizing, nitriding and other hardening processes such as induction hardening, the surface of the part is hardened to typically above HRC 58. Simultaneously, high compressive stresses are created on the surface.

Occasionally some areas may be left unhardened so that soft spots or edges are generated. While the hardened areas have martensitic microstructure, the unhardened spots or edges are more ferritic and have lower hardness. Since the unhardened areas take less volume than the surrounding hardened surface, the residual stress in these two areas will be different and tensile stresses may be generated. Both decrease in hardness and in compressive stress will increase the level of Barkhausen noise in the area of heat treat defects.

Similar effects will be realized in local decarburizing/denitriding or overtempering.

Comparison of the Barkhausen noise level generated by different heat treat defects is given in Figure 12. It is clear that unhardened areas give high Barkhausen noise levels compared to the ground material, which in this case is quenched and tempered. Retempering, decarburizing and overtempering also generate noise levels higher than those from the sound material but less than from the unhardened areas. It is noted that the level of this noise increases with the severity of the defect. In nondestructive testing, Barkhausen noise method can thus be used to assess the severity of the defect.

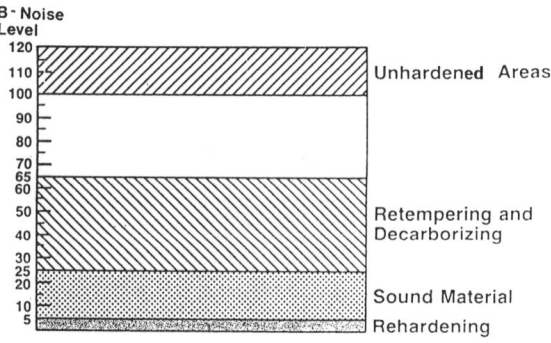

Fig. 12 - Effect of various heat treat defects on the level of Barkhausen noise on case hardened parts.

Example 1. - Noncarburized areas in carburizing of automotive U-joint trunnions, Figure 13. Properly carburized areas, having high hardness and high compressive stress, yielded low Barkhausen noise valves, areas which have been left uncarburized exhibited 5-7 times higher signal levels.

Example 2. - Unhardened (soft) edges of valve lobes of camshafts in induction hardening, Figure 14. The edges left soft due to the misalignment of coils in induction hardening generated about ten times higher signals than the properly hardened valve lobes.

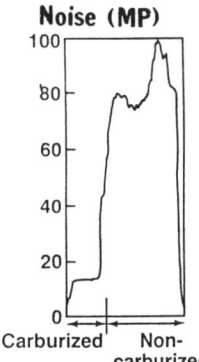

Fig. 13 - Effect of defective carburizing on the level of Barkhausen noise.

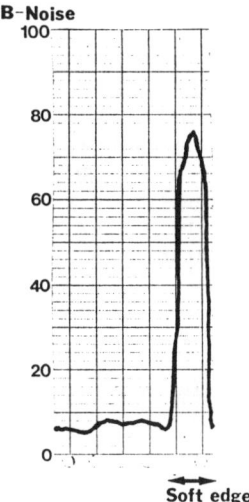

Fig. 14 - Effect of soft edge on Barkhausen noise from an induction hardened camshaft valve lobe.

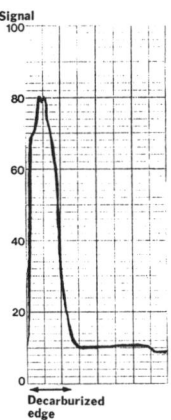

Fig. 15 - Effect of decarburized edge on Barkhausen noise from a carburized bearing race.

case-carburized or nitrided parts, the amount of heat and plastic deformation can be so high that defects called grinding burns are created.

There are two different types of grinding damages: retempering burn and rehardening burn. In retempering type of burn, the tempered surface of the part to be ground is exposed to temperatures which will retemper, overtemper, the surface, creating shallow surface defects with decreased hardness very often below the requirements in specifications. Also, since overtempered material takes less volume than the sound tempered surface, the high compressive stresses are reduced locally and possibly turned into tensile stress. The decrease in hardness and compressive stress will increase Barkhausen noise in the grinding burn area as compared to the sound surface.

In rehardening type of burn the surface temperature in abusive grinding has exceeded the transformation temperature and the grinding coolant has quenched the surface back to fresh, brittle martensite. Often these burns are associated with cracks. The level of Barkhausen noise from rehardened burns is comparable to that from the quenched structure and less than obtainable from quenched, tempered and properly ground surface.

Example 1. - Grinding burn on valve lobes of camshafts, Figure 16. Erratic and high signal levels were obtained from grinding burns on camshaft valve lobes, whereas sound lobes yielded low signals. Rejection limit was set at signal level of 40, meaning that the valve lobe in (a) would be rejected and the lobe in (b) accepted.

Example 2. - Grinding burns on outer diameter of a split ball bearing, Figure 17. A high Barkhausen noise level was associated with a grinding burn measuring approximately 4 mm in diameter on outer diameter of a split ball bearing.

Example 3. - Decarburized edges on bearing races, Figure 15. Up to eight times higher signals were obtained from edges which had experienced decarburizing than from the sound hardened surface.

GRINDING BURNS - Since as-hardened parts are usually brittle, they are tempered before final grinding in order to gain toughness. In tempering both hardness and compressive stress will decrease slightly, which means a somewhat higher Barkhausen noise when compared to the as-heat treated part. Most grinding operations generate heat and plastic deformation in the surface of the part being ground. In abusive grinding of induction hardened and

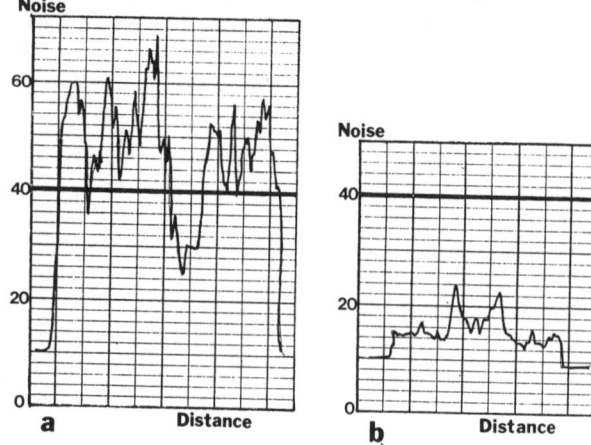

Fig. 16 - Effect of grinding burns on Barkhausen noise from steel camshaft valve lobes.

(a) Heavy grinding burns generating signals above rejection limit.

(b) Sound lobe, signals below rejection limit.

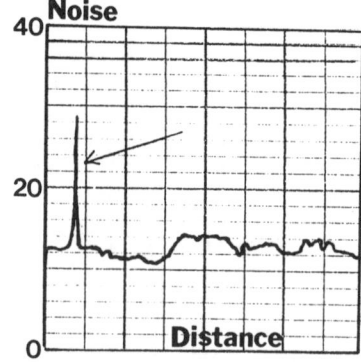

Fig. 17 - Effect of grinding burn (arrow) on Barkhausen noise from outer diameter of a split ball bearing.

Example 3. - Grinding burns on gear tooth flanks, Figure 18. Each tooth flank was evaluated separately, and the Barkhausen noise was recorded and plotted against tooth number. Close to five times higher signals were obtained from teeth number 8 to 16 with grinding burns than from sound teeth.

PREDICTING FATIGUE - Experimental tests have shown that properly hardened, tempered and ground parts have longer service life than parts with heat treat defects or grinding burns. There are two obvious reasons: (i) microstructural softening and (ii) decrease in beneficial compressive stress associated with these defects.

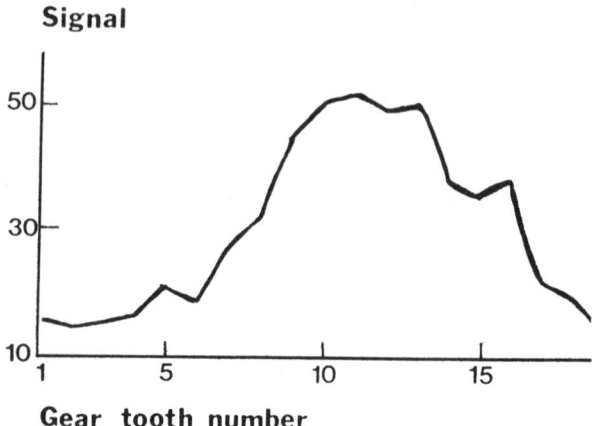

Fig. 18 - Effect of abusive grinding on Barkhausen noise from gear tooth flanks. Gear teeth number 8 to 16 have grinding burns.

Figure 19 gives results of fatigue testing under constand stress of hardened parts. The variation of MP values (Barkhausen noise) was caused by abusive grinding which changes the stress and microstructure of the surface. The time required for rupture was recorded and plotted against initial MP values ranging from 40 to 154. It is clear that there is a good correlation: the higher the MP number, the shorter the time to rupture. On hardened parts, the Barkhausen noise technique can thus be used to predict life expectancy under fatigue loading.

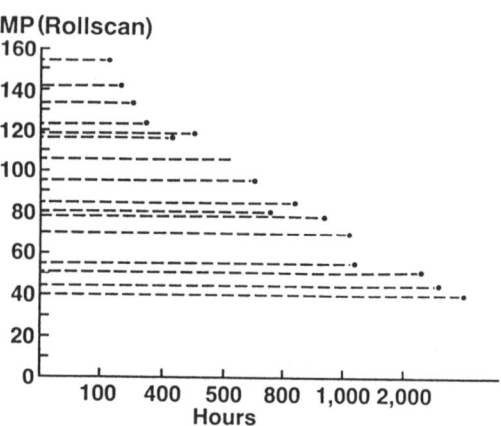

Fig. 19 - Correlation between MP number (Barkhausen noise) and time to rupture in a fatigue test.

DETECTING FATIGUE SOFTENING AND HARDENING - According to Karjalainen et al. (13-15), Barkhausen noise technique can be utilized to give information on the softening and hardening processes during fatigue. Figure 20 exhibits the essential results of their fatigue data on a mild steel with a yield strength of 280 N/mm^2 and a fatigue limit of 200 N/mm^2: The change of Barkhausen noise as a function of cycles depends whether the stress amplitude is below or above the fatigue limit. Below fatigue limit, the Barkhausen noise remains practically unchanged, whereas high changes are observable with stress above 200 N/mm^2. It is interesting to note that differencies can be observed as early as after the first stroke. To explain the above data, the authors compared the curves to fatigue softening and hardening, and came to the conclusion that Barkhausen noise seems to offer an efficient method for detecting softening and hardening processes during bending fatigue.

The above data implies the feasibility of using Barkhausen noise technique to predict the fatique behavior after the few initial strokes. Also, the technique might be used to follow from time to time the material changes due to fatigue on actual components and thus provide a means to prevent catastrophic failures.

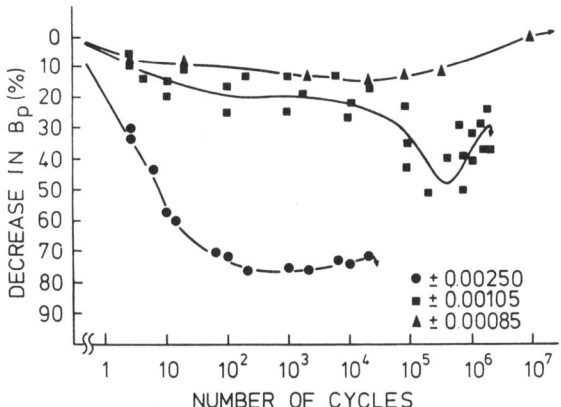

Fig. 20 - Barkhausen noise data against number of cycles from a fatigue test conducted at different strain amplitudes: (i) 0.00085 (170 N/mm^2), (ii) 0.00105 (210 N/mm^2), (iii) 0.00250 (330 N/mm^2).

CONCLUSIONS

Since Barkhausen noise is influenced by the state of stress and microstructure, it is important to note the following:

1. Barkhausen noise can be used to evaluate the state of residual stress, when microstructure is controlled.

2. If residual stress is controlled, Barkhausen noise gives valuable information on hardness.

3. The microstructural and stress contributions to Barkhausen noise can be separated by evaluating the maximum Barkhausen noise amplitude against field strength.

4. Barkhausen noise can be used to detect material defects which include a simultaneous change in stress and microstructure, such as heat treat defects and grinding burns.

5. Since fatigue is affected by residual stress and microstructure, Barkhausen noise can be used to predict life expectancy under fatigue loading.

6. Barkhausen noise offers a means to follow the softening and hardening processes due to fatigue.

REFERENCES

(1) Barkhausen, H., Phys. Zeitschrift 20, 401-403, (1919)

(2) Tiitto, S., Acta Pol. Scand., Ph 119, 19 (1977)

(3) Gerlach, H. and P. Lertes, Phys. Zeitschrift 22, 568 (1921)

(4) Zschiesche, K. Phys. Zeitschrift 23, 201 (1922)

(5) Pashley, R.L., Materials Evaluation 28, No. 7, 157-161 (1970)

(6) Tiitto, S., Schweissen und Schneiden 32, No. 11, 449-452 (1980)

(7) Tiitto, K., "Solving internal stress measurement problems by a new magnetoelastic method", Nondestructive methods for materials property determination, ASM, pp. 105-114 (1984)

(8) Buttle, D.J., Briggs, G.A.D., Jakubovics, J.P., Little, E.A. and C.B. Scruby, Phil. Trans. R. Soc. Lond. A320, 363-378 (1986)

(9) Tiitto, K. and R. Pro, "Detection of heat treat defects and grinding burns by measurement of Barkhausen noise", 2nd International Symposium on the Nondestructive Characterization of Materials, Toronto (1986)

(10) Theiner, W.A., Altpeter, I. and R. Kern, "Untersuchungen von Gefugeparametern mit zerstorungsfreien magnetischen und magnetoelastischen Verfahren", p. 52-61, Sonderbande der Praktischen Metallographie, Dr. Riederer-Verlag GmbH, Stuttgart (1985)

(11) Karjalainen, L.P. and R. Rautionaho, "Detection of machining stresses by the Barkhausen noise method", Quality Test I, Pittsburgh (1982)

(12) Tiitto, K., "Measuring stresses in rolls by magnetoelastic method", Mechanical

Working and Steel Processing XXIII. The Iron and Steel Society, Warrendale (1986)

(13) Karjalainen, L.P. and M. Moilanen, NDT International <u>12</u>, No. 2 51-57 (1979)

(14) Karjalainen, L.P., Moilanen M. and R. Rautioaho, Materials Evaluation <u>37</u>, No. 9, 45-51 (1979)

(15) Karjalainen, L.P. and M. Moilanen, IEEE Transactions on Magnetics <u>MAG-16</u>, No. 3, 514-517 (1980)

BENDING STRESS RELAXATION OF AISI 1095 STEEL STRIP

U.P. Sinha, D.W. Levinson
Department of Civil Engineering, Mechanics and Metallurgy, University of Illinois at Chicago
Chicago, Illinois USA

ABSTRACT

The objective of the present research is to investigate the effects of variable initial applied stress, temperature and heat treatment on the stress relaxation of AISI 1095 spring steel strip under static bending stress, and to evaluate the effects of these variables on the kinetics of stress relaxation.

The test method employed was the one which is described in ASTM Standard Recommended Practice for Stress Relaxation Tests for Materials and Structures, Part C (E328-78). The method measures elastic springback upon unloading at the end of the test interval. A cylindrical mandrel was used for stressing the strip by wrapping the strip around the mandrel and holding at temperature for specified times. Stress relaxation tests were carried out on spring steel strips of 0.127 mm (0.005 in.) thickness.

Test data were obtained for both mill received steel strip and heat treated (normalized) steel strip which had 0.2% offset yield strengths of 1620 MPa (235 ksi) and 669 MPa (97 ksi), respectively. Stress relaxation data for the mill received material were obtained at temperatures of 100, 175, 250°C over a time period of 1000 hours, and 23°C up to 10,000 hours. Test data for heat treated material were obtained at these same temperatures over a time period of 1000 hours.

An individual sample was used for each test period. At each test temperature four different levels of initial stresses were applied to individual samples. These initial stresses were 21, 32, 43, and 64% of the yield stress for the mill received samples, and were 52 and 77% of the yield stress for the normalized material. Additional normalized samples were subjected to initial stresses above the yield stress.

Experimental data indicate that a high yield strength had a beneficial effect in lowering the stress relaxation rate as compared to a lower yield strength temper in this material. The higher resistance to stress relaxation can be attributed to higher activation energy required for the higher yield strength material. Rapid stress relaxation of the steel subjected to initial applied stress in excess of the yield strength may be due to strain recovery in the plastic region.

IN GENERAL, STRESS RELAXATION [1]* is defined as a phenomenon where time and temperature dependent reduction in stress level takes place in an elastically strained solid member. This reduction in stress involves a gradual transformation of elastic strain into inelastic strain. This inelastic strain [2] is due to anelasticity, plasticity, microplasticity and creep. The strength and relaxation of a solid are interrelated because relaxation data can provide a direct measure of the ability of a solid to resist deformation with time. The relation between stress relaxation and creep has long been recognized. The performance of materials in a number of types of stressed condition is affected by stress relaxation. Stress relaxation data are useful for the development of stress relief heat treatments to control the residual stress level and also for providing design data for such mechanical elements as bolts, gaskets, joints and springs. For such applications, a stress level above a minimum allowable working stress must be maintained during service life. The elastic strain is used to maintain a desired stress level. The most familiar example is the use of bolts to counteract the load tending to force apart two members

---

*The number in the brackets refers to the list of references appended to this paper.

such as at a flanged joint. The load-deflection characteristics of springs can also be altered by stress relaxation during service.

In this paper, we report: (1) comprehensive experimental data obtained during stress relaxation of AISI 1095 spring steel strip under various static bending stress (elastic) levels and temperatures (between room temperature and 0.2 T_m where T_m is the melting temperature of this steel in $°K$), (2) effect of a normalizing heat treatment on the rate of stress relaxation and finally show that (3) higher resistance to stress relaxation of high yield strength steel is due to a higher activation energy requirement in the steel of higher yield strength, and (4) for some test results where initial applied stress was in excess of yield strength, the rapid stress relaxation may be due to strain recovery in the plastically deformed region.

MATERIALS AND SPECIMEN

Materials tested were 0.127 mm (0.005 in.) thick strip of mill-received AISI 1095 spring steel and the same mill-received steel strip austenitized at 820°C for 2 min. in a lead bath and cooled in air. The chemical composition of the steel strip is given in Table 1.

Table 1 - Chemical Composition of Steel Strip

Element content (% by weight)

Material	C	Mn	Si	P	S	N
Mill Received	1.00	0.38	0.21	0.018	0.006	0.006

No change in the composition of steel was observed after heat treatment. The 0.2% offset tensile yield strengths of the mill received and heat treated steel in the strip rolling direction were 1620 MPa (235 ksi) and 669 MPa (97 ksi), respectively. The corresponding values of tensile strengths were 1916.8 MPa (278 ksi) and 972.2 MPa (141 ksi), respectively. The test specimens were cut 57.15 mm (2.25 in.) long by 7.11 mm (0.28 in.) wide so that the stress relaxation test could be carried out on the sample deformed parallel to the strip rolling direction.

TEST METHOD

The test specimens were wrapped around a cylindrical mandrel made of steel and held there with a suitable clamp. The mounted specimens were exposed to the test temperatures of 23, 100, 175 and 250°C. An electrically heated furnace was used at temperatures higher than room temperature. At 175 and 250°C a neutral salt bath was used to avoid any decarburization of the test specimens during their exposure. An individual specimen was used at each test temperature and stress relaxation period. Four different mandrels of O.D. 76.2 mm (3 in.), 50.8 mm (2 in.), 38.1 mm (1.5 in.) and 25.4 mm (1 in.) were used at each of the above mentioned test temperatures.

The initial bending stress to be applied on the test specimen was calculated by using elastic flexure equation given in ASTM Recommended Practice [1] for Stress Relaxation Test for Materials and Structure Part (E328-78), paragraph 32.9.1. The elastic flexure equation is given by

$$\sigma = \frac{Mc}{I} \tag{1}$$

where

σ = nominal flexure stress at outer fiber
M = bending moment
c = distance from outer fiber to centroidal axis of the cross section being investigated
I = moment of the inertia about the centroidal axis of the cross section being investigated

For rectangular cross section

$$\frac{c}{I} = \frac{6}{bt^2} \tag{2}$$

where

b = width of the strip
t = thickness of the strip

Combining Eqs. (1) and (2), we get

$$\sigma = \frac{6}{bt^2} \tag{3}$$

The bending moment can be calculated directly from the bend radius using elastic theory [3] with the equation given by

$$M = \frac{bt^3 E}{12}\left(\frac{1}{R_o} - \frac{1}{R}\right) \tag{4}$$

where

M = bending moment
b = strip width
t = strip thickness
E = modulus of elasticity
R_o = mandrel radius
R = strip radius of curvature after stress relaxation (t > 0)

Combining Eqs. (3) and (4) we get the expression, for the stress remaining (residual stress) in the specimen after stress relaxation, given by

$$\sigma = \frac{Et}{2}\left(\frac{1}{R_o} - \frac{1}{R}\right) \tag{5}$$

At time $t = 0$, the strip radius of curvature $R = \infty$, under this condition Eq. (5) reduces to

$$\sigma = \sigma_0 = \frac{Et}{2}\frac{1}{R_0} \qquad (6)$$

where

σ_0 = initial bending stress applied on the specimen

E = Young's modulus of steel = 206.8×10^4 MPa (30×10^6 psi at 23 C)

The initial strain [1] on the specimen was taken as equal to the distance from the neutral axis of the bending to the outer fiber, divided by the radius of curvature of the outer fiber. For the test specimen, the initial strain was calculated by the following (Eq. (7))

$$\text{Strain } \varepsilon_0 = \frac{t}{2R_0 + t} \qquad (7)$$

where

t = specimen thickness
R_0 = radius of the concave surface of the specimen.

The initial calculated values of the stress and strain on the specimens are given in Table 2.

Table 2 - The Calculated Initial Stress and Strain on the Specimen

Material	Mandrel Size O.D. mm (in.)	Initial Stress Eq. (6) MPa (ksi)	Initial Strain Using Eq. (7)	Initial Stress 0.2% Offset Y.S. %
Mill Received	76.2 (3)	344.8 (50)	0.0020	21
	50.8 (2)	517.1 (75)	0.0025	32
	38.1 (1.5)	689.5 (100)	0.0033	43
	25.4 (1)	1034.3 (150)	0.0050	64
Heat Treated	76.2 (3)	344.8 (50)	0.0020	52
	50.8 (2)	517.1 (75)	0.0025	77
	38.1 (1.5)	689.5 (100)	0.0033	103

The schematic diagram of the fixture designed for the purpose of strip curvature measurement is shown in Fig. 1. The strip curvature radius, R, was calculated by using the following Eq. (8)

$$R = \frac{L^2 + h^2}{2h} \qquad (8)$$

where

$2L$ = chord length
h = chord height

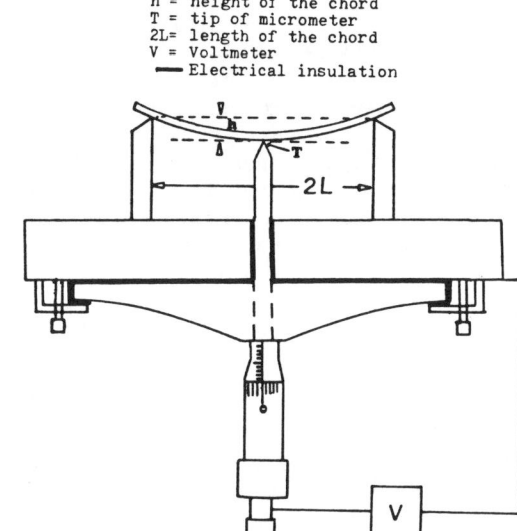

Fig. 1 Schematic diagram of the fixture for the measure of strip curvature

While turning the stem of the micrometer, contact between the tip, T, of the micrometer with the convex surface of the specimen was indicated by the deflection of the needle of suitably connected voltmeter (see Fig. 1). By this device, the minimum value of the chord height up to 0.0025 mm (0.0005 in.) could be reproducibly determined. The loss in stress after stress relaxation was calculated by subtracting Eq. (5) from Eq. (6). The resulting expression is given by

$$\Delta\sigma = \sigma_0 - \sigma = \frac{Et}{2R} \qquad (9)$$

Tests were run on the mill received strips at temperatures of 100, 175 and 250°C over a time period of 1000 hours and at 23°C up to 10,000 hrs. Test data were obtained at these same temperatures over a time period of 1000 hrs for the normalized steel strips. The values of the initial stresses applied to the specimens are shown in Table 2. The specimens stress relaxed at temperatures higher than room temperatures were cooled to room temperature, removed from the mandrel and radii of curvature were measured within 30 minutes of unloading.

The value of the modulus of the steel at room temperature was used for calculating the remaining stress or loss in initial stress on the strip after stress relaxation at elevated temperatures. There is a less than 2 percent difference in the values of the modulus [4] in the temperature range of the present investigation.

RESULTS

The loss in stress for mill-received steel strip during stress relaxation at 23, 100, 175 and 250°C for various specified times and stress levels was calculated by using Eq. (9). The calculated values of loss in stress expressed as percentage of initial stress is defined as % $(\sigma_o - \sigma)/\sigma_o$ where, σ_o = initial applied stress and σ = stress remaining (residual stress) in the strip after stress relaxation at specified test temperature and time interval. The calculated values of loss in stress expressed as percentage of initial stress are given in Tables 3 to 6. The calculated values of $\%[(\sigma_o - \sigma)/\sigma_o]$ when plotted against the stress relaxation time intervals on a log-log scale exhibit a linear relationship and are shown in Figs. 2 to 5. The values of the slopes (B) and intercepts (A) of the curves (lines) are summarized in Table 7.

Similarly, the values of loss in stress expressed as percentage of initial stress (as defined above), for the heat treated steel strips were calculated by using Eq. (9). These calculated values are reported in Tables 8 to 11. These calculated values of loss in stress expressed as percentage of initial stress were plotted against stress relaxation time intervals on a log-log scale and are shown in Figs. 6 to 9. The ** symbol shown in the columns of Tables 9 to 11 indicate the stress relaxation time intervals beyond which stress relaxation curves (Figs. 6 to 9) apparently become asymptotic to the time axis. The values of slopes (B) and intercepts (A) for the heat treated strips were calculated by considering only the linear portion of the curves shown in Figs. 6 to 9. The slopes (B) and intercepts (A) of all the curves were determined by linear regression analysis of the calculated test data.

The calculated values of the stress relaxation parameter (B) pertaining to different test conditions, for both mill received steel strips (Table 7) and heat treated strips (Table 12) are plotted against the reciprocal of the test temperature (°K) and are shown in Figs. 10 and 11, respectively. The same parameters (B) of these materials are also plotted against the initial stress (applied on the specimen) expressed as percentage of 0.2% offset yield strength of these steels and shown in Fig. 12. The calculated values of % (loss in stress/initial stress) for different test temperature and 1000 hours of stress relaxation period are plotted against % (initial stress/0.2% offset y.s.) and are shown in Fig. 13. In the same Fig. 13, the above mentioned calculated values for mill received steel strips stress relaxed at 23°C and 10,000 hrs are shown for the purpose of comparison.

The average activation energy of the process (100-250°C), for the mill received steel strip was obtained by employing the following Arrhenius type equation

$$\frac{1}{t_f} = Ce - (Q/RT) \qquad (10)$$

where

t_f = time required for completion of given fraction of the process during stress relaxation
C = a constant
Q = average activation energy for the process
R = universal gas constant and
T = absolute temperature K

The values of the stress relaxation parameters given in Table 7 were employed in Eq. (11) to calculate, for example, time required for 1 and 5% loss in the initial stress (σ_o = 50 ksi) applied to the mill received steel strip. Figure 14 shows a typical Arrhenius type plot between log of $(1/t_f)$ and $1/T$ (reciprocal of absolute test temperature). The slopes of these lines, Q/2R, then defines the average activation energy. From the slopes of the lines, the average activation energies (100-250°C) for the process were determined. Similarly, the activation energies for 1 and 5% loss in the initial applied stresses of 517 MPa (75 ksi), 690 MPa (100 ksi) and 1034 MPa (150 ksi) were evaluated.

Table 3 – Stress Relaxation in Mill Received Steel Strip at 23°C

Calculated Values of percentage $\dfrac{\text{Initial Stress }(\sigma_o) - \text{Remaining Stress }(\sigma)}{\text{Initial Stress}}$

Initial Stress % 0.2% Offset Y.S.	Stress Relaxation Times (hrs)					
	1000	2000	3000	5000	7000	10000
21	---	---	0.58	0.86	1.30	1.75
32	0.30	0.51	0.76	1.11	1.62	2.20
43	0.35	0.58	0.84	1.41	1.80	2.40
64	0.40	0.76	1.12	1.42	2.20	2.72

Table 4 – Stress Relaxation in Mill Received Steel Strips at 100°C

Values of loss in stress (calculated by using Eq. (9)) expressed as percentage of initial stress = [(initial stress, σ_o – remaining stress, σ)/(initial stress, σ_o)]

Time (hrs)	Initial Stress % / 0.2% Offset Y.S.			
	21	32	43	64
1	0.44	1.20	1.50	7.78
3	0.60	1.85	2.10	9.34
5	0.84	2.32	3.00	10.37
10	1.00	3.00	3.60	11.20
25	1.44	3.75	4.80	13.00
50	1.80	4.20	5.70	15.60
100	2.04	5.80	6.20	18.85
300	3.40	6.91	8.60	22.60
500	4.00	9.40	10.50	24.80
1000	5.00	11.90	12.74	27.20

Table 5 – Stress Relaxation in Mill Received Steel Strips at 175°C

Values of loss in stress (calculated by using Eq. (9)) expressed as percentage of initial stress = [(initial stress, σ_o – remaining stress, σ)/(initial stress, σ_o)]

Time (hrs)	Initial Stress % / 0.2% Offset Y.S.			
	21	32	43	64
1	1.32	3.40	3.89	8.18
3	1.80	4.00	4.98	10.77
5	2.20	4.80	5.99	11.80
10	2.51	5.50	6.50	13.60
25	3.34	6.60	7.50	15.42
50	4.00	7.80	9.60	16.80
100	4.50	9.00	10.50	20.00
300	6.00	11.00	13.10	22.30
500	7.20	12.40	14.82	25.11
1000	8.70	14.30	16.75	28.32

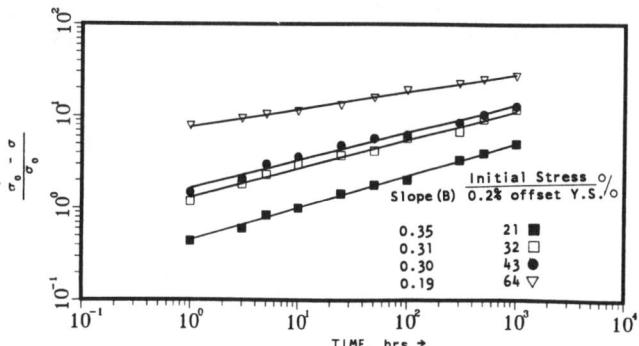

Fig. 3 Stress relaxation in a mill received steel strip at 100°C (212°F)

Table 6 – Stress Relaxation in Mill Received Steel Strips at 250°C

Values of loss in stress (calculated by using Eq. (9)) expressed as percentage of initial stress = [(initial stress, σ_o – remaining stress, σ)/(initial stress, σ)]

Time (hrs)	Initial Stress % / 0.2% Offset Y.S.			
	21	32	43	64
1	5.09	8.80	9.52	13.93
3	6.30	10.80	11.70	16.50
5	7.50	11.60	12.90	18.25
10	8.70	13.42	13.60	19.70
25	11.30	16.20	16.74	22.12
50	12.00	18.50	19.30	24.23
100	13.20	21.00	22.16	29.15
300	16.80	25.00	25.80	31.72
500	18.90	27.80	28.85	33.88
1000	23.50	31.20	32.68	37.38

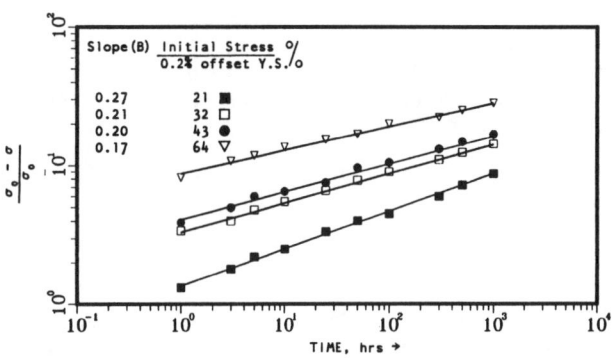

Fig. 4 Stress relaxation in a mill received steel strip at 175°C (347°F)

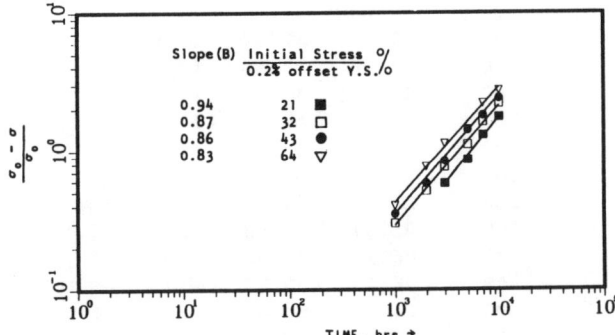

Fig. 2 Stress relaxation in a mill received steel strip at room temperature, 23°C (73°F)

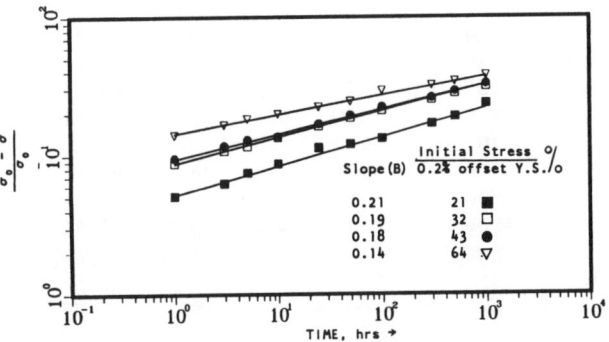

Fig. 5 Stress relaxation in a mill received steel strip at 250°C (482°F)

Table 7 - Stress Relaxation Parameters in Mill Received Steel Strips

Calculated values of slopes (B) and intercepts (A) of the curves (lines) shown in Figs. 2-5.

Initial Stress MPa (ksi)	Test Temperature C (F)	Log A	B	Test Durations (hrs)
344.75 (50)	23 (73)	-3.51	0.94	10000
	100 (212)	-0.35	0.35	1000
	175 (347)	0.13	0.27	1000
	250 (482)	0.71	0.21	1000
517.00 (75)	23 (73)	-3.14	0.87	10000
	100 (212)	0.12	0.31	1000
	175 (347)	0.52	0.21	1000
	250 (482)	0.94	0.19	1000
689.50 (100)	23 (73)	-3.04	0.86	10000
	100 (212)	0.22	0.30	1000
	175 (347)	0.61	0.20	1000
	250 (482)	0.97	0.18	1000
1034.25 (150)	23 (73)	-2.86	0.83	10000
	100 (212)	0.88	0.19	1000
	175 (347)	0.94	0.17	1000
	250 (482)	1.15	0.14	1000

Table 8 - Stress Relaxation in Heat Treated Steel Strips at 23°C

Values of loss in stress (calculated by using Eq. (9)) expressed as percentage of initial stress = [(initial stress, σ_o - remaining stress, σ)/initial stress, σ_o)]

	Initial Stress % 0.2% Offset Y.S.			
Time (hrs)	52	77	103	155
1	--	0.58	14.72	41.10
3	--	1.00	17.50	42.00
5	--	1.42	18.12	42.5
10	--	2.00	19.00	43.16
25	--	3.10	21.50	44.00
50	1.20	4.25	23.00	44.70
100	1.60	5.67	25.72	46.10
200	2.52	ND	ND	ND
300	3.66	7.43	26.50	46.40
500	5.80	10.30	29.15	47.20
1000	10.20	13.20	32.14	48.12

ND = Not Determined

Table 9 - Stress Relaxation in Heat Treated Steel Strips at 100°C

Values of loss in stress (calculated by using Eq. (9)) expressed as percentage of initial stress = [(initial stress, σ_o - remaining stress, σ)/(initial stress, σ_o)]

	Initial Stress % 0.2% Offset Y.S.		
Time (hrs)	52	77	103
1	1.80	9.50	22.50
3	2.50	15.80	28.00
5	3.20	18.10	38.00
10	3.50	21.50	40.00
25	4.20	25.30	43.65
50	5.62	28.23	46.77
100	6.46	33.11	53.70
200	8.22	33.88	54.30
300	9.10	40.74	56.23**
400	9.54	45.54	56.81
500	10.10	46.77	57.00
600	10.82	47.00**	57.00
700	11.31	47.50	57.00
800	12.10	47.50	57.00
900	12.50	47.50	57.00
1000	13.00	47.50	57.00

Table 10 - Stress Relaxation in Heat Treated Steel Strips at 175°C

Values of loss in stress (calculated by using Eq. (9)) expressed as percentage of initial stress = [(initial stress, σ_o - remaining stress, σ)/(initial stress, σ_o)]

	Initial Stress % 0.2% Offset Y.S.		
Time (hrs)	52	77	103
1	4.75	20.00	47.35
3	6.50	22.90	49.98
5	7.40	24.40	52.00
10	8.30	26.60	55.10
25	10.81	29.85	57.70
50	12.20	32.54	59.01
100	15.00	35.48	61.17
200	18.41	36.68	62.17**
300	20.60	42.10	62.40
400	22.00	45.12	62.40
500	23.50	48.00	62.40
600	24.39	49.20**	62.40
700	25.37	50.00	62.40
800	26.25**	50.00	ND
900	26.25	50.00	ND
1000	26.25	50.40	ND

Table 11 – Stress Relaxation in Heat Treated Steel Strips at 250°C

Values of loss in stress (calculated by using Eq. (9)) expressed as percentage of initial stress = [(initial stress, σ_o − remaining stress, σ)/(initial stress, σ_o)]

	Initial Stress %		
	0.2% Offset Y.S.		
Time (hrs)	52	77	103
1	25.75	38.43	55.00
3	31.11	41.54	59.30
5	32.90	43.00	61.13
10	34.70	44.56	66.00
25	35.63	48.00	66.56
50	37.66	51.00	67.00**
100	40.60	52.20	67.00
200	42.08	53.00	68.10
300	43.00	53.60**	68.10
400	43.90**	53.60	ND
500	43.90	53.60	ND
600	45.00	54.00	ND
700	45.00	ND	ND

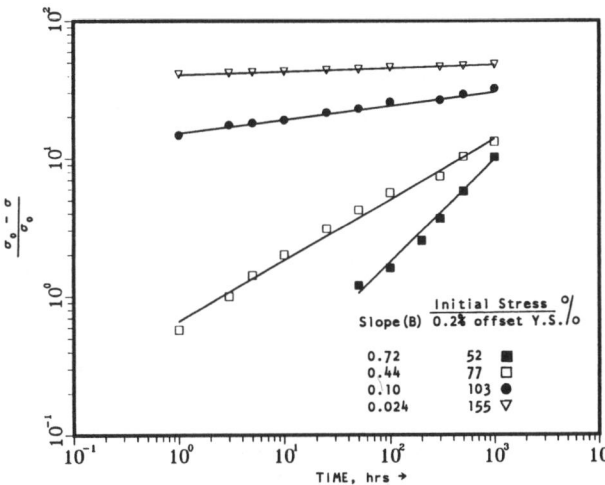

Fig. 6 Stress relaxation in a normalized steel strip at room temperature 23°C (73°F)

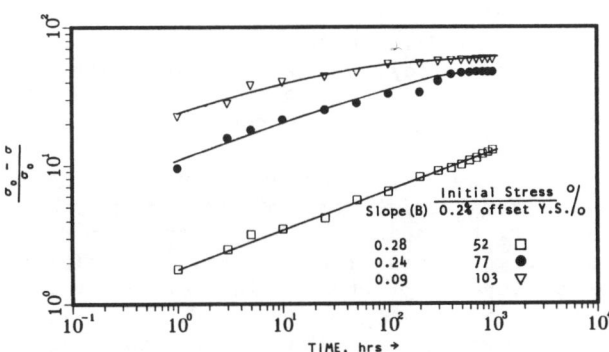

Fig. 7 Stress relaxation in a normalized steel strip at 100°C (212°F)

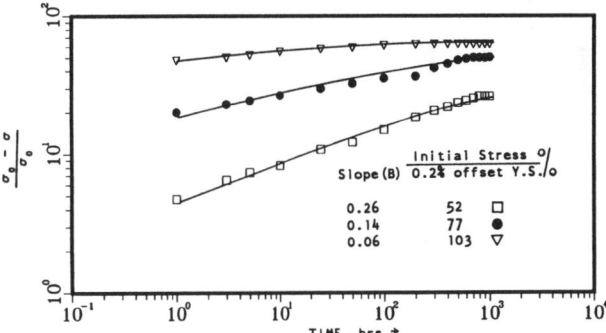

Fig. 8 Stress relaxation in a normalized steel strip at 175°C (347°F)

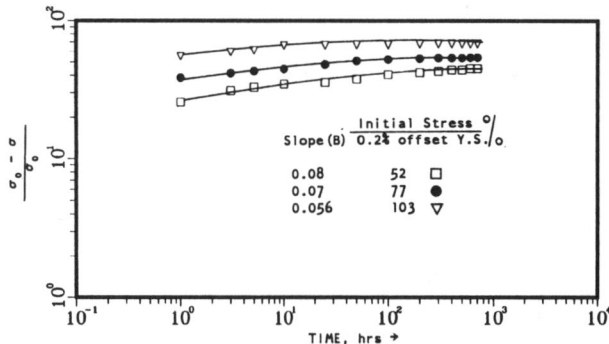

Fig. 9 Stress relaxation in a normalized steel strip at 250°C (482°F)

Table 12 – Stress Relaxation Parameters in Heat Treated Steel Strips

Calculated values of slopes (B) and intercepts (A) of the curves (lines) shown in Figs. 6–9.

Initial Stress MPa (ksi)	Test Temperature C (F)	Log A	B	Test Durations (hrs)
344.75 (50)	23 (73)	−1.20	0.72	1000
	100 (212)	0.26	0.28	1000
	175 (347)	0.89	0.26	1000
	250 (482)	1.44	0.08	1000
517.00 (75)	23 (73)	−0.18	0.44	1000
	100 (212)	1.04	0.24	1000
	175 (347)	1.29	0.14	1000
	250 (482)	1.58	0.07	1000
689.50 (100)	23 (73)	1.18	0.10	1000
	100 (212)	1.38	0.09	1000
	175 (347)	1.67	0.06	1000
	250 (482)	1.74	0.056	1000
1034.25 (150)	23 (73)	1.61	0.024	1000

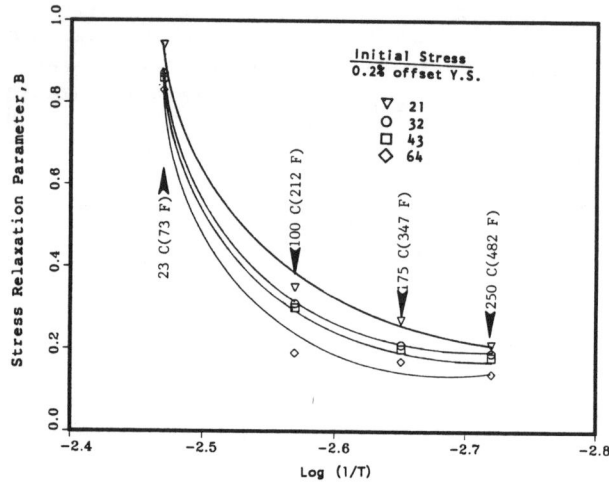

Fig. 10 Variation of stress relaxation parameter (B) in a mill received steel strip with initial applied stress and reciprocal of absolute test temperature

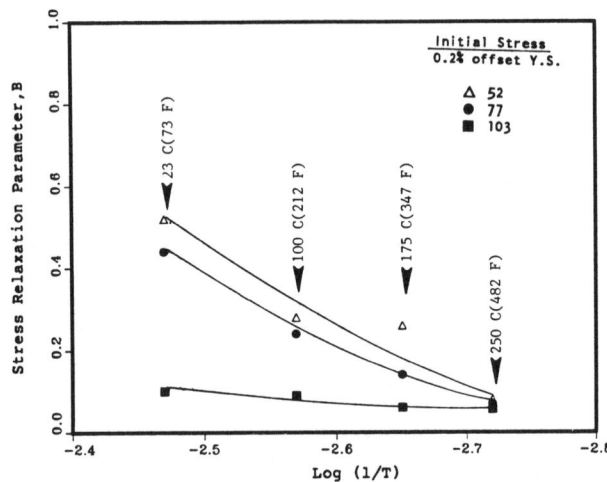

Fig. 11 Variation of stress relaxation parameter (B) in a normalized steel strip with initial applied stress and reciprocal of absolute test temperature

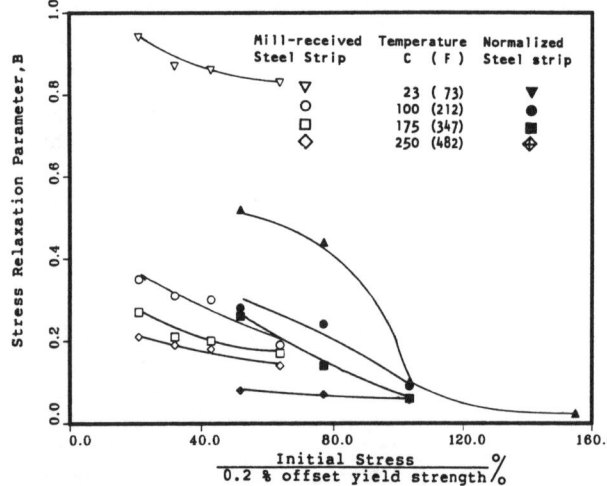

Fig. 12 Comparison of stress relaxation parameters (B) between mill received and a normalized steel strip

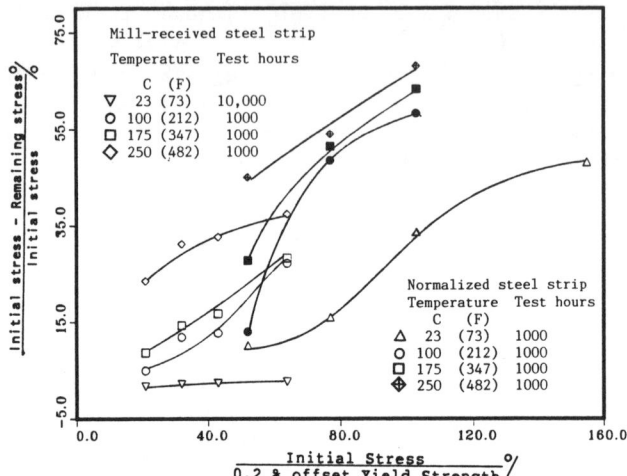

Fig. 13 Effect of the ratio of (Initial Stress/Y.S.) on the ratio of (Loss in Stress/Initial Stress) at various test temperatures

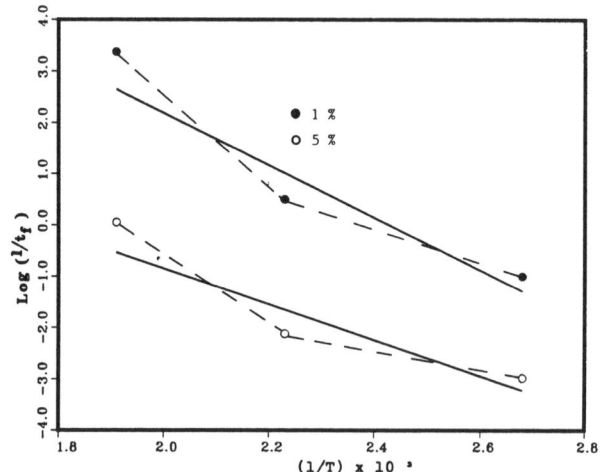

Fig. 14 A typical Arrhenius type plot for stress relaxation in a mill received steel strip at an initial stress 345 MPa (50 ksi)

DISCUSSION

It is well-known that stress relaxation in metals and alloys is a thermally activated deformation process [6-22]. The rate of this deformation process is a function of temperature, stress, strain rate and structure of the metals. During the stress relaxation thermal fluctuations assist the applied stress only in overcoming the resistance to plastic flow due to short-range stress field and not the resistance due to long-range stress field in the metals. As the temperature is increased, the resolved shear stress, τ, on the slip planes necessary to overcome the resistance to plastic deformation is decreased, until this shear stress, τ, reaches a limiting value, τ_μ, which represents the maximum value of the opposing long-range internal stress field. Thus the applied stress is considered [22] to consist of two stress components: (1) an athermal component, τ_μ (2) a thermal component τ^* (also called effective stress). Many experimental observations have shown that τ_μ can change with temperature through the change in shear modulus of the metals and τ^* depends upon both temperature T and shear strain γ. Thus

$$\tau = \tau^*(T,\gamma) + \tau_\mu$$

It is seen in Figs. 2 through 9 that at a given temperature, the higher the initial stress applied to the strip, the greater the stress relaxation during a given time interval. Similarly, it is also observed from these figures that at a given initial applied stress, the higher the temperature the more rapid the stress relaxation during a given time interval. The extent of these stress relaxation rates is manifested by the values of stress relaxation parameters (B) given in Tables 7 (for mill received steel) and 8 (for the same steel in a normalized state). The variation of this parameter (B) with the reciprocal of absolute temperature is shown graphically in Figs. 10 and 11, respectively. The linear relationship between the logarithm of the loss in stress expressed as a percentage of the initial stress (as defined earlier) and the log of stress relaxation time can be represented by

$$\log \frac{\sigma_o - \sigma}{\sigma_o} = A + B \log t \qquad (11)$$

where

σ = initial stress
σ_o = remaining stress at any time
t = stress relaxation time interval
B = slope of stress relaxation curve
A = intercept of stress-relaxation curve at log t = 0 (1 hr).

The above mentioned relationship has also been reported in other materials [5] when stress relaxation is carried out over a long period of time. It can be observed from Tables 7 and 12 and corresponding stress relaxation curves in Figs. 2 to 9 that the smaller the values of B, the faster the stress relaxation. It is also observed from Figs. 7 to 9 that in the normalized steel a fully stress relaxed condition is possible under a suitable combination of initial stress and temperature. This 'fully relaxed flow stress' can be achieved when the flow stress after a particular stress relaxation interval is nearly balanced by the opposing internal long-range stress field. This difference in the stress relaxation behavior of this steel in two conditions (mill received and normalized) may be due to differences in the stability of their structures against the plastic deformation. Figure 12 shows how differences in the stress relaxation resistance of this steel in two conditions can be represented when the slopes of stress relaxation curves, represented by B, are plotted against the ratios of initial stress and their respective 0.2 percent yield strength values. This figure shows that for a given ratio of the above mentioned parameters, the larger the values of B, the more resistant the steel under a given stress relaxation condition. Figure 13 shows how the stress relaxation strength of the mill received steel strip changes with the temperature for a given ratio of an applied initial stress to the yield stress.

Examination of Fig. 14 shows that stress relaxation in this steel follow an Arrhenius-type behavior over the temperature range of investigation (100 to 250°C). However, the data points also indicate the change in the slope of the curves shown by dotted lines. This may indicate that two different mechanisms may be operative in this temperature range. It has been observed in this steel [23] that the first stage of tempering ceases to operate at about 175°C and the second stage of tempering follows thereafter. This indicates that different stages of tempering in this steel can also be reflected by their stress relaxation behavior in this temperature range. A more detailed study of stress relaxation behavior in this temperature range may explain the existence of different slopes of the plot shown in shown in Fig. 14. It follows from Fig. 15 that average activation energy of the process determined over the temperature range (100-250°C) first increases with initial stress applied to specimen, reaches a maximum value and then decreases slowly. In Fig. 15 data of Grachev, et al., [24] are also included with the data for the variation of average activation energies with initial stress. They studied the stress relaxation behavior of quenched high carbon steel at various temperatures up to 100°C. They have attributed the decrease in the activation energy after reaching a maximum value due to passing from elastic to elastoplastic strain. In the present study, although the initial bending stress applied to the specimen was within the elastic limit, it is quite possible that decrease in activation energy after a certain stress level may be due to the actual flow stress value exceeding the elastic limit (it is to be noted that yield stress decreases with increase of temperature).

It is difficult to compare the stress relaxation behavior of this steel due to unavailability of any published data in the literature on this steel. However, a comparison of stress relaxation characteristics of other grades of high carbon steel [25] with the present one at room temperature is given in Fig. 16. In Fig. 16, the 'values of loss in stress expressed as percentage of initial stress' extrapolated to 10,000 hrs based on their experiments conducted up to 170 hrs, are shown for the purpose of comparison. The importance of the 'ratio of initial stress to 0.2 percent offset yield strength' can be realized from Fig. 16, when this ratio is used to characterize the stress relaxation behavior of the steel. It is also clear from this figure that high carbon steels of different grades but comparable yield strength can exhibit similar stress relaxation behavior.

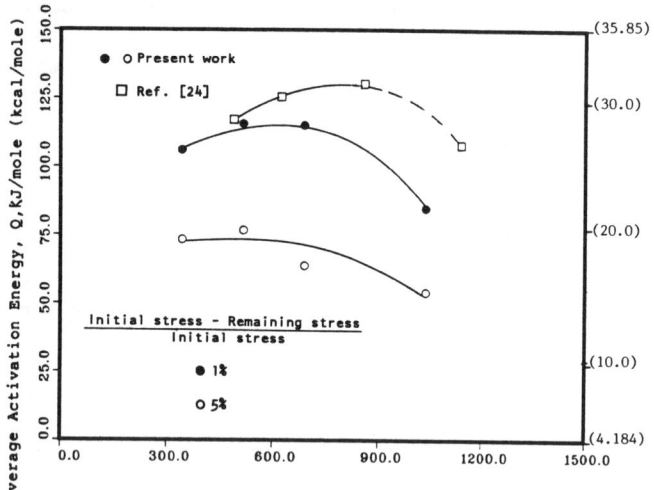

Fig. 15 Effect of an initial applied stress on the variation of average activation energy in a mill received steel strip

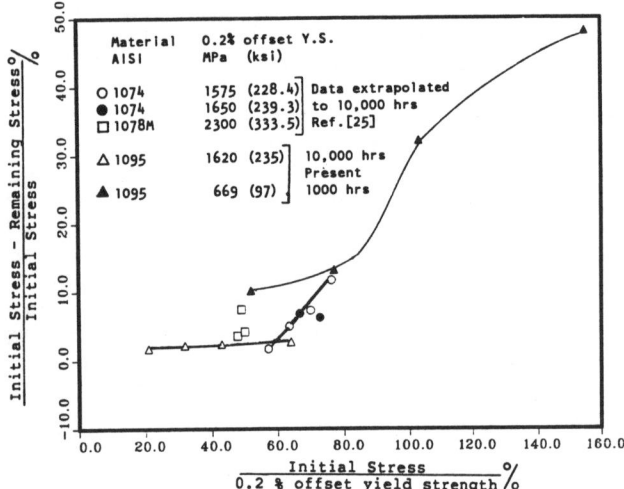

Fig. 16 Comparison of stress relaxation strength of different grades of high carbon spring steels at room temperature

CONCLUSIONS

Based on the present investigation and available experimental data on the other grades of high carbon steels, the following conclusions can can be drawn:

1. The higher yield strength is beneficial in lowering the stress relaxation rate of the steel.
2. In low yield strength steel, a rapid stress relaxation is observed when initial stress applied to the steel is in the plastic range.
3. The average activation energy of the steel increases with increase in applied stress in the elastic region and decreases when the applied stresss exceeds the elastic limit.
4. The different grades of spring steels with comparable yield strengths can exhibit the same stress relaxation behavior provided the 'ratio of initial stress to 0.2 percent yield strength' is the same. A further study on the different grades of high carbon steels, other than the one employed in the present investigation, for longer time interval is required.

REFERENCES

1. ASTM Standard Recommended Practice for Stress Relaxation Tests for Materials and Structures (E-328), ASTM Annual Book of Standards, Part 10.
2. Manjoine, M. J. and H. R. Voorhees, 'Compilation of Stress Relaxation Data for Engineering Alloys,' ASTM Data Series Publication DS 60, (1982).
3. Austen, A. R. and W. Taylor, 'Stress Relaxation in Beryllium Copper and Beryllium Nickel Springs,' in Proceeding, 1971 Fourth Annual Annual Connector Symposium, Electronic Connector Study Group, Inc., Cherry Hill, N.J., Oct. (1971).
4. Carlson, H., 'Spring Designer's Handbook,' p. 312, Marcel Dekker, Inc., N.Y. (1978).
5. Fox, A., J. Testing and Evaluation, Vol. 8, $\underline{3}$, pp. 119-26, May (1980).
6. Seeger, A., Z. Naturforsch, $\underline{9a}$, pp. 758, 819, 856 (1954).
7. Conrad, H., Journal of Metals, pp. 582-88, July (1964).
8. Conrad, H. and W. Hayes, Trans. ASM, Vol. 56, pp. 125-34, 229-62, (1963
9. Conrad, H., Canadian Journal of Physics, Vol. 45, pp. 581-90, (1967).
10. Conrad, H., JISI, pp. 364-75, (1961).
11. Becker, R., Z.Physik, Vol. 26, p. 919, (1925).
12. Wilson, J.F. and F. Garafalo, Materials Research and Standard, pp. 85-92, Jan. (1966).
13. Wilson, J. F. and N. K. Wilson, Trans. of Soc. of Rheology, Vol. 10 $\underline{1}$, pp. 399-418, (1966). 14. Krausz, A. S. and H. Eyring, 'Deformation Kinetics,' John Wiley and Sons, Inc., New York (1975).
15. Conway, J. B., R. H. Stentz and J. T. Berling, 'Fatigue, Tensile, Relaxation Behavior of Stainless Steels,' pp. 228-63, Technical Information Center, U.S. Atomic Energy Commission (1975).
16. Freudenthal, A. M., Proc. ASTM, pp.986-1000, (1960).
17. Feltham, P., Phil. Mag., $\underline{6}$, pp. 847-50, (1961).

18. Feltham, P., Journal of Institute of Metals, Vol. 89, pp. 210-14 (1960-610.
19. Fox, A., edited, 'Stress Relaxation Testing,' ASTM Special Technical Publication, p. 676 (1979).
20. Rohde, R. W. and T. V. Nordstron, Mater. Sc. and Eng., 12, pp. 179-85, (1973).
21. Gupta, I. and J. C. M. Li, Met. Trans., Vol. 1, pp. 2323-30, Aug. (1970).
22. Conrad, H., Mater. Sc. and Eng., 6, pp. 265-73, (1970).
23. Antia, D. P., S. G. Fletcher and M. Cohen, Trans. ASM, Vol. 32, pp. 290-332, (1944).
24. Grachev, S. V. and V. YA. Zubov, Fiz. metal. metalloved, Vol. 15, 6, pp. 854-59, (1963).
25. Idermark, S. U. V. and E. R. Johansson, in 'Stress Relaxation Testing,' A. Fox edited, pp. 61-77, ASTM Special Technical Publication 676.

SELECTING QUENCHANTS TO MAXIMIZE TENSILE PROPERTIES AND MINIMIZE DISTORTION IN ALUMINUM PARTS

Charles E. Bates
Southern Research Institute
Birmingham, Alabama USA

Abstract

Quenching refers to the rapid cooling of metal parts from the solution treating temperature which is typically in the range of 465 to 565°C (870 - 1050°F) for aluminum alloys. Several factors including the kind of quenchant, the quenchant use conditions, the section thickness of the part, and the transformation rate of the alloy being quenched determine whether a part can be successfully quenched and then aged to produce the desired strength.

A quench factor, Q, has been devised that interrelates quenching variables (velocity, concentration, temperature effects, etc.) the section size of parts, and transformation data of specific alloys to provide a single number indicating the extent to which a part can be through hardened or strengthened.

The quenchant solution and quenchant operating conditions should be selected to provide the proper quench factor without being excessively severe so as to produce high thermal gradients that can cause distortion.

Cooling curves, interface heat transfer coefficients, quench factors, and thermal gradients produced by a variety of quenchants are presented. The concept of quench factors is then related to the problem of distortion during quenching.

QUENCHING REFERS TO THE RAPID cooling of metal from the solution treating temperature, typically between 465 and 565°C (870 - 1050°F) for aluminum alloys, and is practiced to retain hardening elements and compounds in solid solution. Quenched aluminum parts may then be aged to promote controlled precipitation hardening and strengthening.

A balance must be obtained between the need to quench sufficiently fast to retain most of the hardening elements and compounds in solution and the need to minimize residual stress and distortion in the parts being quenched. Precipitation during quenching can lead to localized overaging, loss of grain boundary corrosion resistance, and in extreme cases, an inadequate response during the age hardening treatment.

Highly agitated cold water is an excellent quenchant in terms of retaining hardening elements and compounds in solution. Unfortunately, the high cooling rates associated with quenching in cold water can produce large differences in temperature between thick and thin sections which can cause localized plastic flow. Plastic flow results in the distortion observed after quenching or during machining.

If distortion is to be minimized, the temperature differences between different areas of a part must be minimized consistent with cooling all regions of the part sufficiently fast to avoid excessive precipitation during quenching. Usually distortion is controlled in aluminum parts by adding polymers to water quenchants to reduce the convective or film coefficient between the part and the water. Reducing the film coefficient retards heat transfer from the part surface and improves the temperature uniformity from surface to center of thick sections and between thick and thin sections.

Aluminum alloys have a relatively high thermal conductivity, between 1.4 and 2.38 W cm/cm^2°K (975-1650 Btu·in./hr ft^2°F), compared to a conductivity of about 0.14 to 0.29 W cm/cm^2°K (100 to 200 Btu·in./hr ft^2°F) for austenite in most carbon and low alloy steels. The high conductivity of aluminum can be both a blessing and a problem. If heat is being rapidly extracted at the part surface by the quenchant, the high conductivity results in rapid temperature losses in thin sections and large temperature differences between thick and thin sections. If heat is being extracted more slowly, the high metal conductivity aids in maintaining temperature uniformity within the part. The practical difficulty lies in establishing just how fast a part of a particular alloy needs to be quenched to retain sufficient hardening elements and compounds in solid solution to obtain an acceptable age hardening reaction while not cooling so fast that plastic deformation occurs that causes distortion of the part.

FACTORS INFLUENCING QUENCH SEVERITY - Several material and quenchant characteristics influence the rate of heat removal from a part being quenched. An infinite quench is one that instantly decreases the skin of the part to the bath temperature. The rate of cooling in the part is then a function only of the diffusivity of the metal, i.e., its ability to diffuse heat from the interior to the surface. In practice, however, quenchants never provide the idealized "infinite" quench.

The cooling rates, in practice, are controlled by the vapor blanket formation, boiling characteristics, and the velocity, temperature, specific heat, heat of vaporization, conductivity, density, viscosity, and wetting characteristics of the quenching fluid. Practically, the cooling rates are controlled by the fluid selected, the type and amount of polymer put in water based quenchants, and the bath temperature and velocity.

Mathematically, heat transfer from parts can be described using Newton's law of cooling:

$$q = hA(T_1 - T_2) \qquad (1)$$

where
  q = rate of heat transfer
  A = surface area of the part in contact with fluid
  T_1 = surface temperature
  T_2 = fluid temperature away from surface
  h = interfacial or film coefficient

If this equation is rearranged, h, the film coefficient, can be defined in terms of the part area, difference in temperature between the part and the quenchant, and the heat being transferred. An analytical determination of the interface coefficient, h, requires that the properties of the fluid moving past the part be examined.

The properties of the quenchant, including boiling temperature, viscosity, density, thermal conductivity, and specific heat combine to make the quenchant an important, if not the most important, variable affecting quench severity. Increases in quenchant velocity generally increase the quench severity. Increasing the bath temperature puts the quenchant nearer its boiling point, decreases the temperature difference between the part and bath, and decreases the quench severity.

The actual heat flow from the interior of a part being quenched to the surface can be described with Fourier's equation:

$$q = k \cdot A \cdot dT/dx \qquad (2)$$

where
  q = amount of heat transferred
  k = thermal conductivity of the alloy
  A = area of the part
  dT/dx = thermal gradient in the part

The expression for heat transfer from a bar, neglecting axial flow is

$$\frac{d^2T}{dr^2} + \frac{1}{r}\frac{dT}{dr} = \frac{1}{\alpha}\frac{dT}{dt} \qquad (3)$$

where
  r = bar radius
  α = thermal diffusivity
  dT/dr = thermal gradient

The thermal conductivity (k) of the part can be related to the film coefficient (h) of the bath using Biot's number in equation (4)

$$\text{Biot's number} = B_i = \frac{hX}{k} \qquad (4)$$

where X is a characteristic size of the part.

A similar ratio, more widely used in quenching, is the Grossman number defined in equation (5):

$$H = h/2k \qquad (5)$$

The Grossman number has been reported to equal approximately one for one-inch sections quenched in still water.

The Grossman number for various experimental conditions evaluated in the current study was determined by first solving Eq. (3) with a finite difference heat transfer program which allowed specific Grossman numbers to be used as input values.

Cooling curves at the center of various-sized bars and plates were calculated and the cooling rate between 425 and 150°C determined. A polynomial least squares fit was then obtained to relate the average cooling rate between 425 and 150°C to the Grossman number. The Grossman number, H, under experimental conditions used in this study, was determined by recording a cooling curve using a thermocouple located in the center of cylindrical (bar) test probes, determining the cooling rate between 425 and 150°C from the cooling curve, putting this value in the polynomial expression relating the cooling rate to the Grossman number, and solving for the H value.

QUENCH FACTOR CONCEPTS

Evancho and Staley introduced a method of interrelating the cooling rate in a part, the section thickness of the part, and the precipitation kinetics of the alloy being solution treated.(1) The general technology for heat treating aluminum alloys using quench factor concepts is described in Vol IV of the 9th Edition Metals Handbook (2) and Aluminum Properties and Physical Metallurgy.(3) The philosophy of using quench factors to evaluate quenchants is reviewed in this paper, and the coefficients defining the C curve for 7075-T73 presented. The concept of quench factors is then related to minimizing distortion during quenching.

There exists for each age hardenable aluminum alloy a C curve or time-temperature curve that describes the precipitation of solute elements and compounds as a function of time and temperature. Long times are usually required for a given amount of solute precipitation at elevated temperatures in the range of about 370 to 425°C because the solute supersaturation is low, and consequently the thermodynamic driving force for precipitate formation is low. At temperatures in the range of 275 - 350°C, however, the undercooling and thermodynamic driving force is high, and the time required to achieve a particular amount of solute precipitation is relatively low-- typically a few seconds in precipitation hardenable aluminum alloys. In this temperature range the diffusion coefficients for the solute species are reasonably high, and precipitates can nucleate and grow in short times. Nucleation and precipitation usually occurs first along grain boundaries because less lattice strain is necessary in grain boundaries and diffusion rates are higher than within grains.

At still lower temperatures, below about 200°C, the time required for a given amount of solute precipitation increases to hundreds or thousands of seconds because solute diffusion coefficients are low. The thermodynamic potential for precipitate formation is high because of the high degree of solute supersaturation, but the rate of precipitate formation is low because of the inability of atoms to move around and react to form the precipitating species.

Evancho and Staley developed an equation of the form

$$C_T = -K_1 * K_2 * \exp(\frac{K_3 * K_4^2}{R*T(K_4 - T)^2}) * \exp(\frac{K_5}{R*T}) \quad (6)$$

to mathematically represent C curves or time temperature curves.(1) The terms in this expression are defined as follows:

C_T = critical time required to precipitate a constant amount of solute (the locus of the critical times is the C-curve),

K_1 = constant which equals the natural logarithm of the fraction untransformed formed during quenching i.e., the fraction defined by the C-curve, typically ln 0.99 or ln 0.995

K_2 = constant related to the reciprocal of the number of nucleation sites,

K_3 = constant related to the energy required to form a nucleus,

K_4 = constant related to the solvus temperature,

K_5 = constant related to the activation energy for diffusion,

R = 8.3143 $J \cdot K^{-1} \cdot mol^{-1}$ (gas constant),

T = temperature °K

Numerical values for each of these constants define the C curve for the particular alloy composition and temper condition.

QUENCH FACTOR CALCULATION - The quench factor, Q, interrelates the cooling curve (time-temperature history) in a part produced by a quenchant and the precipitation kinetics of the alloy which is described by the C_T function. The quench factor can be calculated from time-temperature data and the equation describing the precipitation of a specified amount of solute. In aluminum alloy metallurgy, the amount of solute precipitated is typically chosen as an amount that will cause a loss of 0.5% of the attainable yield strength in the alloy.

Experimentally, the average temperature between data points in the cooling curve is first calculated. The C_T value is then calculated at that temperature using equation (6). The ratio of the time step length, Δt, divided by the C_T value provides the incremental quench factor, q, which is calculated using equation (7)

$$q = \frac{\Delta t}{C_T} \quad (7)$$

where

q = incremental quench factor, and
Δt = time step.

The incremental quench values are progressively summed as the part is cooled through the precipitation range, normally about 425 to 150°C in aluminum alloys, to produce the cumulative quench factor, Q according to equation (8):

$$Q = \sum_{150}^{425} q \quad (8)$$

The integration or numerical summation process is schematically illustrated in Figure 1. Published coefficients for the C_T function for 7075-T6, 7050-T76, and 2024-T851 are given in Table I.

The cumulative quench factor reflects the heat extraction characteristics of the quenchant (as a function of velocity, polymer concentration, temperature and other quenchant variables) over the precipitation range of the alloy, as well

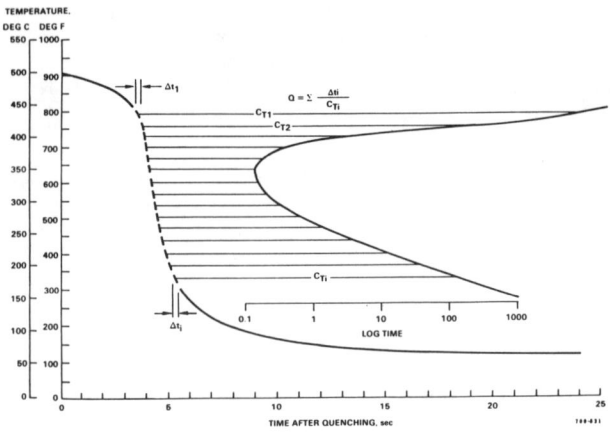

Figure 1. Schematic illustration of the method for calculating a quench factor.

as the section thickness of the part and the precipitation kinetics of the alloy.

An alloy with a low rate of precipitation will produce a lower Q value at a given cooling rate compared to an alloy with a high precipitation rate. Quench factors calculated for different alloys might be quite different even if similar section sizes are cooled in the same quenchant since quench factors take into account individual alloy precipitation kinetics by means of the equation describing the C curve for each alloy. This method of describing quench severity is different from that used with Grossman numbers (H) which are related solely to the ability of a quenchant to extract heat and not to the transformation kinetics of the alloy being heat treated.

The quench factor provided by a particular quenchant can be determined using parts or probes instrumented with thermocouples, a quenchant system in which the concentration, temperature and velocity can be controlled, and a data acquisition system for recording the temperature in instrumented parts or probes as a function of time after quenching. In operation, a part or probe is solution treated at the proper temperature for the alloy and quenched into a bath with the quenchant being evaluated at the desired velocity and temperature. Cooling curves are recorded and the quench factors calculated using the mathematical expression for the C_T function for the alloy.

The tensile strength of the alloy after a proper temper cycle can be predicted using equation (9).(1)

$$\sigma_y = \sigma_{max} e^{K_1 Q} \qquad (9)$$

where

σ_y = predicted yield strength
σ_{max} = yield strength after an infinite quench (and temper cycle)
e = base of the natural logarithm
K_1 = ln (0.995) = -0.00501
Q = quench factor

Low values of Q are associated with high quench rates, minimum precipitation during cooling and with high yield strengths. Conversely, higher Q values are obtained with slower quench rates and lower strength values. Solute elements and compounds are precipitated during cooling from the solution treating temperature at high Q values. As a consequence, an improperly quenched alloy may not properly harden during aging, and it may be susceptible to intergranular corrosion, stress corrosion or exfoliation.

The relationship between quench factor and yield strength for 7075-T73 aluminum extrusions is illustrated in Figure 2.(5) Forty five tensile specimens were removed from 5 lots of bar extrusions ranging in diameter from 12.7 to 127 mm (0.5 to 3.0 inches). Sections of the bar stock were solution treated at 465°C and quenched in water ranging in temperature between 25 and 100°C, a 20% polyalkalylene glycol (UCON A) solution, a 25% solution of polyvinyl pyrrolidone 60, and a 25% solution of polyvinyl pyrrolidone 90, fast oil, and a fluidized sand bed.

The standard deviation in yield strength about the regression line was 2.1 MPa (0.3 ksi) and ± 3 standard deviations was 11.0 MPa (1.6 ksi). This implies that the yield strength in a part can be predicted rather accurately if the C_T expression for the alloy being quenched is known and if a valid cooling curve is available. The coefficients describing the C curve for 7075-T73 are given in Table I.

Since there is a predictable functional relationship between the quench factor and the yield strength of alloys where the C curve is available, it is possible to select an upper limit value of Q, above which specified yield strength values cannot be statistically met. Data relating the quench factor to predicted yield strength in 7075-T73 is given in Table II.

Table 3.7.3.0 from Military Handbook V, reproduced in part in Table III, specifies minimum longitudinal yield strength "A" and "B" design allowable values of

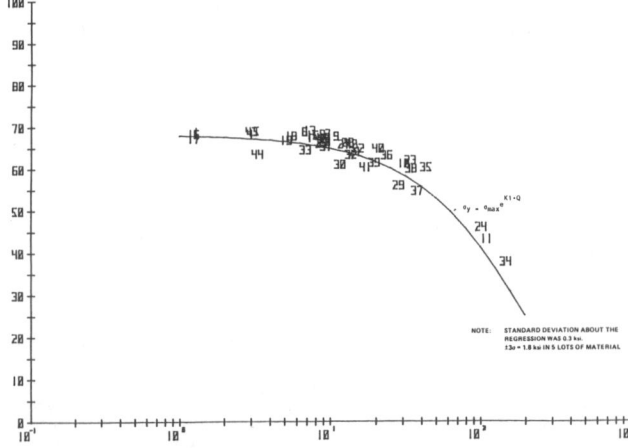

FIGURE 2. YIELD STRENGTH IN 7075-T73 ALUMINUM PLOTTED AS A FUNCTION OF QUENCH FACTOR, (5 LOTS OF MATERIAL)

Table I

Coefficients for Calculating C_T at 99.5% of Yield Strength

Alloy	σm	K_1	K_2	K_3	K_4	K_5	Calculation Range °C	Reference
7050-T76	--	-0.00501	2.2x10^{-19}	5190	850	1.8x10^{+5}	425 - 150	
7075-T6	70.42	-0.00501	4.1x10^{-13}	1050	780	1.4x10^{+5}	425 - 150	1
2024-T851	66.6	-0.00501	1.72x10^{-11}	45	750	3.2x10^{+4}	445 - 150	4
7075-T73	68.9	-0.00501	1.37x10^{-13}	1069	737	1.37x10^{+5}	425 - 150	

Alloy heat treatment and temper designations:

1. 7050-T76 Solution treated at 880-890°F, quenched, and over aged to obtain exfoliation resistance
2. 7075-T6   Solution treated at 860-880°F, quenched, and aged at 240-260°F for 22-24 hours
3. 2024-T851 Solution treated at 925°F, slack quenched, strained 2.25-2.5%, aged at 375°F for 12 hours
4. 7075-T73 Solution treated at 860-880°F (460-471°C), quenched and aged at 215-235°F (100-112°C) for 6-8 hours and 340-360°F (170-182°C) for 8-10 hours.

413.7 MPa (60 ksi) and 434.4 MPa (63 ksi) respectively in 19 to 36.8 mm (0.750 to 1.449 inch) thick extruded rods, bars and shapes.(6) The data in Table II indicates that to meet the A allowables, the quench factor in 7075-T73 must not exceed 28.0, and to meet the B allowables, the quench factor must not exceed 18.0. These values represent upper limit values. In order to provide some safety margin and accommodate other variables in the quench system such as racking differences, a quenchant capable of providing somewhat lower quench factors in the section thicknesses being heat treated must be used in the quench tank.

QUENCHANT SELECTION - Now that an upper limit quench factor value can be defined above which properties in parts cannot be expected to meet specified minimum values, the question becomes how to select quenchants that will provide appropriate quench factors in section thicknesses of interest. Quenchant and quenchant operating conditions can be evaluated by instrumenting a variety of probes, solution treating each one, quenching the probes into quenchant solutions of interest, recording the cooling curves, and calculating film coefficients, Grossman numbers and quench factors. In the current study, 7075 probes were solution treated at 465°C and quenched in a variety of solutions under controlled conditions.

TABLE II

Relationship Between Quench Factor and Yield Strength in 7075-T73 Aluminum

Quench Factor "Q"	% Attainable Yield Strength	Predicted Yield Strength (MPa)	Predicted Yield Strength (ksi)
0.0	100.0	475.1	68.9
2.0	99.0	470.2	68.2
4.0	98.0	465.4	67.5
6.0	97.0	461.3	66.9
8.0	96.1	456.5	66.2
10.0	95.1	451.6	65.5
12.0	94.2	447.5	64.9
14.0	93.2	442.7	64.2
16.0	92.3	438.5	63.6
18.0	91.4	434.4	63.0
20.0	90.5	429.6	62.3
22.0	89.6	425.4	61.7
24.0	88.7	421.3	61.1
26.0	87.8	417.2	60.5
28.0	86.9	413.0	59.9
30.0	86.0	408.9	59.3
32.0	85.2	404.7	58.7
34.0	84.3	400.6	58.1
36.0	83.5	396.5	57.5
38.0	82.7	393.0	57.0
40.0	81.8	388.9	56.4
42.0	81.0	384.7	55.8
44.0	80.2	381.3	55.3
46.0	79.4	377.2	54.7
48.0	78.6	373.7	54.2
50.0	77.8	369.6	53.6

TABLE III

Design Mechanical and Physical Properties of 7075 Aluminum

Specification	QQ-A-200/11											
Form	Extrusion (rod, bars, and shapes)											
Temper	T73, T73510, T73511											
Cross-sectional area, in.²	≤ 20				≤ 25				≥ 20		≤ 32	
Thickness, in.	0.062–0.249		0.250–0.499		0.500–0.749		0.750–1.499		1.500–2.999		3.000–4.499	3.000–4.499
Basis	A	B	A	B	A	B	A	B	A	B	A	B
Mechanical properties:												
F_{tu}, ksi												
L	68	72	70	74	70	73	70	73	69	74	68	71	65	70
LT	66	70	68	72	67	70	66	69	62	67	58	61	56	60
F_{ty}, ksi												
L	58	61	60	63	60	63	60	63	59	65	57	62	55	60
LT	56	59	57	60	57	60	56	58	51	56	46	50	44	48

Military Standardization Handbook, 5D, Metallic Materials and Elements for Aerospace Vehicle Structures, Vol. 1, June 1983.

Table IV, published in 1940, provides Grossman numbers for several quenchants.(7) This table has provided valuable guidance to heat treaters since its publication in spite of the fact that quenchant velocities were not well defined.

Table V provides experimental data on water under a range of velocity and temperature conditions, and Table VI provides similar data on two polymers under selected conditions. The Grossman number, H, and the film coefficient, h, provide useful information about the rate of heat removal from the surface of a part. The thermal conductivity of 7075 aluminum is about 1.70 W cm/cm²°K (1150 Btu·in/ ft²hr°F). Unagitated water at 25°C has a Grossman number of about one as reported by Grossman and a film coefficient of about 3.55 W/cm²°K. The film coefficient increased to 4.78 and 5.14 W/cm²°K at velocities of 0.25 and 0.50 m/sec, respectively. These high film coefficient values can therefore cause cold water quenching to create high thermal gradients from surface to center of a part and high temperature differences between thick and thin sections. The film coefficients generally decreased with increasing water temperature until at temperatures of 90-100°C, the film coefficients were in the range of 0.13 to 0.30 W/cm²°K.

Table VI provides similar data on two polymer solutions under selected conditions. The 25% solution of polyalkalene glycol (UCON A) produced Grossman numbers of 0.19 to 0.23 and film coefficients of 0.66 to

TABLE IV

APPROXIMATE QUENCHING SEVERITY (H) OF VARIOUS MEDIA (IN PEARLITE TEMPERATURE RANGE)

Circulation or Agitation	Grossman number for			
	Brine*	Water*	Oil* and salt	Air†
None	2	0.9-1.0	0.25-0.30	0.02
Mild	2-2.2	1.0-1.1	0.30-0.35	--
Moderate	--	1.2-1.3	0.35-0.40	--
Good	--	1.4-1.5	0.4-0.5	--
Strong	--	1.6-2.0	0.5-0.8	--
Violent	5	4	0.8-1.1	--

*After M.A. Grossman, Trans. Am. Inst. Mining Metal. Eng., 1942.

†After M.A. Grossman and M. Asimow, Iron Age, 145, 1940.

TABLE V
GROSSMAN NUMBERS AND FILM COEFFICIENTS
FOR WATER UNDER SELECTED CONDITIONS

Type	Temp °C	Temp °F	Velocity m/sec	Velocity ft/min	Grossman Number (H=h/2k)	Effective Film Coefficient (h) (W/cm²°K)	Effective Film Coefficient (h) (Btu/(hr)(ft²)(°F))
water	27	80	0.00	0	1.07	3.55	2460
			0.25	50	1.35	4.78	3105
			0.50	100	1.55	5.14	3565
water	38	100	0.00	0	0.99	3.28	2275
			0.25	50	1.21	4.01	2785
			0.50	100	1.48	4.91	3400
water	49	120	0.00	0	1.10	3.65	2530
			0.25	50	1.29	4.29	2970
			0.50	100	1.60	5.31	3680
water	60	140	0.00	0	0.86	2.85	1980
			0.25	50	1.09	3.62	2510
			0.50	100	1.33	4.41	3060
water	71	160	0.00	0	0.21	0.70	485
			0.25	50	0.57	1.89	1310
			0.50	100	0.79	2.62	1815
water	82	180	0.00	0	0.11	0.36	255
			0.25	50	0.21	0.69	485
			0.50	100	0.27	0.89	620
water	93	200	0.00	0	0.06	0.20	138
			0.25	50	0.08	0.27	184
			0.50	100	0.09	0.30	207
water	100	212	0.00	0	0.04	0.13	92
			0.25	50	0.04	0.13	92
			0.50	100	0.04	0.13	92

TABLE VI
GROSSMAN NUMBERS AND FILM COEFFICIENTS
FOR WATER SOLUBLE POLYMERS, OIL AND AIR
UNDER SELECTED CONDITIONS

Type	Temp °C	Temp °F	Velocity m/sec	Velocity ft/min	Concentration (%)	Grossman Number (H=h/2k)	Effective Film Coefficient (h) (W/cm²°K)	Effective Film Coefficient (h) (Btu/(hr)(ft²)(°F))
UCON A	30	85	0.00	0	25	0.19	0.63	429
			0.25	50	25	0.21	0.70	475
			0.50	100	25	0.23	0.77	529
PVP90	30	85	0.00	0	25	0.44	1.49	1012
			0.25	50	25	0.40	1.34	912
			0.50	100	25	0.42	1.41	966

0.77 W/cm²°K when the bath was operated at 30°C and with velocities from 0 to 0.5 m/sec. Polyvinyl pyrrolidone 90 produced Grossman numbers of 0.40 to 0.44 and film coefficients of 1.34 to 1.49 under these conditions. The higher film coefficient of PVP 90 suggests that it can be used to quench heavier sectioned parts.

If the Grossman number, H, or the effective interface heat transfer coefficient, h, between the part and the quenchant is established, the quench factor in commercial shapes can be calculated using finite element or finite difference heat transfer programs. The results of calculations on sheets and plates made using constant film coefficients are illustrated in Figures 3 and 4 respectively. These figures illustrate interrelationship between aluminum sheet or plate thickness, film coefficient and quench factor. The calculations for these graphs were made using film coefficients indicated at the end of each diagonal line. The diagonal lines represent lines of constant film coefficient.

The important feature of these figures is that when used in combination with Table II, which relates quench factor to yield strength, and data on film coefficient such as that presented in Tables V and VI, estimates can be made about the ability of specific quenchants and operating conditions to provide cooling rates sufficiently high to meet minimum mechanical properties in parts of various thicknesses. Using a previous example, Military Handbook V specifies a "B" allowable minimum strength in 7075-T73 in 19-38.1 mm thick (0.75- 1.5 inch) extrusions of 434.4 MPa (63 ksi). The upper limit Q value capable of meeting this strength is 18.0.

Boiling water without agitation has a film coefficient of 0.20 W/cm²°K and could not be expected to provide an acceptable quench factor in a 20-38 mm thick section. This can be determined by locating the upper limit Q value of 18 on the ordinate of Figure 4, following this value horizontally until it intersects the diagonal line representing a film coefficient of 0.21 w/cm²°K and then reading the abscissi value for the maximum thickness that can be expected to be properly quenched. In this case, the maximum hardenable thickness is expected to be about 5.5 mm (0.22 in.). In order to produce an acceptable quench factor in a 38 mm (1.50 in.) thick plate, a quenchant providing a film coefficient of about 2.16 W/cm²hr°K (1500 Btu/ ft²hr°F) or higher is required. A film coefficient of 2.88 W/cm²°K (2000 Btu/ft²hr°F) provides a higher margin of safety, and will produce a quench factor of about 6 in a 19 mm (0.75 inch) thick plate and a quench factor of 13 in a 38.1 mm (1.50 inch) thick plate. High film coefficients and Grossman numbers can be obtained by immersion quenching in agitated cold water, brine, or with high pressure spray quenching.

QUENCH SEVERITY AND DISTORTION - Quenchants must be selected to provide film coefficients and quench factors capable of producing acceptable properties in the section thicknesses of interest, as previously discussed. However, it is desirable not to use quenchants with excessively high film coefficients if distortion is to be minimized consistent with meeting the required properties. Excessively high film coefficients result in higher temperature gradients across thick sections and large temperature differences between thick and thin

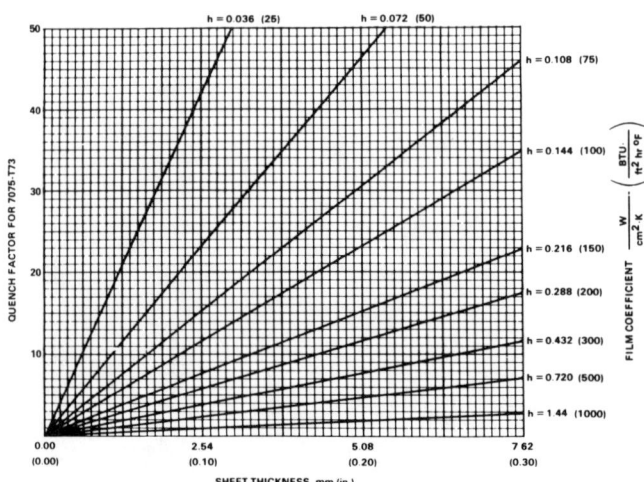

Figure 3. Effect of sheet thickness and film coefficient on quench factor in 7075-T73 aluminum

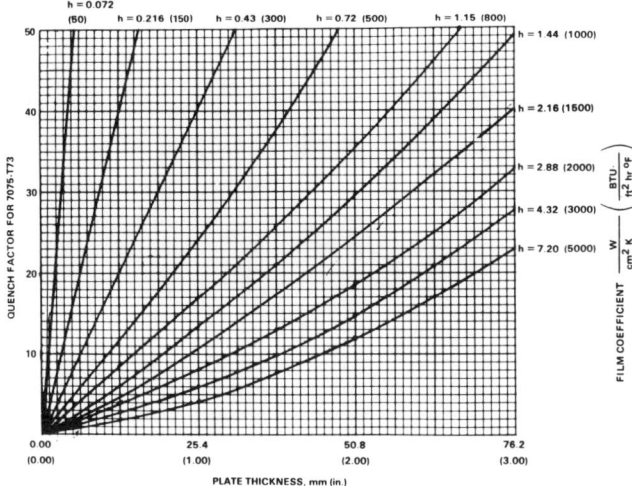

Figure 4. Effect of plate thickness and film coefficient on quench factor in 7075-T73 aluminum.

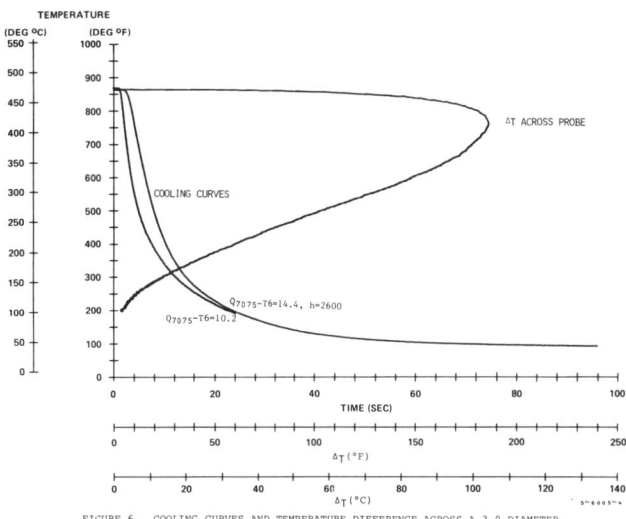

FIGURE 6. COOLING CURVES AND TEMPERATURE DIFFERENCE ACROSS A 3.0 DIAMETER 7075 PROBE QUENCHED IN 90°F WATER FLOWING AT 50 fpm.

sections. This in turn aggravates residual stress and distortion problems.

Figure 5 illustrates a 76.2 mm (3 inch) diameter bar probe instrumented with two thermocouples, one on the centerline and the second 6.35 mm (0.25 in.) from the surface. Cooling curves obtained at these thermocouple positions when this bar was quenched in 32°C water flowing at 0.125 m/sec are illustrated in Figure 6.

The temperature difference between the two thermocouples calculated as the center temperature minus the surface temperature is also illustrated in Figure 6 and plotted as a function of the center temperature (on the ordinate). It is observed that the maximum temperature difference across the bar was approximately 110°C and the maximum value occurred shortly after quenching was begun while the center temperature was approximately 425°C.

Similar temperature-time histories and the temperature differences across the probe when quenching in 60°C water and 25% polyvinyl pyrrolidone 90 flowing at 0.125 m/sec past the probe are illustrated in Figures 7 and 8 respectively. The data in both cases are superimposed on the 32°C water data previously shown in Figure 6. The higher water temperature of 60°C reduced the maximum temperature difference to 61°C and the 25% solution of polyvinyl pyrrolidone 90 reduced the difference still further to a value of 39°C.

The Grossman numbers and film coefficients associated with these quenchants were presented in Tables V and VI. The temperature difference across the section decreased as the Grossman number and film coefficient decreased.

High film coefficients can produce even larger differences in temperature between thick and thin sections of a part. Cooling curves associated with a 76.2 mm section and a 12.7 mm section quenched in 32°C water flowing over the part at a velocity of 0.125 m/sec are illustrated in Figure 9. Water at this temperature and velocity had a film coefficient of about 4.78 W cm/cm²°K (3100 Btu/ft²hr°F) and produced a maximum temperature difference between the two sections of approximately 290°C (520°F).

The thermal stress in a part which causes distortion during quenching, is a function of the alloy thermal expansion coefficient, elastic modulus, and temperature difference within the part. Minimizing the thermal stress requires that the temperature differences be minimized since

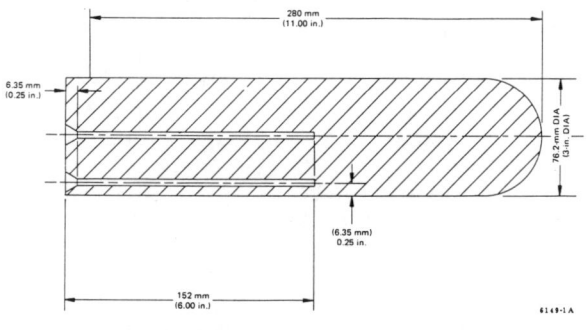

Figure 5. Cross section of 76.2-mm (3-inch) diameter 7075 Aluminum probe.

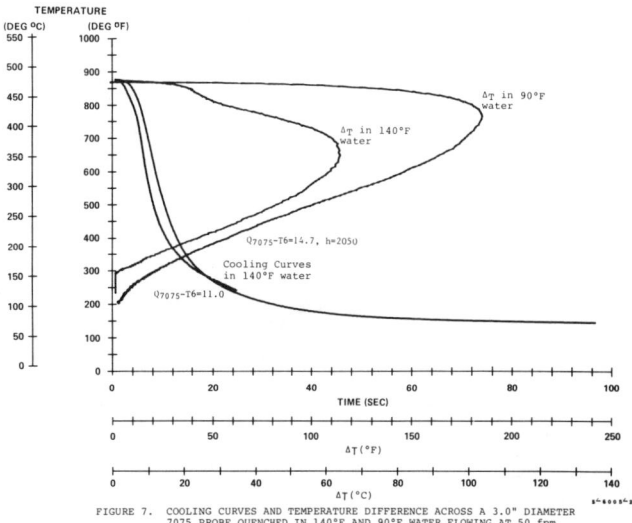

FIGURE 7. COOLING CURVES AND TEMPERATURE DIFFERENCE ACROSS A 3.0" DIAMETER 7075 PROBE QUENCHED IN 140°F AND 90°F WATER FLOWING AT 50 fpm.

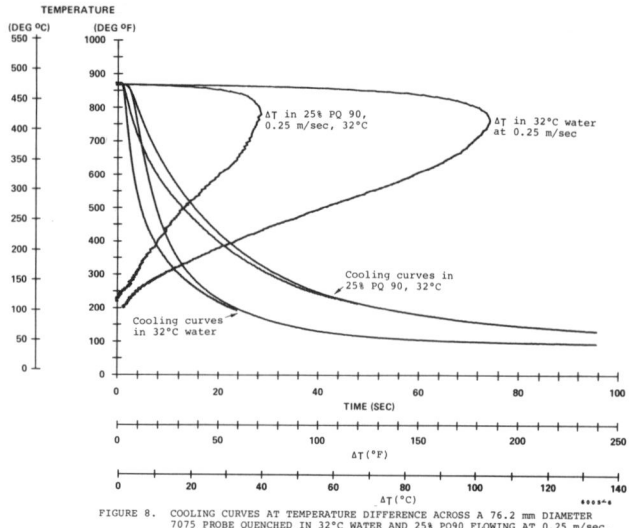

FIGURE 8. COOLING CURVES AT TEMPERATURE DIFFERENCE ACROSS A 76.2 mm DIAMETER 7075 PROBE QUENCHED IN 32°C WATER AND 25% PQ90 FLOWING AT 0.25 m/sec.

no control can be exercised over either the expansion coefficient or elastic modulus.

The temperature differences across plates up to 76 mm thick were calculated using a finite difference heat transfer program in which the film coefficient was used as an input value. The results are illustrated in Figure 10. The temperature difference is the maximum calculated value between the plate center and a location 1.60 mm (1/16 in.) beneath the surface. The temperature difference across a given section thickness progressively increased as the film coefficients increased. In order to minimize the temperature difference across a part to reduce the thermal stress, the film coefficient must be minimized consistent with cooling fast enough to guarantee the minimum yield strength can be met. This can be done by selecting and using the most appropriate quenchant and quenchant operating conditions.

Quenching real parts is usually more complicated than quenching sheets and plates because most real parts do not have a uniform cross section. Many aircraft

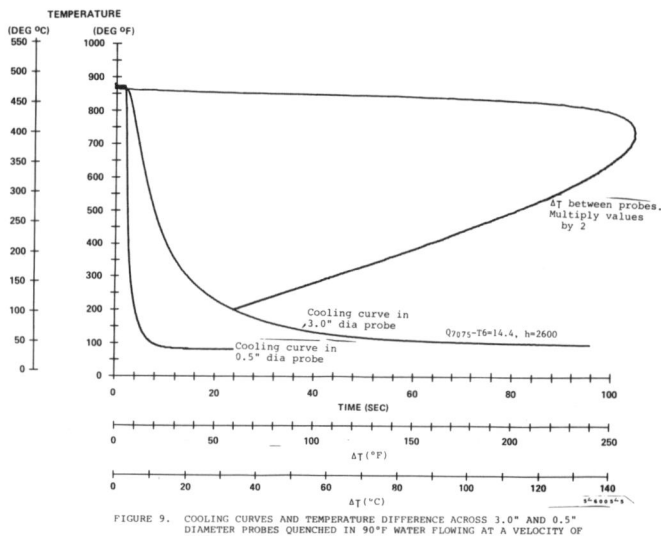

FIGURE 9. COOLING CURVES AND TEMPERATURE DIFFERENCE ACROSS 3.0" AND 0.5" DIAMETER PROBES QUENCHED IN 90°F WATER FLOWING AT A VELOCITY OF 50 fpm.

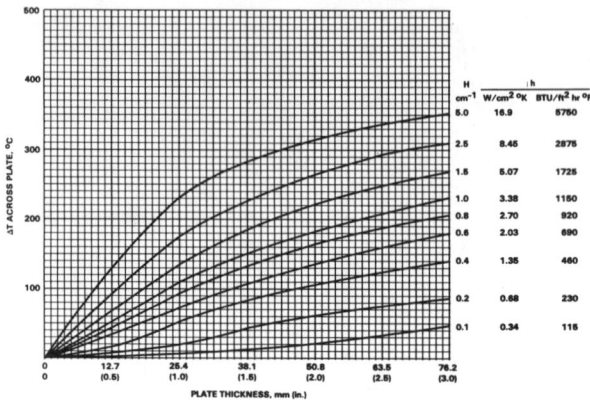

Figure 10. Effect of film coefficient on the temperature difference between surface and center of 7075 aluminum plates.

structural numbers consist of flanges reinforced by thin web sections. When such a part is quenched in a solution such as water that produces a high film coefficient, the web section temperature drops far more quickly than the temperature in the heavier flange. The large temperature difference creates a high thermal stress which often causes plastic deformation in the web. When the heavy flange cools, it contracts onto the thin web and often causes buckling or the "oil-can" effect. Efforts to reduce web thicknesses in order to remove weight from structural members aggravates the buckling problem by producing thinner sections that cool even faster than the flanges. The only feasible approach to reducing plastic flow in web sections with conventional quenching technology is to reduce the heat transfer coefficient at the web--quenchant interface while not reducing the film coefficient so much that the quench factor in the flange is too high to allow properties to be met.

SUMMARY - The function of a quenchant is to remove heat from age hardenable aluminum parts sufficiently rapidly to insure that elements and compounds used to harden the alloy do not precipitate during the quenching process. The quenchant must also minimize, as far as possible, thermal gradients within the part that may cause plastic deformation and residual stress. Deformation and residual stress may cause problems when the part is machined or put into service.

Quench factor analysis can be used to predict the strength in aluminum alloys from experimental and analytical cooling curves. Quenchants should be selected to produce an adequately high interface heat transfer coefficient, h, and sufficiently low quench factor, Q, to insure that the design minimums can be achieved in the heaviest or critical sections. Quenchants that provide excessively high values of h or H should not be used, however, if distortion is of concern, because they introduce high thermal gradients that induce plastic flow, deformation, and residual stress. The optimum quenchant is one that will provide a film coefficient slightly higher than that required to produce the required Q factor in the critical section thickness.

Grossman numbers are presented for a variety of quenchants and the general methodology is outlined for selecting quenchants that will provide sufficiently high heat transfer rates to meet specified mechanical properties without producing excessively high residual stresses.

REFERENCES

1. Evancho, J.W. and J.T. Staley, "Kinetics of Precipitation in Aluminum Alloys During Continuous Cooling," Met. Trans., Vol. 5, (Jan 1974), pp. 43-47.

2. Metals Handbook, Ninth Edition, Vol. 4, pp. 689-695, Heat Treating, (1981).

3. John E. Hatch, ed., Aluminum-Properties and Physical Metallurgy, pp. 157-175, American Society for Metals, Metals Park, OH, 1984, 397 p.

4. L. K. Ives, et al., Processing/Microstructure/Property Relationships in 2024 Aluminum Alloy Plates, U.S. Dept. of Commerce, National Bureau of Standards Technical Report NBSIR 83-2669, January 1983.

5. Charles E. Bates, "Technical Support for PQ90 Aluminum Heat Treatment Specification," Southern Research Institute Report SoRI-EAS-86-1147 to Park Chemical Company, Southern Research Institute, Birmingham, AL, 1986, 30 p.

6. Military Standardization Handbook, 5D, Metallic Materials and Elements for Aerospace Vehicle Structures, Vol. 1, June 1983.

7. John H. Holloman and Leonard D. Jaffe, "Ferrous Metallurgical Design," p. 176, John Wiley & Sons, New York, NY (1947)

RESIDUAL STRESS IN QUENCHED STEEL CYLINDERS

Mark E. Todaro, Mark A. Doxbeck, George P. Capsimalis
U.S. Army Armament, Munitions, and Chemical Command
Benet Weapons Laboratory
Watervliet, New York USA

ABSTRACT

Measurements were made on high strength, low alloy steel cylinders to determine the residual stress distributions resulting from various heat treatments. Cylinders of 239 mm outer diameter and 94 mm inner diameter were austenitized at 843 °C and quenched at various rates to 93 °C. Residual stress measurements were made on cylindrical cross sections that had been cut from the larger cylinder at least one foot from the nearest end. Using ultrasonic and X-ray diffraction techniques, we measured tangential and radial components of stress as a function of radial position.

IN THE HEAT TREATMENT of steel, a fast quench rate is often necessary to produce the desired martensitic phase but can result in cracking and undesirable stress states. This paper presents various measurements of residual stress distributions in quenched steel cylinders. The stress measurements were made with plane shear ultrasonic waves using the acoustoelastic effect and with X-ray diffraction using single-exposure and multiple-exposure techniques.

SPECIMEN PREPARATION

Residual stress measurements were made on cylindrical specimens cut from a long hollow cylinder of ASTM A723 steel (MIL-S-46119A) with an outer diameter of 239 mm and an inner diameter of 94 mm. Two different heat treatments were chosen. In the first treatment, the entire cylinder was austenitized at 843 °C and quenched to 93 °C in 12 minutes using an outer-diameter quench. A 3-inch thick cylindrical section was cut from near one end, 1 foot from the end itself, taking care not to heat the specimen to the point where stresses might be relieved by annealing. Both faces of the section were then wet-ground, leaving them parallel and 2 inches apart. The same large cylinder was again austenitized at 843 °C, but then quenched to 93 °C in 25 minutes using an outer-diameter quench. A 2-inch thick section was taken from one end, 1 foot from the end, as before. The faces were then wet-ground, leaving them parallel and 1.11 inches apart. The ultrasonic data was taken after each specimen was prepared in this way.

After taking the ultrasonic data, we prepared each specimen for X-ray diffraction by electropolishing a portion of its face, removing a 0.1-mm thick layer of surface material that might contain residual stress due to machining and grinding.

ULTRASONIC TECHNIQUE

THEORY. The change in velocity of an acoustic wave in a solid due to stress is known as the acoustoelastic effect. If the relation between stress and velocity is known, measurement of the velocity of an acoustic wave can, at least in principle, be used to determine the stress.

Hughes and Kelly,[1] Bach and Askegaard,[2] and Husson and Kino[3] have derived expressions for the velocities of acoustic waves in homogeneously stressed solids. The velocity, V, of a plane shear wave propagated along the 1-axis and polarized along the 2-axis is given by the following expression, adapted from those authors:

$$\rho_0 V^2 = \mu + \frac{1}{3\lambda + 2\mu} \Big[(3\lambda + 2\mu) \sigma_1 \\ + \Big(3\lambda + 2\mu + \frac{3n\lambda}{4\mu} + \frac{n}{2}\Big)(\sigma_1 + \sigma_2) \\ + \Big(m - \frac{n}{2} - 2\lambda - \frac{n\lambda}{2\mu}\Big)(\sigma_1 + \sigma_2 + \sigma_3) \Big] \quad (1)$$

Here, λ and μ are the Lame constants (second-order elastic constants), while m and n are two of the three Murnaghan constants (third-order elastic constants). ρ_0 is the density of the undeformed medium. The σ's are the triaxial principal stresses along the 1-, 2-, and 3-axes.

For the type of steel encountered in our research, the factor $(m - n/2 - 2\lambda - n\lambda/2\mu)$ is essentially zero, as shown by Frankel et al.[4] For situations where σ_1 is negligible as well, eq. (1) can be further simplified to

$$\frac{\Delta V}{V_0} = \frac{1}{2\mu(3\lambda+2\mu)}\left(3\lambda + 2\mu + \frac{3n\lambda}{4\mu} + \frac{n}{2}\right)\sigma_2 \quad (2)$$

where $\Delta V/V_0$ is the relative change in velocity due to the stress.

Although these solutions are derived for homogeneous stress, they may also be used for an inhomogeneous stress that does not vary greatly over a distance of one wavelength.

EXPERIMENTAL DETAILS. We measured shear wave velocities using a pulse-echo technique and a computer-controlled phase-detection system (Matec Instruments, Model MBS-8000), shown schematically in Fig. 1. The system uses phase detection methods that involve interactive computer control of the frequency and measurement of phase relationships. As indicated by the figure, the transducer converts an electrical pulse into a shear wave that travels through the specimen. Each time an echo strikes the front face, it is converted into an electrical pulse that is then amplified and sent to a pair of phase detectors. From the phase detectors, the computer receives signals proportional to the sine and cosine of the echo pulse's phase with respect to a reference wave. The computer can then calculate the amplitude and phase of the echo pulse. Because the phase can only be calculated as an angle between $-\pi$ and π, the system varies the frequency slightly and measures the corresponding phase shifts for one or more echoes. From this phase data, we may calculate the transit time of the shear wave through the specimen and deduce its velocity.

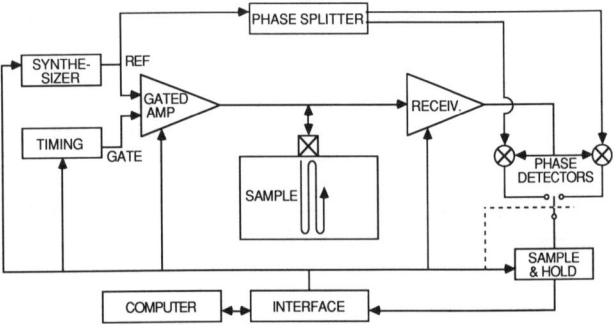

Fig 1. Schematic diagram of computer-controlled phase detection system to measure ultrasonic velocities.

The measurement of stress, however, requires an accuracy that can only be achieved by measuring changes in transit time as the system is perturbed. For mapping stress distributions, that perturbation is the relocation of the transducer on the specimen's face. By measuring the phase shift due to a change in transducer position, the computer can calculate the corresponding time change.

The system was used with a pulse frequency of 5 MHz and duration of several microseconds. A normal incidence shear wave transducer (Panametrics, V156) introduced shear waves into the specimen and received echoes. The transducer was acoustically coupled to the specimen with a viscous resin (KB-Aerotech, SLC-70) and held in place by hand. Velocity variations were mapped at 0.1-inch intervals along 4 radial directions as shown in Fig. 2. To calculate the tangential (hoop) component of stress through eq. (2), we polarized the shear wave tangentially. To calculate the radial component, we polarized the wave radially.

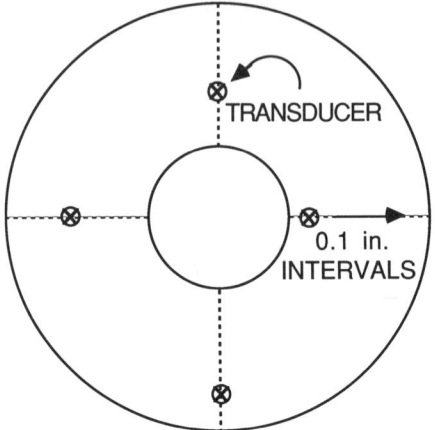

Fig. 2. Transducer positions on face of specimen.

X-RAY DIFFRACTION TECHNIQUES

THEORY. In the single-exposure technique, two diffraction patterns are obtained simultaneously in a single X-ray exposure, as drawn in Fig. 3. The incident X-ray beam strikes the specimen at an angle β with the surface normal and is diffracted by atoms in the lattice. In the plane of the incident beam and the surface normal, two diffraction peaks are observed at angles α_1 and α_2 with respect to the incident beam. The component of stress parallel to the surface and in the plane of the incident beam, σ, and the surface normal can then be related to the positions of the diffraction peaks with the expression[5,6]

$$\sigma = \left(\frac{E}{1+\nu}\right) \frac{\alpha_2 - \alpha_1}{4\sin^2\theta_0 \sin 2\beta} \quad (3)$$

In this equation, θ_0 is the Bragg diffraction angle for the unstrained material, E is Young's modulus, and ν is Poisson's ratio.

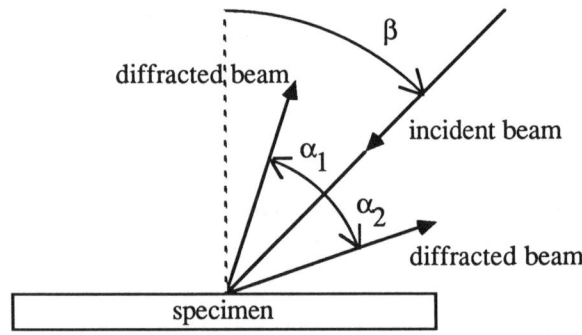

Fig. 3. Single-exposure X-ray diffraction technique.

The multiple-exposure technique,[5,7] also known as the $\sin^2\psi$ technique, relies on the relation

$$\frac{d_\psi - d_0}{d_0} = \left(\frac{1+\nu}{E}\right)\sigma \sin^2\psi \quad (4)$$

where d_ψ is the distance between diffracting planes whose normal makes an angle ψ to the surface normal; d_0 is the distance between diffracting planes parallel to the surface. The term $(d_\psi - d_0)/d_0$ is plotted for several values of ψ, and σ is obtained from the slope of the resulting line.

EXPERIMENTAL DETAILS. Diffraction peaks for the single-exposure technique were obtained and analyzed with a Denver X-ray Instruments Model D-1000-A system. The system uses a miniature X-ray tube and collects diffraction peak profiles using two Ruud-Barrett position sensitive scintillation detectors.[8,9] Each detector consists of a fiber optic bundle with its end covered by a CdZnS scintillation coating. The coating converts the diffracted X-rays to light pulses that then travel along the fibers, are amplified by an image intensifier, and directed to a 512 photodiode array. Each photodiode charges a capacitor to a level proportional to the amount of incident light.

The X-ray tube used a Chromium target and a Berylium window, providing Cr Kα_1 radiation, with a wavelength of 2.29 Å. The system was used with β at 20° and with the two detectors centered on the 211 peak of iron (θ_0 approximately 78°).

Data for the multiple-exposure technique was taken using a computer-controlled system based on a Siemens stress goniometer,[10] which also used Cr Kα_1 radiation and scanned the 211 peak of iron.

RESULTS AND DISCUSSION

The change in shear wave velocity data was converted to stress using a simplified form of eq. (2),

$$\sigma = A \frac{\Delta V}{V_0} \quad (5)$$

where σ refers to the stress component in the direction of polarization. For the type of steel used in this study, it has been found that $A = -2630$ MPa (ref. 4). Figs. 4 and 5 show the results for the 12- and 25-minute quench specimens. Both cases show a compressive tangential stress at the outer diameter. The tangential stress becomes tensile with increasing depth. In the fast quench specimen, it again becomes compressive toward the inner diameter.

The preliminary results of the X-ray diffraction analysis show qualitative agreement with the ultrasonic results. Figs. 6 and 7 show the tangential stress distributions for the 12-minute quench as obtained with the single-exposure and multiple-exposure techniques. Despite a high level of scatter in the data, one can observe that the stress is compressive at the outer diameter, changes to tensile with increasing depth, and again becomes compressive toward the inner diameter. The large scatter in the data is likely due to the broadness of the diffraction peaks, a direct result of quenching. Furthermore, we might expect the X-ray data to differ somewhat from the ultrasonic data because the radiation only penetrates the specimen to a depth of about 10 µm. The ultrasonic technique uses a wave that travels through the thickness of the specimen, averaging out velocity variations along the wave's path.

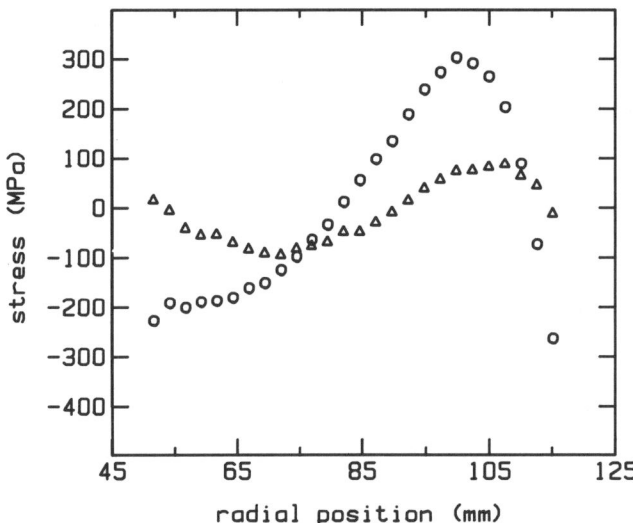

Fig. 4. Tangential (circles) and radial (triangles) stress for 12-minute quench specimen using ultrasonic technique.

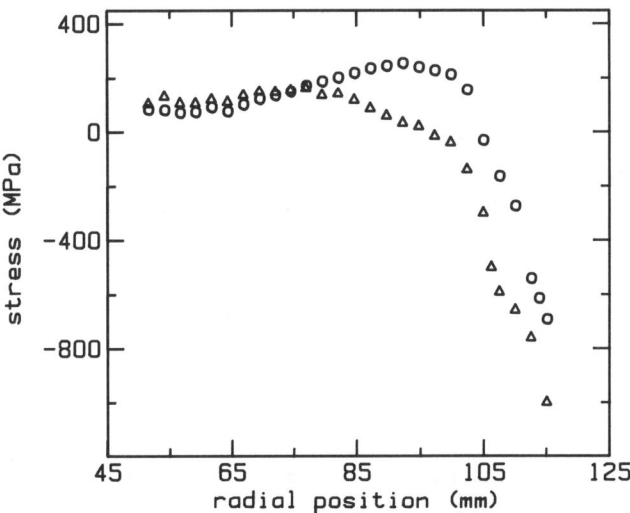

Fig. 5. Tangential (circles) and radial (triangles) stress for 25-minute quench specimen using ultrasonic technique.

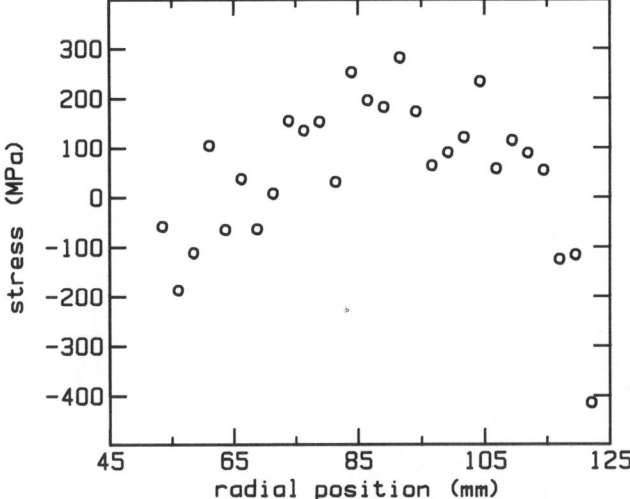

Fig. 6. Tangential stress for 12-minute quench specimen using single-exposure X-ray diffraction technique.

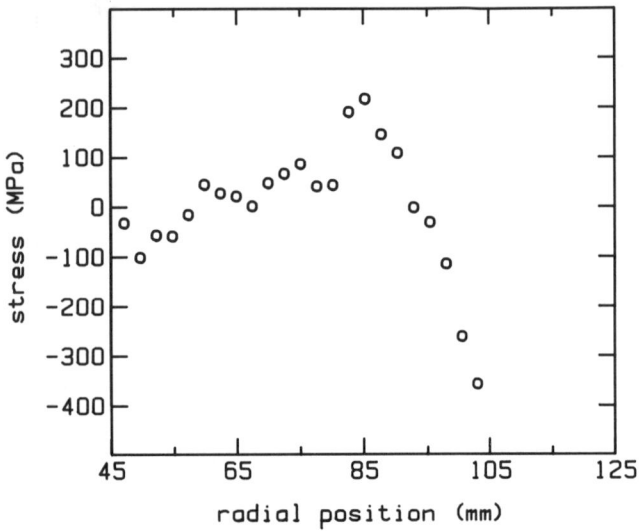

Fig. 7. Tangential stress for 12-minute quench specimen using multiple-exposure X-ray diffraction technique.

One problem with the radial stress distribution for the 25-minute quench tube, Fig. 5, is that it shows a high compressive stress at the outer diameter, where a radial stress would not be expected. A possible explanation is that we are not actually sampling the stress at the surface because of the finite width of the transducer.

One of the major problems with using the ultrasonic technique to measure residual stress is that a preferential alignment of grains results in an elastic anisotropy that can cause velocity variations at least as large as those due to applied or residual stress.[11] Although the ultrasonic technique is relatively fast and easy to use, the possibility of velocity changes due to preferred grain orientation reduces confidence in the results.

Future work will concentrate on verifying the remainder of the ultrasonic data using X-ray diffraction and on calculating theoretical stress distributions based on available models of the quench process.

REFERENCES

1. D. S. Hughes and J. L. Kelly, Phys. Rev. 92, 1145 (1953).

2. F. Bach and v. Askegaard, Exp. Mech. 19, 69 (1979).

3. D. Husson and K. S. Kino, J. Appl. Phys. 53, 7250 (1982).

4. J. Frankel, W. Scholz, G. Capsimalis, and W. Korman, Proc. Ultrasonics Symp., 1009 (1983).

5. H. P. Klug and L. E. Alexander, "X-Ray Diffraction Procedures for Polycrystalline and Amorphous Materials," 2nd Ed., John Wiley and Sons, New York (1974).

6. J. T. Norton, Adv. X-Ray Anal. 11, 401 (1968).

7. Society of Automotive Engineers, "Residual Stress Measurement by X-Ray Diffraction," Society of Automotive Engineers, Inc., Warrensdale, Pennsylvania (1971).

8. C. O. Ruud, Ind. Res. and Dev. 25, 84 (1983).

9. C. O. Ruud, P. S. DiMascio, and D. J. Snoha, Advances in X-Ray Analysis 27, (1984).

10. G. P. Capsimalis, R. F. Haggerty, and K. Loomis, Technical Report WVT-TR-77001, Watervliet Arsenal, Watervliet, New York (1977).

11. D. I. Crecraft, Ultrasonics 6, 117 (1968).

RESIDUAL STRESS IN LASER HARDENED GEAR MATERIAL STUDIES

Richard L. Frohlich
Westinghouse Marine Division
Sunnyvale, California USA

ABSTRACT

A contrast between successful long-running roller contact fatigue tests and abrupt surface failures on laser surface hardened rollers was attributed to procedural changes immediately after laser processing that developed tensile residual surface stresses in the shorter running tests. The x-ray diffraction technique was used to measure the residual stress profiles in both test programs. This report describes the roller fatigue test program, a failure analysis made on a short running test and the impact this investigation had on later gear hardening material and residual stress studies.

INTRODUCTION

Residual stresses in finished parts have often been regarded as factors contributing to premature part failure, geometric distortion or general nonconformance that was unexpected and unplanned for. Tensile residual stresses created by welding, machining, or grinding

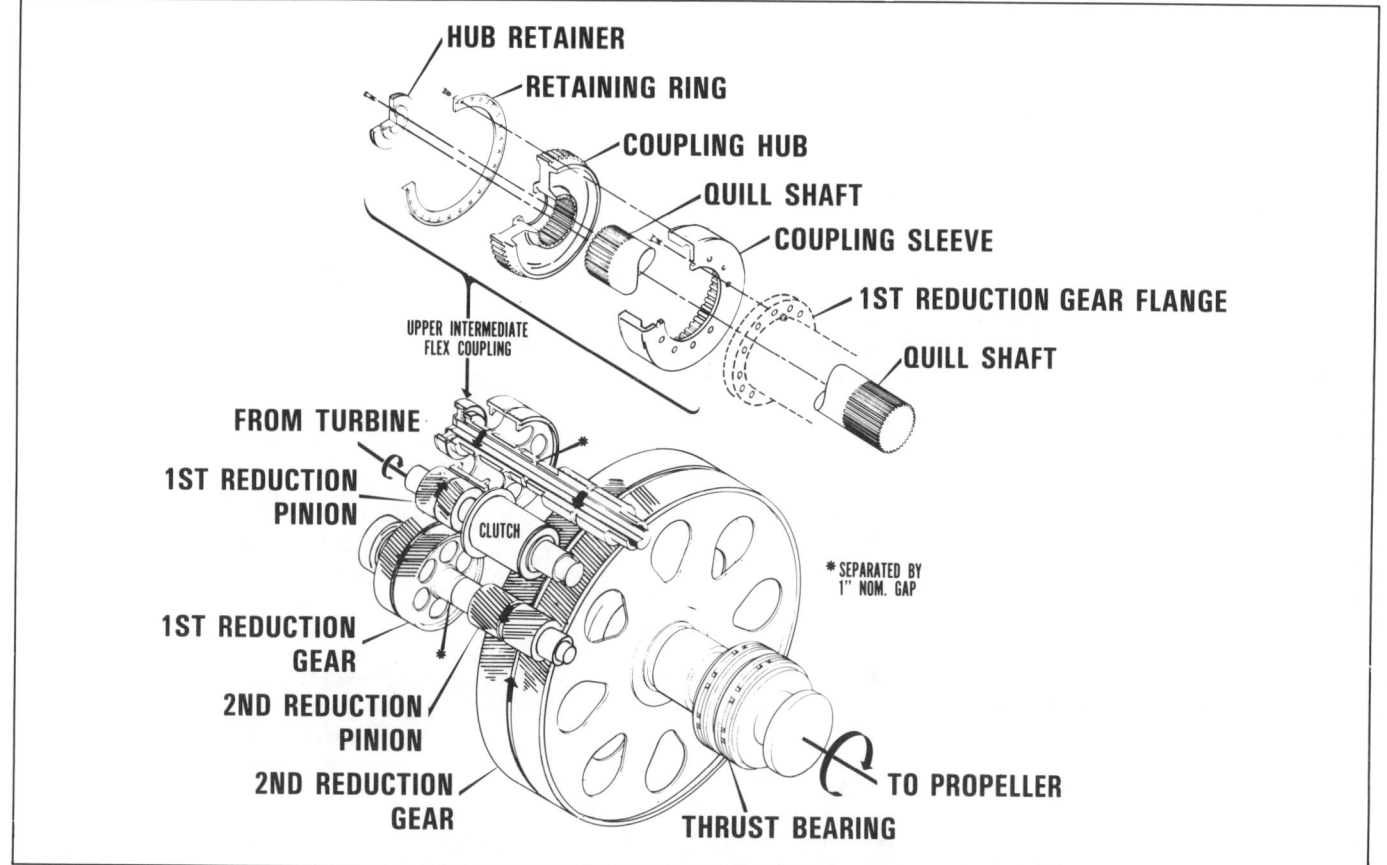

Figure 1 - Laser Gear Hardening Objectives

operations have contributed to the premature failure of numerous parts. In contrast, compressive residual stresses can improve fatigue and stress corrosion properties in many applications. This paper discusses two laser development programs, roller tests and spline hardening, that were influenced by early residual stress investigations. Our results to date have shown that residual stress measurements can be an important diagnostic tool in the development of improved gear hardening processes.

The investigation of laser transformation hardening for surface hardening large marine gears, Figure 1, has been a long-term objective of the Westinghouse Electric Corporation (WEC) Marine Division in Sunnyvale, California. Laser technology offered a very flexible alternative to other options that are costly in terms of capital equipment, production space, manufacturing cycle time, and energy utilization. While laser hardening of large marine gears remains an overall program objective, early development work was performed on small test gears and roller fatigue test parts to examine materials, establish program guidelines, and to foster a better understanding of the laser process. Little thought was given to investigating residual stress values in the earliest tests until unexpected failures occurred in roller fatigue tests. Residual stress measurements by x-ray diffraction methods helped diagnose the problem. In a later spline hardening program, residual stress measurements were performed to assure that laser process variables and techniques produced the desired compressive residual stresses in critical locations. This paper discusses events in those two programs that led to residual stress studies in laser heat treated gear materials.

LASER TRANSFORMATION HARDENING

In transformation hardening iron-carbon alloys, the alloy must be austenitized then rapidly cooled to form martensite. Martensite forms by a diffusionless phase transformation with specific crystallographic relationships to the parent phase characterized by an acicular microstructure, a significant volumetric increase, and accompanying lattice strains that result in locally high levels of residual stresses.[1] During laser transformation hardening, a CO_2 laser heat treating beam is used to raise the surface temperature of ferrous alloys above critical transformation temperatures while the base material immediately beneath the surface remains substantially colder to provide the thermal gradient necessary for a rapid cool to produce martensite.

The depth of a hardened zone depends upon the austenitizing temperature reached at locations beneath the surface and the rate of heat extraction or self-quenching that results from the much colder base material. A balance of energy absorption with thermal conduction must be maintained to prevent overheating and possible surface melting. Residual stresses result from thermal contraction of the alloy as it rapidly cools on which is superimposed an expansion of the metal as the surface zones transform from austenite to martensite.[2] The magnitude of residual stresses in low alloy steels depends primarily upon temperature, quench severity, carbon content, and previous microstructure of the base material.

EARLY LASER MATERIALS STUDIES

The WEC gear hardening program used continuous wave CO_2 lasers of 10.6 μm wave length at power settings ranging from 2.5 to 17.5 kW in its research. In that output power range, the laser beam can be defocused over a relatively broad zone for general heat treating applications by using special optics such as integrating mirrors, scanning mirrors, geometrically shaped mirrors, or combinations of mirrors and oscillators. Travel speed and beam scanning or oscillation were techniques used in this program to control the time the heat treating beam remains trained on a finite zone, thus preventing surfaces from overheating and possibly melting.

Residual stress investigations were not performed originally in these studies, instead it was assumed that stress values would be similar in magnitude and sign to that obtained by the tooth by tooth induction gear hardening process in which gear tooth surfaces are rapidly austenitized and then quenched by the cool core. Highly compressive residual stresses were expected at the surface due to the volume expansion created by the martensitic reaction. The roller fatigue tests described in this report demonstrated that acceptable case depth and hardness ranges could not necessarily assure compressively stressed surfaces nor accompanying good fatigue performance. Based on this study, recommendations were made to have residual stress profiles made on future laser transformation hardened surfaces. Residual stress determinations were felt to be a powerful diagnostic tool in evaluating fatigue failures and as a means of improving the general understanding of the laser process.

ROLLER TEST PROGRAM

The roller test program was initiated to rapidly evaluate laser treated gear materials for hardening characteristics and rolling fatigue performance while controlling program costs. In a gear hardening program, numerous process variables, material chemistries and hardening techniques must be evaluated before implementing bending and pitting fatigue endurance tests. The simple geometry of the roller made it much less costly to prepare than hobbed gears and it provided a means of acquiring surface contact fatigue data well before fatigue studies would normally occur in a more tradition gear fatigue program.

In roller wear tests, surface endurance limits can be established by rotating two highly loaded test rolls against each other until surface failures initiate. The contact stress developed between the two rollers simulates the stresses generated by highly loaded gear sets rotating at their respective pitch diameter. Hardened and ground gear reduction systems must exhibit good surface endurance characteristics in order to avoid surface degradation such as pitting and spalling failures which can contribute to excessive noise during operation. The Hertzian surface contact stresses can be calculated from the test loads placed on the rollers using Equation 1.[3] The roller tester, operating at 1800 RPM, can reach a test objective of 10^7 cycles in approximately 4 days if surface pitting does not initiate. Roller surfaces are periodically inspected for micro-pitting using a microscope built into the equipment during scheduled shutdowns of the system.

creating surface slippage at the point of contact. This action simulates the sliding motion of gear teeth as they mesh just before and after the pure rolling condition at the pitch diameter. Using the Hertzian equation for

Figure 3 - Typical Roller Test Specimen

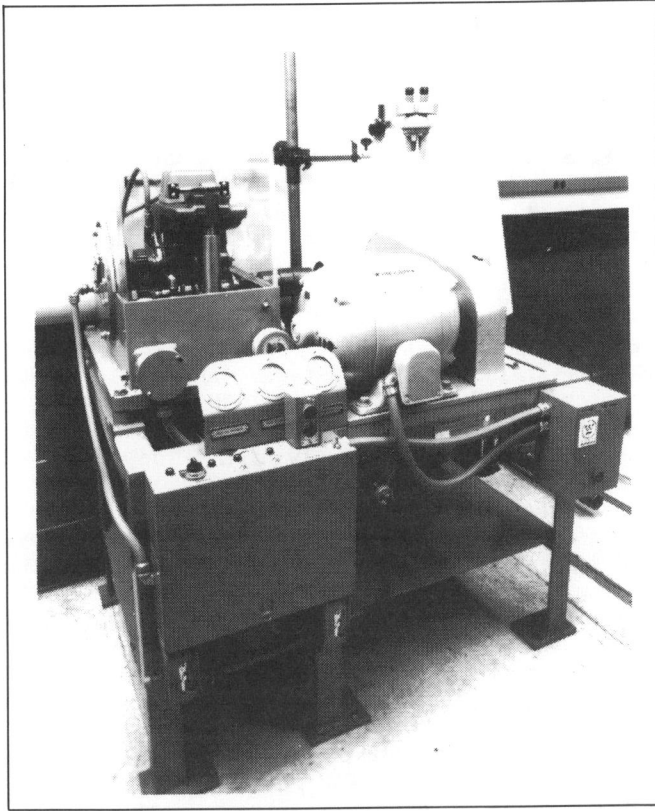

Figure 2 - Overall View of Roller Testing Machine

Figures 2 and 3 shows the laboratory roller testing machine and a sample roller that was used to evaluate gear materials laser hardened in two programs. In a roller fatigue setup, two 3.010 inch diameter by 0.5 inch wide test rollers, are shown aligned and mounted onto parallel shafts that can be hydraulically loaded, Figures 4 and 5. Special changing gears mounted on the two shafts force the roller to rotate at slightly different speeds thereby

Figure 4 - Rollers, Support Bearings, and Lubrication System Aligned and Positioned in Roll Tester

surface contact pressure, the Hertzian load, W, is given as follows:

$$W = \frac{L \left(\frac{1}{E_1} + \frac{1}{E_2}\right)}{0.35 \left(\frac{1}{R_1} + \frac{1}{R_2}\right)} (S_c^2) \quad \text{Eq.(1)}$$

or when L = 0.5 inch (face width of rollers)
E_1 and E_2 are taken to be equal at 30×10^6 psi
R_1 and R_2 are equal to 1.5 inches (radii of rolls)

$$W = 715 \times 10^{-4} (S_c)^2$$

$$S_c = (W/715 \times 10^{-4})^{1/2} \quad \text{Eq.(2)}$$

where S_c is the Hertzian contact pressure given in ksi
and W = Hertzian load in lb.[4]

Figure 5 - Microscopic Inspection of Roller Surfaces

LASER HARDENING TESTS AND RESULTS - The laser roll testing program was conducted in phases using two medium carbon, high hardenability alloy steels whose nominal chemistries are shown in Table 1. Test results from earlier roller fatigue studies performed with furnace hardened gear materials had shown that increased surface hardnesses resulted in higher pitting threshhold limits, consequently, high surface hardness was a primary objective for laser hardening the test rollers.

In the first phase of roller tests, a medium carbon, high alloy steel, designated PDS 10325GZ, was selected because of its mechanical and chemical properties, and its availability at the Westinghouse Marine Division. The alloy had special chemistry requirements for use in the manufacture of production thru-hardened pinions which required a very high hardenability alloy. The material specification permitted a maximum carbon content of 0.45, however, the test materials available for use in the laser studies contained a relatively low 0.28 to 0.34 carbon. The low carbon content in the PDS 10325GZ alloy was a program concern because information extrapolated from carbon-martensite formation curves indicated that a program hardness objective of Rc60 could not be obtained under optimum laser conditions. The second phase of

TABLE 1 - COMPARISON OF CHEMISTRIES FOR TWO ROLLER FATIGUE STUDIES

		10325 GZ	MIL-S-19434A CLASS 5
CARBON	C	0.45/Max	0.50 Max*
MANGANESE	Mn	0.40/.80	0.60/.90
SILICON	Si	.15/.35	0.15 Min
NICKEL	Ni	2.75/Min	1.65 Min
CHROMIUM	Cr	0.40/1.25	0.50 Min
MOLY	Mo	0.30/.55	0.13/.50

*Ordered C = 0.48 -.53

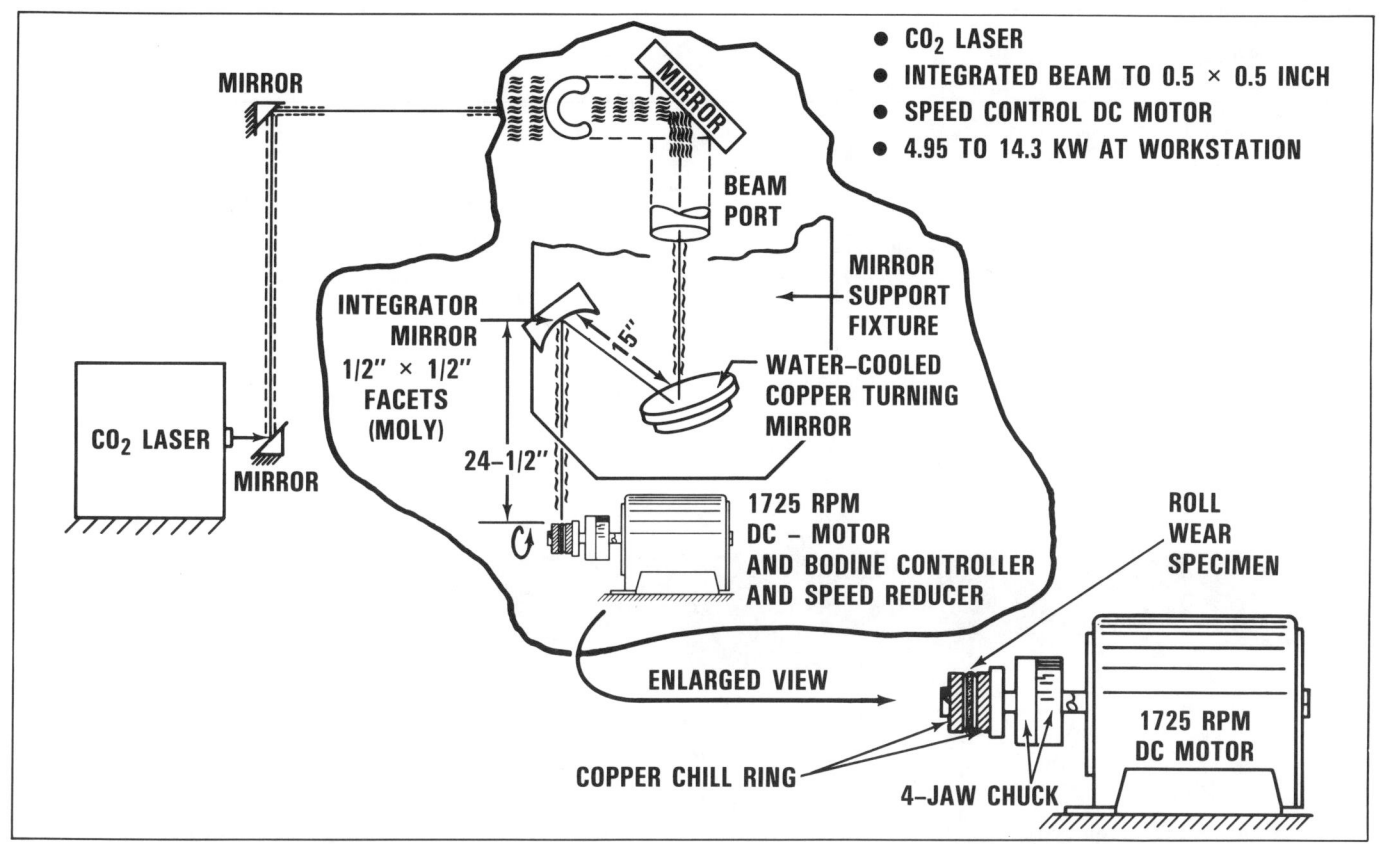

Figure 6 - Laser and Roller Fixture Arrangement for Slow and High Speed Tests

roller tests used an AISI 4350 alloy steel that was purchased to MIL-S-19434A CLASS 5. This alloy approached the chemistry and hardenability characteristics of the Westinghouse alloy and was readily available from steel suppliers. The alloy contained a carbon content range that would satisfy surface hardness objectives using available laser facilities. The Rc60 hardness would permit comparison against roller fatigue data from earlier programs on furnace nitrided samples. Both furnace nitriding and case carburizing practice can typically develop gear tooth surface hardnesses of at least Rc60.

Laser heat treated roller samples were prepared on a fixture shown schematically in Figure 6. The beam entered an enclosed work station via a series of mirrors maintained in ducts purged with dry air or nitrogen. The final mirror was a reimaging integrator mirror which focused the beam to a final 0.5 inch square heat treating geometry. The fixture used to rotate the rollers beneath the laser featured special speed controls which permitted running the first phase of roller tests at a rotation speed comparable to travel speeds used in early gear hardening investigations; the second phase of this study used the same fixture replaced with a high speed motor and digital tachometer.

The most serious problem with heat treating rollers with the slow rotation technique was the local overlap or tempered zone. The termination of the heat treating beam on a previously hardened surface after one revolution effectively softens or tempers a small zone immediately ahead of the beam. In Figure 7, a thin temper-overlap zone can be clearly seen in the center of the photographs that was prepared by overlapping the laser heat treating beam on a previously hardened surface. In this test, the material was a flat plate of PDS 10325GZ steel with a carbon content of 0.34 that was normalized, quenched, and tempered to a core hardness of Brinell 311. The zone overlap between the parallel passes on the flat plate closely simulates the overlap on the test rollers. A Knoop microhardness traverse made approximately 0.005 inch beneath the surface showed the change in hardness that occurred across the transformation hardened zone and overlap zone. The surface hardness on either side of the overlap zone ranged between Rc54 -57, however, within the overlap zone, the material was tempered or softened by as much as 10 Rockwell points. The high hardness values outside of the tempered zone indicated that the surface must have transformed to nearly 100% martensite. The micrographs also show the significant grain refinement that occurred in the laser transformation hardening zone and the sharp delineation of microstructure at the case-core interface. A CO_2 laser equipped with an integrating mirror was used in preparing this test (see parameters in Figure 7).

PHASE 1 ROLLER FATIGUE TEST RESULTS - The first phase of roller fatigue tests using the medium

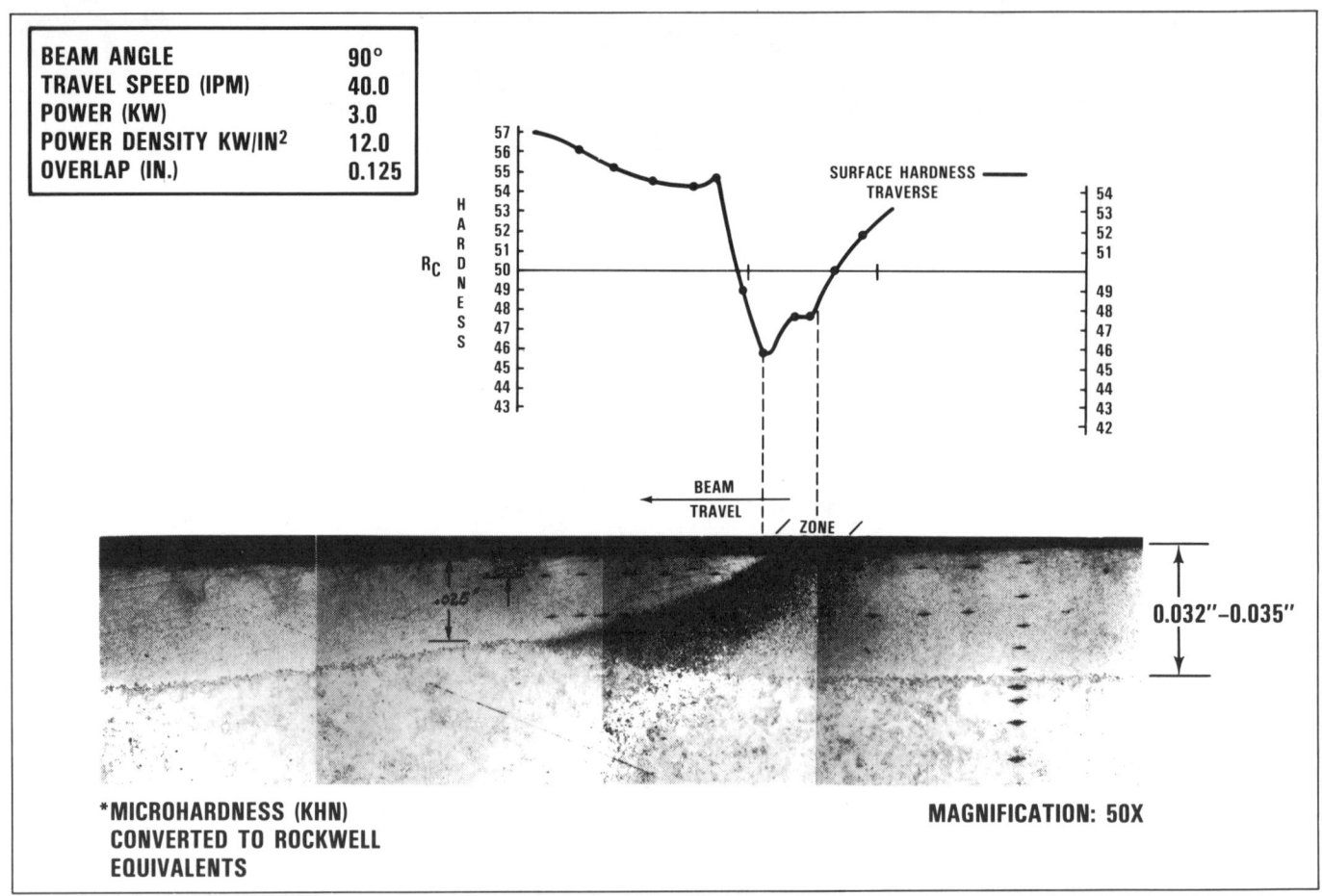

Figure 7 - Surface Tempering of Overlapping Laser Transformation Passes

carbon, laser treated PDS 10325GZ alloy were prepared using parameters listed in Table 2. Excellent surface endurance performance was obtained with this alloy even though the maximum surface hardness of Rc53 was lower than the program objective of Rc60. An S-N curve that plotted Hertzian stress vs. load cycles showed that the PDS 10325GZ alloy could sustain a Hertzian contact stress of 250 ksi for 10^7 cycles without pitting failure. At a higher Hertzian contact stress of 300 ksi, an unacceptable level of micro-pitting occurred in the softened, overlap zone. Laser treated roller test results were comparable to the highest contact pressures recorded in earlier studies on nitrided rollers even though a true measure of the surface endurance properties of a laser treated surface had not been obtained because of the overlap zone.

PHASE 2 ROLLER FATIGUE TEST PREPARATION AND RESULTS - In the second phase of roller tests, much higher rotational speeds, 1550 rpm, and significant increases in laser power over earlier tests were used. The belief was that laser power and rotational speed would circumvent the overlap zone of the earlier tests while sustaining transformation hardening at the surface. AISI4350 alloy steel was selected to increase the surface hardness potential, the heat treating fixture was modified with a high speed motor, and procedural changes were made in laser power levels and couplant to eliminate the soft overlap zone of earlier tests. The test fixture was modified with a water misting system to externally spray quench the roller surfaces after the laser was shutoff. The higher laser power settings were required to maximize surface austenization while minimizing overall roller heatup. During the preparation of roller test samples, the laser power was fixed at 14.3 kW at the work station. In Table 2, a comparison of the laser parameters for the two phases of roller test shows that approximately 10 kW of additional power was used in the second phase of tests. The laser couplant was changed to a phosphate coating to improve coupling (i.e., minimize reflections) the laser beam to the roller surface. Figure 8 shows the general arrangement of the test setup used in these high speed tests.

Case depth was controlled by varying the heat treating time. A change of more that 1 to 1.5 seconds could alter case depth by 0.015 to 0.025 inches. Test samples were hardened and sectioned to verify that a case depth of between 0.040 and 0.060 inch was obtained. An examination of the roller showed that the overlap zone had been successfully eliminated and that a surface hardness objective of Rc60 had been attained on roller surfaces.

TABLE 2 - COMPARISON OF LASER PARAMETERS USED TO SURFACE HARDEN
LOW SPEED AND HIGH SPEED ROLLERS

	PHASE 1	PHASE 2	
	3-10	6-7	6-15
MATERIAL	PDS 10325GZ	AISI4350	AISI4350
CORE HARDNESS	311-352 BHN	SAME	SAME
LASER POWER	4.5 kW	14.3 kW	14.3 kW
POWER DENSITY	18 kW/IN^2	57.2 kW/IN^2	57.2 kW/IN^2
SURFACE HARDNESS	Rc53	Rc59 -60	Rc59 -60
CASE THICKNESS	0.060 IN.	0.040 IN.	0.055 IN.
COUPLANT	PAINT	PHOSPHATE	PHOSPHATE
TRAVEL SPEED	53.5 IPM (5.7 RPM)	1550 IPM	1550 IPM
TIME (SEC.)	10.6	6.5	8.0
COOLING	AMBIENT	MIST	AMBIENT
RESIDUAL STRESS			
o SURFACE	-101 ksi	+12.7 ksi	-46 ksi
o 0.003 INCH	-62 ksi	-58 ksi	-37 ksi
o 0.035 INCH	-23 ksi	+75 ksi	-50 ksi

The second phase of roller fatigue tests began at a contact stress of 300 ksi where earlier GZ tests had failed to reach a 10^7 cycle endurance limit due to micropitting in the overlap zones. The tests were begun with a great deal of optimism, but instead of performing well at this stress level, vibration sensors built into the roll test machine shut down the tests after less than the required number of cycles. An inspection of test rollers revealed that transverse surface cracks had formed on the driver and driven rolls. This was a first instance of surface cracking occurring in a roller test program. Cases of crushing failures had occurred in earlier programs that used softer core hardness materials and thin case depths, however the nature of the transverse cracking in the high speed 4350 alloy tests was completely new. The reason for the failures were felt to be either equipment, materials, or process oriented. There were no indications of problems with the roller samples when inspected by visual, dye penetrant, and magnetic particle methods.

The elimination of equipment and test part geometry as possible contributors to the failures, indicated that procedure changes in the laser heat treating cycle were the probable causes of the failures. An examination of one of the transverse cracked rollers and a review of the laser procedure was initiated.

ROLLER #6-4 FAILURE INVESTIGATION - Roller #6-4 was selected from the high-speed tests for failure analysis because it contained a single transverse surface crack that had occurred after approximately 3 million cycles. Other rollers had shown multiple surface cracks, which was felt to be less valuable for examining the origin of the failure. When the roller containing the transverse surface crack was broken open, the most notable visual observation was a cleavage crack parallel to the roller surface at a depth approximately 0.035 in. This was approximately the location of the case core interface, the transition zone between the hard martensitic case and the ductile core. Low magnification photographs of the fracture face in Figures 9 and 10 show the transverse crack as it progressed through the laser-hardened roller surface into the ductile core material. The fracture surface appeared to have macroscopic beach marks bowing to the surface and towards the inside radius of roller at approximately the location of the case-core interface. At higher SEM magnifications, Figures 11 and 12, the fracture face immediately beneath the surface clearly showed fatigue striations in the hardened zone just beneath the roller contact surface. As the crack progressed into the hardened roller, the striation spacing decreased rather than increased as was expected. The decreasing spacing of the fatigue striations, and the macroscopic beach marks were indicators that the failure may have initiated at the case core interface. However, maximum contact stresses occur at the surface and maximum shear

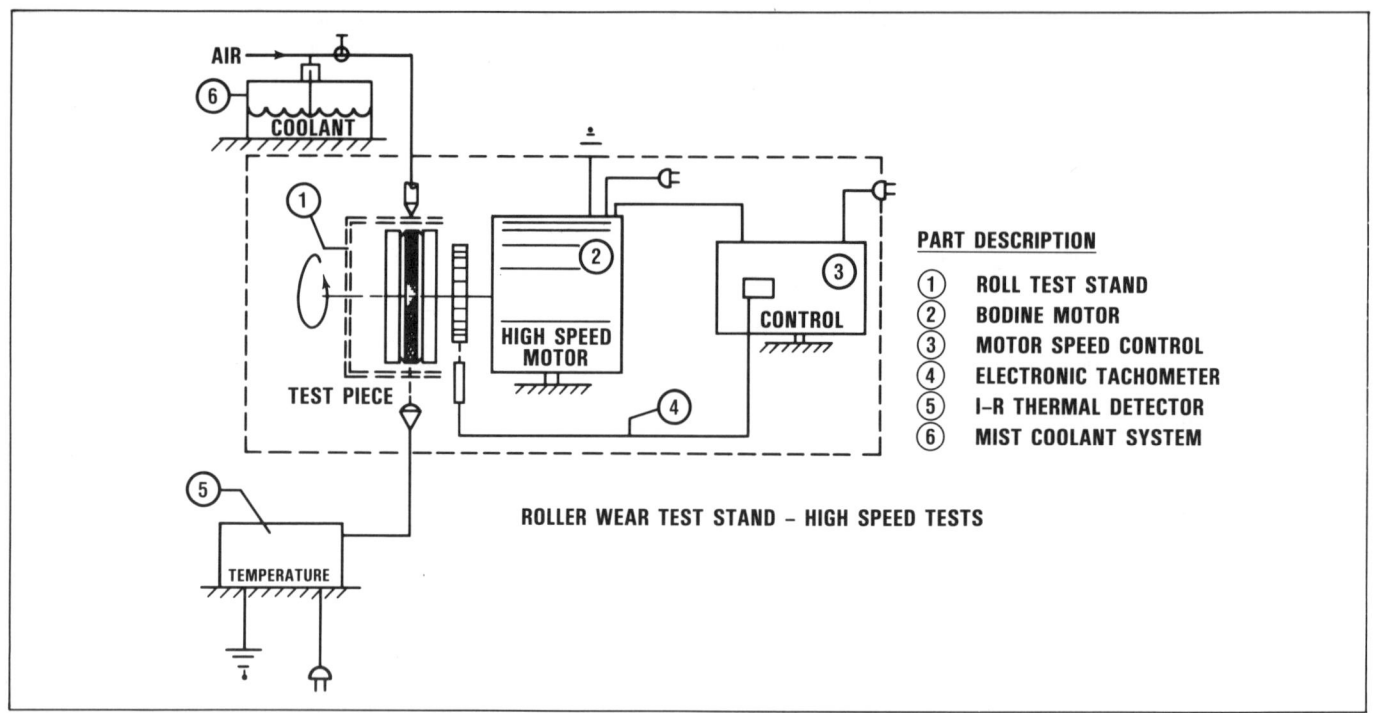

Figure 8 - Schematic of High Speed Roller Test Fixture with
Water Mist Cooling and Infrared Temperature Control Features

stresses occur a few thousands beneath the surface, therefore supporting an alternative opinion that failure should initiate in that zone.[5,6] Because the failure mode and initiation site were unclear, it was decided to use x-ray diffraction to quantify residual stress profiles in both long running roller tests and the high speed rollers which failed by cracking.

Three rollers were selected for residual stress examination using the x-ray diffraction method. Each stress profile included surface and subsurface measurements made after electropolishing to specific depths. The three rollers represented laser conditions performed in the initial slow roll test of PDS 10325GZ alloy, and two conditions taken from the AISI4350 alloy. Roller #6-7 represented an untested roller taken from the same lot of water mist cooled rollers as the one that had failed by transverse surface cracking. Roller #6-15 was also an untested roller that differed from the rest of the high-speed rollers because the external mist quench had not been used to treat the surface. Roller #3-10 represented a tested sample from the first phase of tests that had survived the 2×10^7 cycles without pitting.

A sharp contrast of residual stress values occurred when examining high speed rollers #6-4, #6-15, and low speed #3-10.

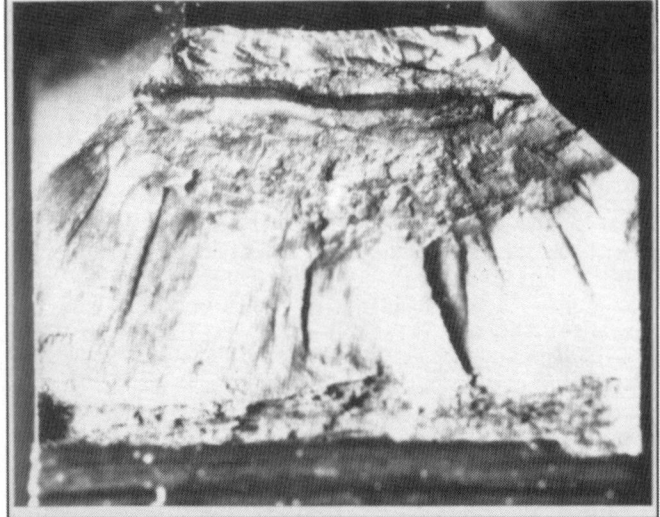

Figure 9 - Fracture Surface of Roller #6-4; Mag: x8

Figure 10 - Fracture Surface of Roller #6-4 with Beach Marks Bowing About Cleavage Crack; Mag: x25

Figure 11 - Roller #6-4: Fatigue Striations in Laser-Hardened Case; Mag: x800

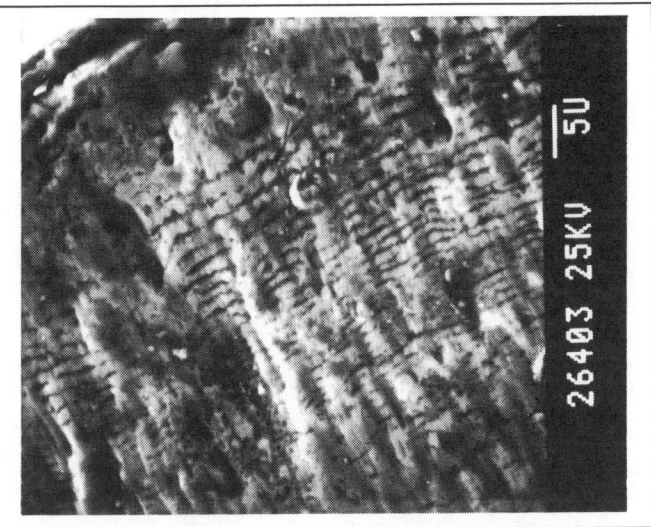

Figure 12 - Roller #6-4: Closeup - Fatigue Striations Becoming Closure Together with Depth into Surface; Mag: x1600

The residual stress profile made on Roller #3-10 showed a surface compressive stress of -101ksi, and subsequent lesser magnitudes of compressive stresses to a depth of 0.038 inch where tensile stresses were first noted. A similar residual stress pattern was exhibited on Roller #6-15 except the surface stress was not as highly compressive as Roller #3-10.

The stress pattern was completely different in Roller #6-7 than the other two test pieces. The surface exhibited a tensile stress of +12.8 ksi at the surface which rapidly changed to compressive stress immediately beneath it. It again changed to tensile residual stress approximately where the critical case core interface occurs. Figure 13 (a, b, & c) shows graphs of residual stress plotted against depth into the roller surface for the three rollers. Changes to the laser heat treating procedure had effected the stress profiles in each of the three tests.

The presence of tensile residual stresses on the surface of water mist quenched rollers could explain the rapid failures in this test series. Superimposing tensile residual stresses onto a surface already under high Hertzian contact stresses with maximum subsurface shear stress immediately beneath the surface could account for a crack initiating at or very near the surface under the cyclic loading of the two rollers. The subsurface compressive residual stress zone located immediately beneath the tensile surface would resist further crack propagation thus providing a possible explanation why the fatigue striation spacing decreased as the crack grew deeper. Eventually, the cyclic loading of the highly stressed rollers propagated the crack towards the case-core interface where residual tensile stresses were measured. The transverse crack in a zone of tensile residual stresses helped propagate the cleavage failure noted earlier.

The unexpected failure of rollers in Phase 2 tests was attributed to changes in the heat treating procedures. The use of residual stress measurements provided a quantitative means of analyzing the results of these changes. An examination of the residual stress profiles, Figure 13 and Table 2, shows that rollers exhibiting continously compressive residual stress profiles had a common element of air or ambient surface quenching after laser hardening. In sharp contrast, the spray mist rollers exhibited tensile residual stresses at critical locations in the case, e.g., the surface and near the critical case core interface. Superimposing cyclic loading conditions onto roller surfaces exceeded the fatigue endurance limit of the material resulting in transverse surface cracking and shearing at case-core interface. The values of residual stress accessible by the x-ray diffraction method clearly showed that laser procedural changes had influenced the outcome of the tests.

Further residual stress evaluations of the (thick/thin) case were unnecessary because of the similarity of failures in both thin and deep case AISI4350 rollers. The roller test program had provided a quantitative measure of surface endurance for two laser treated materials and thus had satisfied early program objectives. The use of spray quenching was discontinued as a means of securing additional surface hardness. A high hardenability alloy with an intermediate carbon level such as AISI 4340 was felt to be a better base metal choice for future development. The use of residual stress measurements to evaluate the roller fatigue program had shown the influence that residual stresses could have on cyclically loaded parts. Residual stress determinations were specified for future laser programs.

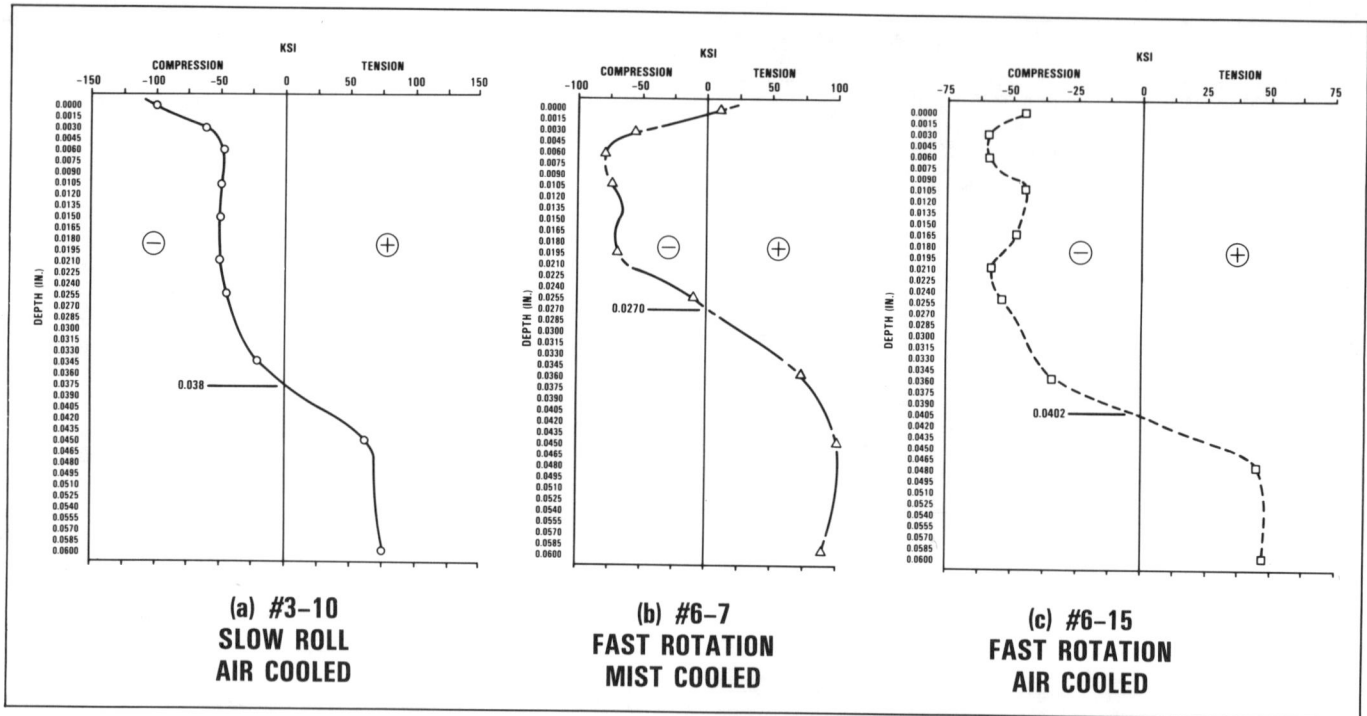

Figure 13 a, b, c - Residual Stress Measurements on Air-Cooled Rollers #3-10 and #6-15 Showing Surfaces in Compression; Water Quenched. Roller #6-7 Showed a Surface in Tension

SPLINE HARDENING

The technology acquired in the roll test program coupled well with a heat treat application that promised a shorter lead time into a production application than laser hardened marine gears. Within locked train double reduction gear boxes are torque transmitting quill shafts, Figure 1, that couple the 1st reduction gear to the 2nd reduction pinion. The ends of the nearly 9-foot shaft of AISI 4340 alloy were machined with 5/10 diametral pitch spline teeth which would eventually be furnace nitrided for 72 hours at 925-975 °F. Since the spline teeth and a local circumferential zone adjacent the teeth were the only portions of the shaft requiring surface hardening, extraordinary measures were required to protect critical shaft dimensions and nonhardened surfaces while the part was in the furnace. The return of the shaft from the furnace subcontractor initiated a series of costly machining and inspection operations that increased cycle time and costs. The decision to investigate the laser transformation hardening process for this application was easily justified.

The qualification procedures for laser hardening splines were successfully performed by progressively heat treating individual spline teeth using either scanning or integrating optics. Figure 14 shows how the flanks and common or shared roots between adjacent spline teeth were simultaneously surface hardened. The 30° pressure angle and the aspect ratio (width to height) of spline teeth allowed direct beam impingement on the common spline flanks and shared root which greatly simplied the hardening sequence. Water cooled copper shields shown just above each spline tooth served two purposes in the qualification. First they protected the tips of the spline teeth from direct coupling with the laser beam thereby avoiding local overheating and possible melting, and secondly, the copper shoes were fitted with orifices to water cool the adjacent flanks on either side of the surface being heat treated. The water cooling served several important functions. It helped maintain the overall shaft temperature below 115 °F, which had been established as an interpass temperature between heat treat passes to assure good mass quenching of the laser. It also minimized back tempering (softening) of previously hardened flanks. Without the external quench, a previously hardened flank could temper below the desired hardness level.

CASE DEPTH AND RESIDUAL STRESS MEASUREMENTS IN SPLINES - The laser spline hardening program did not have a fatigue test requirement because as torque transmitters in power trains, splines do not receive the cyclic bending and sliding loads that occur with gears. Earlier roller fatigue tests had shown that procedural changes in the laser heat treating process could effect the residual stress state, therefore, one of the first tests performed on spline teeth after initial hardness, depth and uniformity requirements had been satisfied was a determination of residual stresses. The x-ray diffraction method was used to measure the magnitude, sign, and depth the residual stress penetrated into the transformation hardened zone before changing signs. Another concern was the

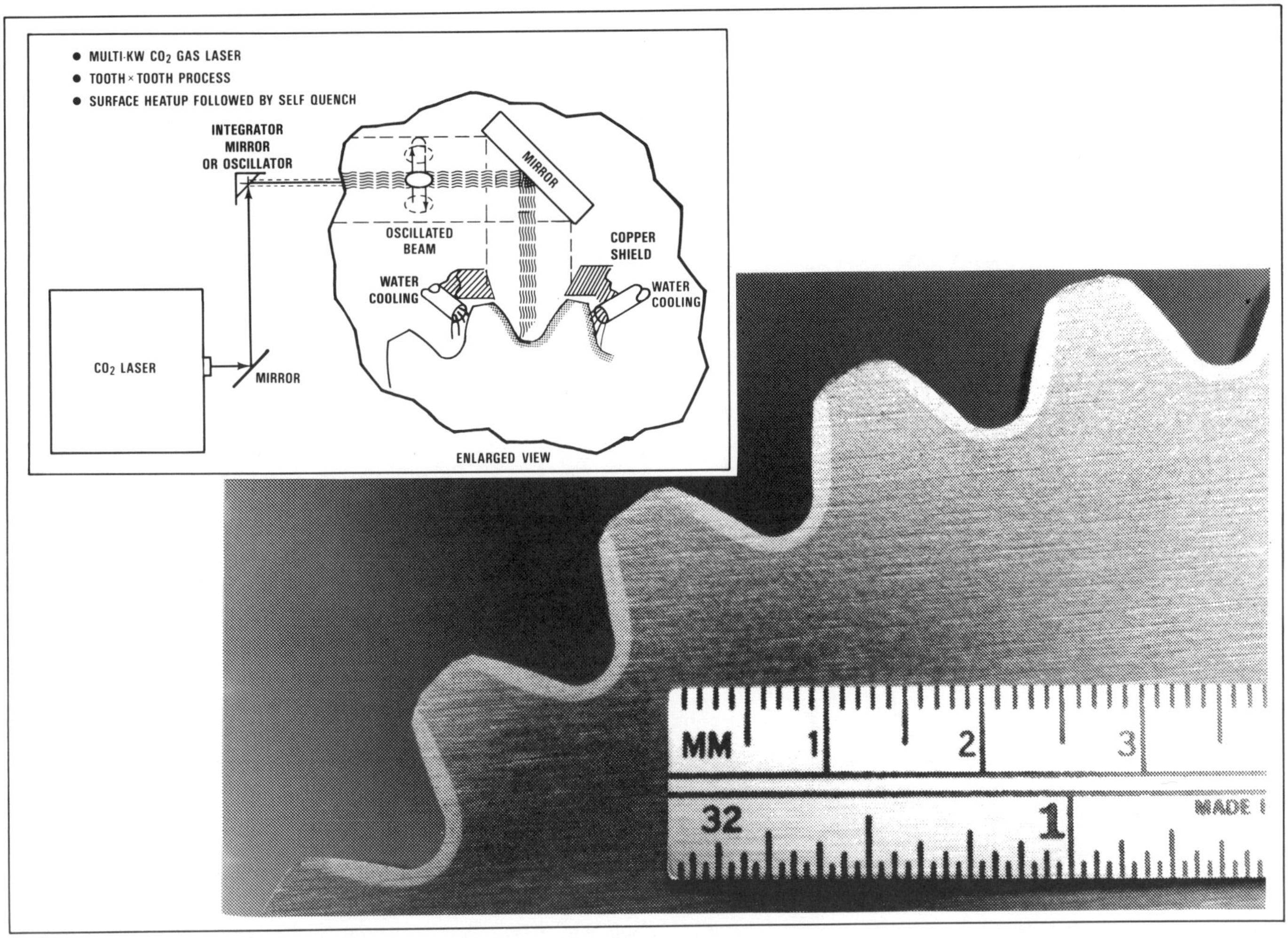

Figure 14 - Schematic Showing Spline Tooth Being Laser Transformation Hardened Using an Oscillating Beam Technique

effect that back tempering would have on residual stress values. A general relaxation of compressive residual surface stresses was expected on the back tempered flanks due to the reduction in surface hardness, but the extent of relaxation required investigation. Questions were raised whether procedures such as shotpeening would benefit previously hardened surfaces.

Figure 15 depicts a range of case hardness for spline teeth made from AISI 4340 alloy that were heat treated to a nominal case depth of 0.045 inch (at Rc50). The lower hardness values result from back tempering of sequentially hardened spline teeth. Microhardness checks on numerous occasions confirmed that an acceptable hardness range of R_C 54-60 was attainable using AISI 4340 alloy with carbon between 0.38 and 0.43.

X-ray diffraction measurements were made to differentiate the residual stress values at selected positions on the flanks and root of splines. Figure 16 shows the locations of residual stress measurements at the surface and at incremental depths beneath the surface at three positions along the involute spline tooth profile: the pitch diameter, root radius 45° and bottom root area. Figures 16, 17, and 18 summarize the results of residual stress profiles in these tests. The maximum compressive residual stress values always occurred on a non back-tempered surface.

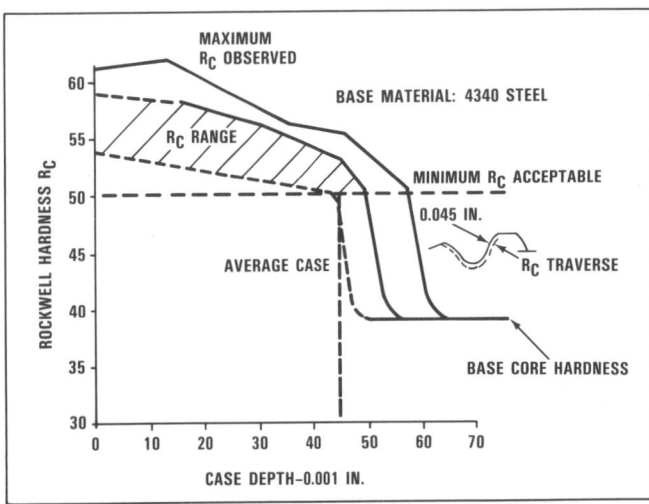

Figure 15 - Typical Hardness Range Measured on Spline Teeth Laser Hardened to a Case Depth of 0.045 Inch

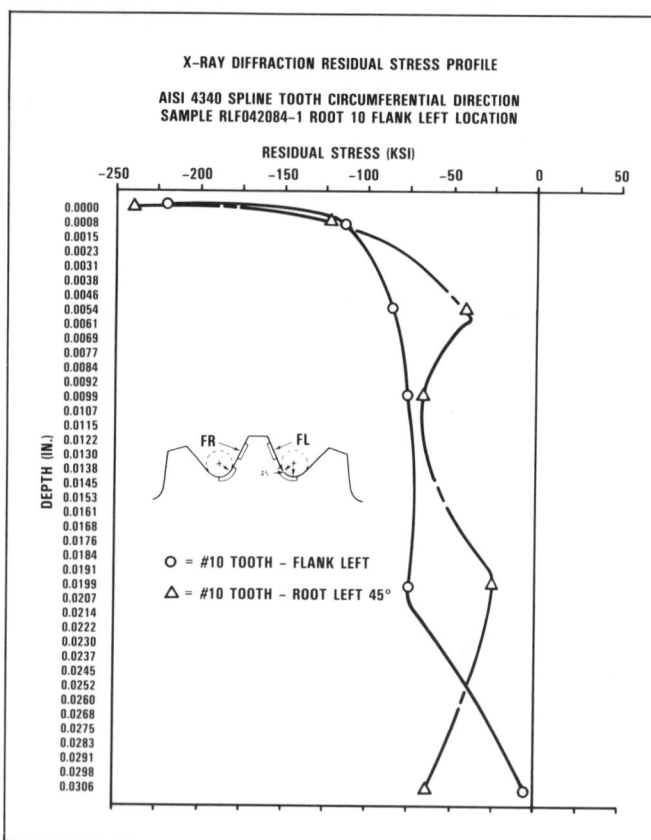

Figure 16 - Residual Stresses in Laser-Hardened Splines

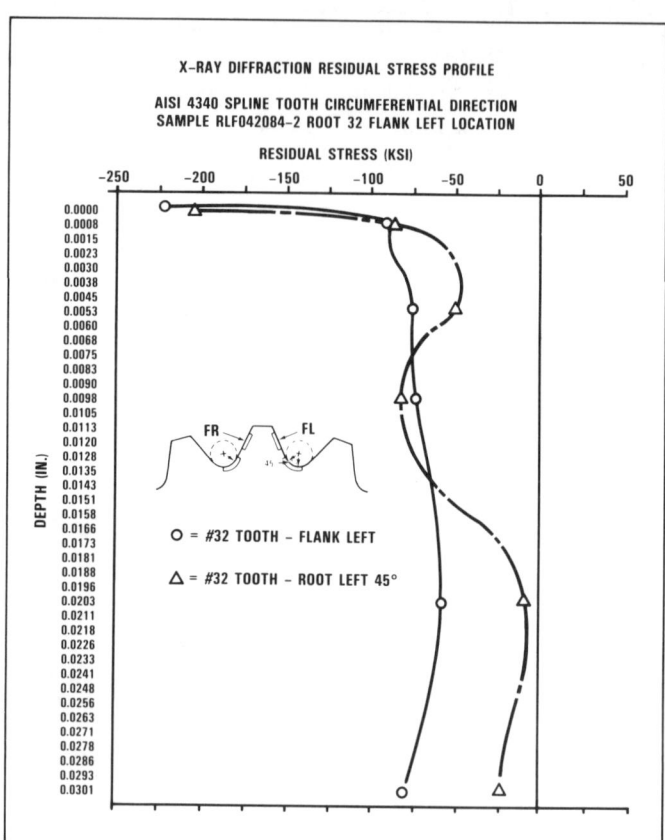

Figure 17 - Residual Stresses in Laser-Hardened Splines

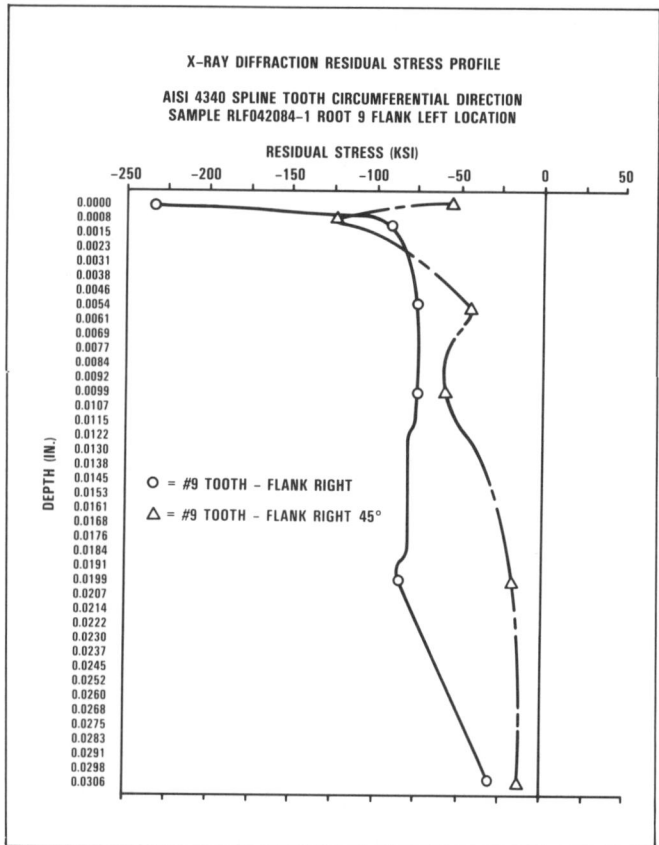

Figure 18 - Residual Stresses in Laser-Hardened Splines - Back-Tempered Flanks

The surface compressive stresses often exceeded -200 ksi, however by approximately 0.0015 to 0.005 inch, the residual stress values stablized between -50 ksi and -100 ksi. The residual stress measurements on the slightly back-tempered flank of the spline tooth were more varied than their counterpart flanks. Residual stress measurements at the surface varied from a low of -58 ksi (-126.5 ksi @ 0.001 in) to a high of -219 ksi (-94.4 ksi @ 0.0012 in). The variation in stress magnitudes may have resulted from slight changes in cooling procedures. More importantly, residual stresses remained compressive well beneath the surface of the spline tooth. Residual stress measurements made at a depth of 0.040 inch, had values that ranged from a very compressive -57 ksi to tensile +5.7 ksi. Residual stress measurements performed in the bottom root between adjacent splines showed compressive surface stresses of -135 ksi and -188 ksi.

SHOTPEENING OF LASER TREATED SPLINES - One last investigation performed on early splines studied the effects of shotpeening a previously laser-hardened surface. Residual stress measurements were made to establish surface stress changes after shotpeening with the parameters shown in Table 3. Little change was anticipated in the surface residual stress values because the shot used was softer than the spline surface.

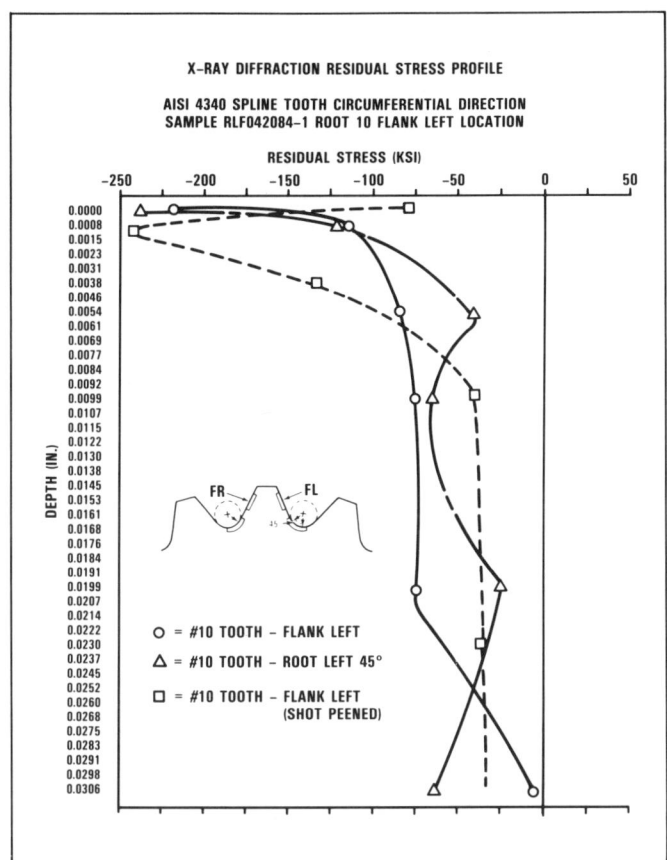
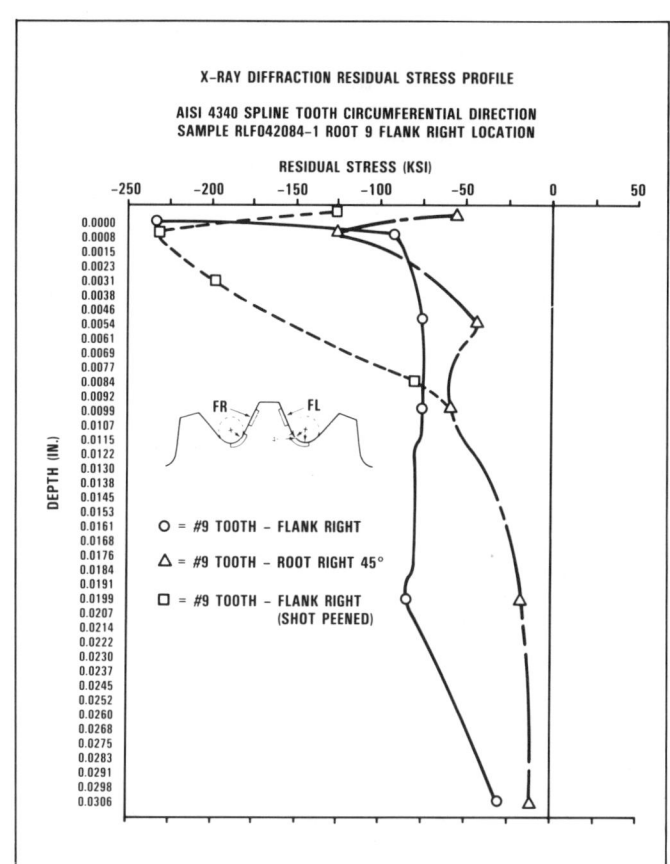

Figure 19 - Shotpeening Laser-Hardened Splines

Figure 20 - Shotpeening Laser-Hardened Splines - Back-Tempered Flanks

TABLE 3 - SHOTPEENING DATA USED TO TREAT LASER-HARDENED SPLINES

SHOT SIZE	110
SHOT TYPE	CAST STEEL R_C 45-48
COVERAGE	200%
INTENSITY	A 9-12 (ALMEN STRIP "A")

Residual stress measurements made on involute spline teeth previously checked for residual stress values showed changes in surface stress magnitudes had occurred after shotpeening. Figures 18-20 show that surface residual stress values were less compressive than earlier studies, however, more importantly, residual stress values were more compressive as measured 0.001 to 0.003 inch beneath the treated surface. This investigation indicated that shotpeening could relocate the maximum compressive stress below the surface. This condition could benefit surfaces requiring grinding for special dimensional requirements, repair zones where softening due to overlap or back tempering could have resulted or applications where subsurface stresses were a concern.

SUMMARY

The roller fatigue program and the spline hardening program used residual stress determinations in the evaluation of laser heat treating procedures and in the resolution of a fatigue failure investigation. The roll test program demonstrated that procedural changes in a laser heat treating program could alter normally compressive surface stresses making them tensile. X-ray diffraction residual stress measurements were felt to be an effective means studying subtle changes in the heat treating process. The investigations performed on roller wear fatigue samples and on involute spline teeth made from AISI 4340 alloy resulted in the following observations:

o Residual stress magnitudes exceeding -200 ksi were recorded on spline shafts of AISI 4340 alloy treated by laser procedures.

o Shotpeening of previously laser hardened surfaces resulted in maximum residual stresses being relocated slightly below the surface. This could potentially be beneficial for hardened gears in which the surface is removed by final grinding operations for quiet running performance.

o Surface residual tensile stresses resulted from water mist quenching roller surfaces immediately after terminating the laser heat treating beam. Tensile surface stresses resulted in the premature failure of highly stressed rollers. Subsurface compressive residual stresses resisted further propagation of surface initiated cracks in highly stressed rollers.

o Slight back tempering of previously hardened surfaces generally results in lower compressive stress values.

The roller fatigue and spline tooth investigations demonstrated that residual stress determinations could be a valuable diagnostic tool in developing laser heat treating procedures and examining fatigue programs. There will continue to be further investigations as this program matures.

ACKNOWLEDGMENTS

The author wishes to acknowledge Dr. J. Nurminen and Mr. J. Smith of the Westinghouse R&D Center (WEC R&D) for helping conduct laser tests on the roller fatigue specimens. Mr. Howard Kaufman, WEC R&D, performed all roller fatigue testing. Mr. Leo Albertin, WEC R&D, and Dr. A. Nakagawa of the Westinghouse Marine Division are acknowledged for providing technical advice during the preparation of this paper.

REFERENCES

1. REED-HILL, R. E., "Physical Metallurgy Principles", pp. 495-505, Van Nostrand Company, Princeton, N.J. (1964).

2. IBID, p.504.

3. KUNSMAN, L. D., "Effect of Material Parameters on Gear Load-Bearing Capacity in Marine Reduction Gearing," Westinghouse Electric Research Report 78-ID9-GERMA-RI (1978).

4. IBID, p.14.

5. IBID, p.45.

6. CASTLEBERRY, G., "Designing for Contact Stresses," Machine Design, 8 August 1985.

DETERMINATION OF THE SOURCE OF VARIANCE IN DISTORTION OF SURFACE HARDENED BEARING RACES

David L. Milam
The Timken Company
Canton, Ohio USA

ABSTRACT

The source of variance in the out-of-roundness (OOR) of surface hardened outer bearing races was sought by evaluation of residual stress, retained austenite, microhardness profile, and constituents of the microstructure. A correlation was found between OOR, residual stress, and microhardness profile. Variation of the heat input to the outer surface during hardening was determined to be the source of the variance in OOR. Residual stress existing in the races prior to hardening was not a factor.

INTRODUCTION

DISTORTION of precision components during their manufacture is compensated for by a metal removal finishing operation. Metal removal from hardened steel precision components such as bearings and gears is costly because the hard material (> 58 HRC) necessitates the use of grinding or ceramic tools. The cost of metal removal is proportional to the extent of distortion existing in the hardened component. A reduction in the extent of distortion will result in a reduction in the cost of finishing and a reduction in the variance of metallurgical quality.

Distortion of bearing races during heat treatment is caused by the following factors: relief of residual stress, nonuniform heating or cooling, and mechanical stress.[1,2] Bearing races are generally produced from rings cut from a tube, forging, or rolled ring. Residual stress created in the ring during manufacture persists in the race after machining. Heating of the race for carburizing or hardening will allow this stress to be relieved because the yield stress decreases with increasing temperature and becomes less than the residual stress. Permanent plastic deformation, resulting in distortion, will accompany this relief of residual stress. The extent of this distortion is a function of process temperature, magnitude of residual stress, and thickness of the race. Distortion attributable to relief of residual stress can be minimized by reduction of process temperature or by elimination of residual stress with a stress relief or normalizing treatment applied to the ring.

The constituents of the microstructure of a bearing race prior to carburizing or hardening are ferrite, carbide, and ferrite plus carbide (pearlite). During austenitizing, the microstructure transforms to austenite and carbide. During quenching, most of the austenite transforms to martensite. An expansion in volume occurs during heating and again during quenching. An expansion in volume occuring at one position in a race will cause distortion if the volume expansion does not occur simultaneously at all positions in the race.

Phase transformation will <u>never</u> occur simultaneously at all locations in a race because of (1) Variation of the concentration of carbon with depth below the surface of carburized races and (2) Variation of temperature with depth below the surface resulting from the finite rate of heat conduction across a surface and between surface and core. The simultaneous phase transformation, that is, transformation at the same time, at <u>constant</u> <u>depth</u> below the surface can be approached by (1) Creation of a uniform carbon concentration profile around the race and (2) Elimination of temperature variation at constant depth during heating and cooling. It has been reported that improved homogeneity of steel hardenability also will reduce the variation in the start of allotropic phase transformation at constant depth.[3] Distortion will decrease as the phase transformation at constant depth tends to occur at the same time.

The yield stress of a bearing race at the carburizing or hardening temperature is 34 to 68 MPa (5 to 10 ksi). An imposed stress--resulting from stacking or material handling within the furnace--could deform the race. Distortion attributable to mechanical stress can be minimized by the use of fixturing and designing furnace and quench systems to prevent creep, relative movement between races, and impact.

Phase transformation hardening by rapid heating and cooling of the surface will produce less distortion of a bearing race than carburizing and quenching because the core is not heated. Phase transformation of the core is not possible. The core of a surface hardened race will have a greater yield stress than that of the core of a carburized race, so it can better resist the nonuniform volume expansion resulting from nonsimultaneous phase transformation of the austenitized layer. The extent of distortion resulting from relief of residual stress during heating, a significant cause of distortion, will depend upon the magnitude of residual stress existing in the race prior to surface hardening.

Consequently, a high energy input surface hardening process will produce hardened precision components requiring less metal removal in the finishing operation. Reduction of finishing cost is possible through use of surface hardening processes. Examples of high energy input surface hardening processes are induction [4], laser [5], electron beam [6], and flame [7].

Process development consists of modifying equipment and process to produce acceptable components. Critical to the development of a surface hardening process is the evaluation of the hardness profile, size and shape, and cost. During an evaluation of bearing race shape produced by surface hardening, an unexpected large variance in out-of-roundness (OOR) was observed.

The objective of this study was to determine the cause of the variance of OOR. The variance of OOR was assumed to be caused by a variance in the magnitude of residual stress existing in the races prior to surface hardening or the inconsistency of the process. Relief of residual stress during heating was considered a factor because the thickness of the race was believed insufficient to prevent heating of the core. It was believed that an inconsistency of heating could be detected by a measurement of residual stress at three locations of the race because an inconsistency of heating will create a nonsimultaneous volume expansion. Nonsimultaneous expansion of volume around the race will create a nonuniform residual stress around the race because the unheated core is resistant to yielding. It was expected that a correlation between OOR and the residual stress state of soft and/or hardened races could be found. This paper describes the residual stress state found existing in soft and hardened races, the correlation made between residual stress and the amount of OOR, and the identification of the source of stress and distortion.

PROCEDURE

Three outer races that had been hardened and two races that had not been hardened were evaluated. The races had been cut from a tube of spherodized AISI 51100 steel and stress-relieved. The OOR of the races is shown in Table I.

Table I. Out-Of-Roundness Of Races, mm (in.)

Code	Prior To Hardening	After Hardening
82	0.013 (0.0005)	0.406 (0.016)
87	0.013 (0.0005)	0.254 (0.010)
98	0.013 (0.0005)	0.076 (0.003)
99	0.013 (0.0005)	Not Hardened
101	0.013 (0.0005)	Not Hardened

The two unhardened races had been stress-relieved and machined along with the other races that had been subsequently hardened. It was assumed that the residual stress state existing in these two unhardened races was identical to that existing in all the races prior to hardening.

The residual stress in the circumferential direction was measured on and below the outside surface (O.D.) and the inside surface (I.D.). The measurements were made on pieces cut from three equidistant positions on each race. The location of these positions relative to the location of the largest diameter of the race is shown in Fig. 1. The stress was measured using a $\sin^2\Psi$ or multiple tilt method.[8] For this work, a chromium x-ray tube was used to analyze the (211) planes. The power level was 35 kV and the current was 1.5 ma. Five Ψ angles were used: -42, -29, 0, 29, and 42 degrees. A 6 degree Ψ angle oscillation was used to minimize the effect of K (alpha) separation on the calculation of stress. The upper 30 percent of the peak was used for calculation of peak location. An elastic stress constant of 3.714×10^{-8} was used in the calculation of stress. This value of elastic stress constant is one that is used for quenched and tempered carburized steel. Because the study focused on relative stress rather than absolute stress, an x-ray elastic stress constant for this material was not measured. The error in stress calculation varied from 103 - 172 MPa (15 - 25 ksi) at the surface to 14 - 28 MPa (2 - 4 ksi) below the surface.

Stress was measured at the surface and below the surface. Material was removed in increments of 0.25 or 0.50 mm (0.010 or 0.020 in.) by electropolishing. Grinding was used to compensate for nonuniform material removal by electropolishing that was encountered at depths greater than 1.5 mm (0.060 in.). At least 0.25 mm (0.010 in.) of material was removed by electropolishing after grinding to eliminate the grind-affected layer. The stress calculations were not corrected for depth of material removed.

Microstructure specimens were cut from each of the three sections of each race following completion of the residual stress measurements. Microhardness traverses using a Knoop indentor under a 500 gram load were made in a direction from the inner and outer surfaces toward the core.

RESULTS AND DISCUSSION

Residual stress measurements of an unhardened race to a depth of 1 mm (0.040 in.) below the O.D. surface are shown in Fig. 2. Each curve

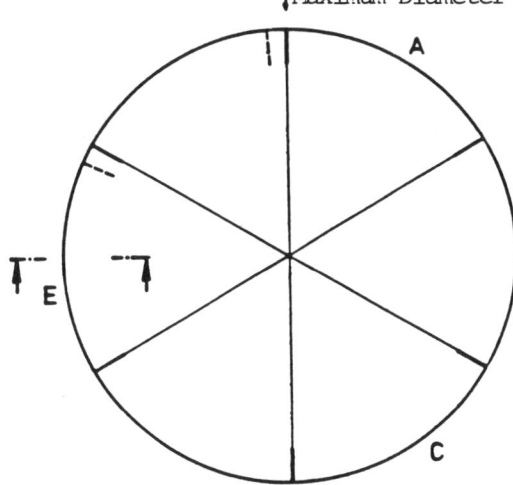

Fig. 1—Location of residual stress measurements relative to position of maximum diameter.

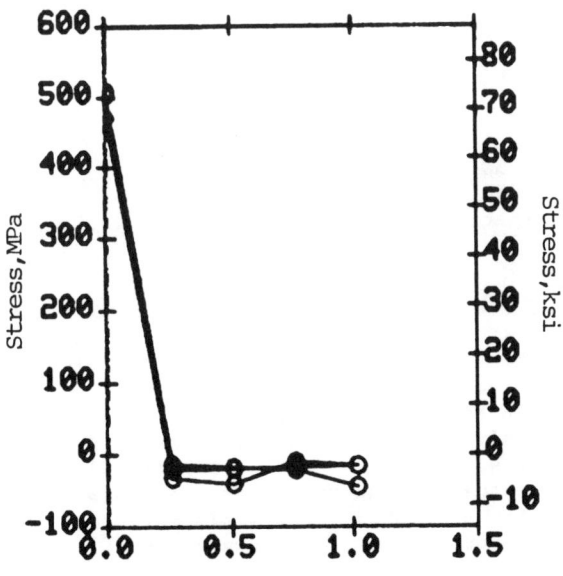

Fig. 2—State of residual stress existing prior to rapid heating and cooling of O.D. surface.

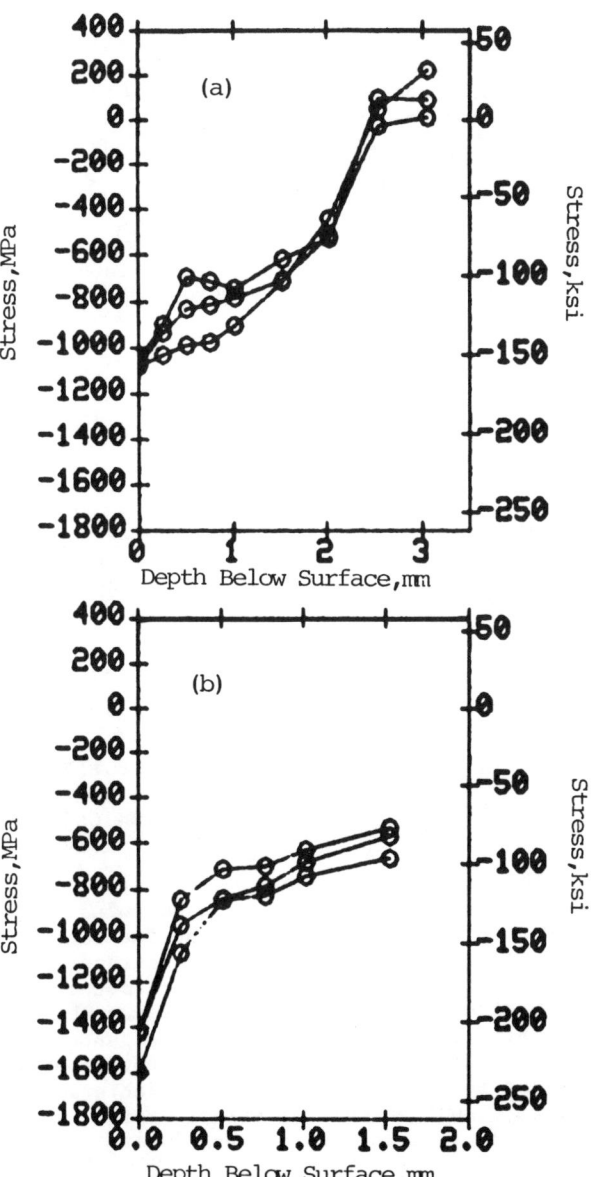

Fig. 3—Residual stress profile existing after rapid heating and cooling of (a) O.D. surface and (b) I.D. surface.

represents one of the three locations on the race. Other than a tensile stress existing at the surface, the race was stress-free. The tensile stress existing at the surface was created by machining of the race from the tube. It can be seen that the depth of machining-affected material is < 0.25 mm (0.010 in.). Beyond 0.25 mm (0.010 in.) depth, the residual stress varied between 0 and -34 MPa (0 and -5 ksi).

Rapid heating and cooling of the surface significantly altered the state of residual stress, Fig. 3. The residual stress at the surface changed from a tensile stress of 482 MPa (70 ksi) to a compressive stress of -1,034 MPa (-150 ksi) on the O.D. and -1,379 MPa (-200 ksi) on the I.D.. Stress calculations at depths > 1.5 mm (0.060 in.) below the I.D. surface are not reported here because of (1) Inability to control the metal removal process and (2) A large error in location of the centroid of the diffraction peak. Surface hardening produced a greater compressive stress than is typical of carburizing and quenching (-345 MPa (-50 ksi)).

The residual stress created by surface hardening decreased continuously below the surface until it became tensile at 2.5 mm (0.100 in.) depth. This is the state of stress in cut pieces. The residual stress state in uncut races were measured and it was determined that the transition from compressive to tensile stress occured at 3.5 mm (0.140 in.) depth below the surface. The depth of zero residual stress produced by surface hardening this high carbon-containing material is greater than that produced in carburized and quenched races to the same aim depth of hardening. The source of this significantly more compressive and deeper stress is in the unaustenitized core. Because the temperature of the core of a race being surface hardened will be less than that of a race being

Table II. Residual Stress Measurements (MPa) of Hardened Races

(a) Outer Diameter

Depth Below Surface, mm

Race/Position	0.00	0.25	0.51	0.76	1.02	1.52	2.03	2.54	3.05
82 A	-1224	-1278	-1142	-1041	-944	-842	-467	-274	-139
C	-1132	-1242	-1054	-978	-1098	-774	-419	17	61
E	-1122	-1122	-1111	-968	-862	-725	-580	156	76
87 A	-1082	-1031	-991	-977	-906	-712	-437	49	227
C	-1078	-935	-832	-812	-784	-709	-512	-23	14
E	-1050	-899	-696	-714	-741	-616	-527	100	89
98 A	-968	-840	-684	-686	-725	-667	-477	-66	52
C	-856	-1014	-1100	-1020	-887	-629	-487	-72	46
E	-898	-994	-794	-853	-771	-471	-387	-54	-8

(b) Inner Diameter

Depth Below Surface, mm

Race/Position	0.00	0.25	0.51	0.76	1.02	1.52
82 A	-1390	-1379	-1275	-1144	-920	-831
C	-1480	-947	-886	-802	-807	-656
E	-1404	-869	-859	-822	-767	-662
87 A	-1595	-1078	-843	-826	-741	-666
C	-1423	-954	-839	-782	-681	-564
E	-1414	-847	-710	-698	-630	-530
98 A	-1379	-1012	-814	-742	-760	-601
C	-1244	-1046	-838	-914	-816	-707
E	-1305	-1068	-806	-834	-749	-594

carburized, both the elastic modulus and the yield stress of the core of a surface hardened race will be greater than that of a carburized race. It is the higher elastic modulus and yield stress of the core in a surface hardened race that permit a more compressive and deeper residual stress to develop in the hardened layer near the surface. Yielding of the core will be limited, so the residual stress of the hardened layer will be more compressive.

The magnitude of residual stress at the surfaces of the three races are not equal, Table II. The average of the stress calculations at the three positions is plotted as a function of depth in Fig. 4. Clearly, the stress is most compressive in race 82, O.D. and I.D.. It is of interest to note that the OOR of race 82 was the greatest of the three hardened races.

If a phase transformation occurs simultaneously at a constant depth around a

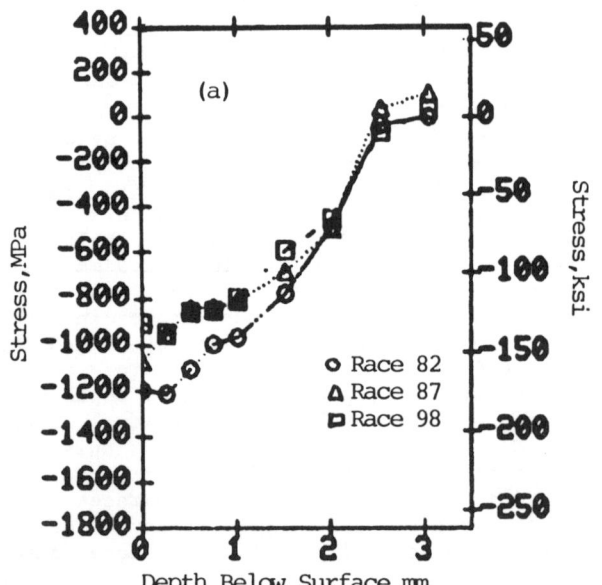
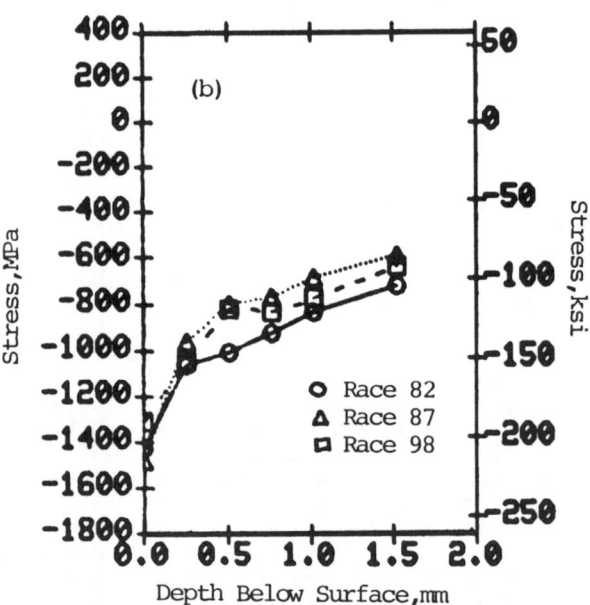

Fig. 4--Residual stress profiles of three hardened races--(a) Outer Diameter and (b) Inner Diameter

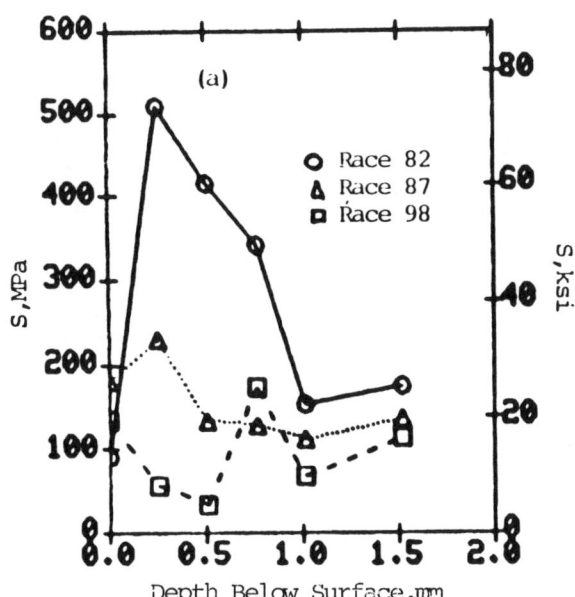
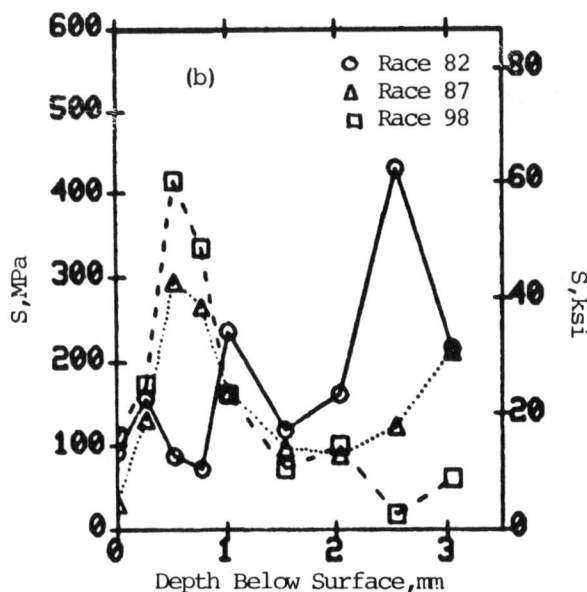

Fig. 5--S, the range of residual stress, as a function of depth below (a) I.D. surface and (b) O.D. surface.

race, the residual stress at that depth around the race will not vary. The range of stress, S, would be zero. If the stress does vary (that is nonzero S), it could only have resulted from nonsimultaneous phase transformation. The range of stress at each depth was calculated and plotted in Fig. 5. With respect to the I.D. surface, S is greatest in race 82 and least in race 98. The relative ranking of S below the I.D. surface is identical to that for OOR: greatest in race 82 and least in race 98. The relative ranking of S below the O.D. surfaces is mixed. There is a transition depth of 1 mm (0.040 in.) below which the ordering of S is the same as that below the I.D. surface. At depths shallower than 1 mm (0.040 in.) the ranking of S is opposite that of the I.D. Thus, there exists a correlation between OOR and S below the ID surface and beyond 1 mm depth below the O.D. surface.

Table III. Microhardness Measurements (knoop) of Hardened Races

(a) Outer Diameter

Depth Below Surface, mm

Race/Position	0.08	0.25	0.51	0.76	1.02	1.52	2.03	2.54	3.05
82 A	659	685	682	666	632	562	332	296	295
C	632	672	662	644	632	552	458	305	289
E	606	632	614	603	565	498	325	285	286
87 A	494	488	492	496	518	518	464	305	285
C	634	634	628	614	594	565	472	313	307
E	666	666	666	632	626	603	511	330	293
98 A	538	541	542	532	511	484	370	285	293
C	500	511	500	498	498	460	353	286	290
E	417	494	490	490	506	478	370	300	296

(a) Inner Diameter

Depth Below Surface, mm

Race/Position	0.00	0.25	0.51	0.76	1.02	1.52	2.03	2.54	3.05
82 A	546	534	518	548	528	494	346	294	301
C	605	617	586	594	584	548	466	293	286
E	573	570	562	550	555	555	480	296	291
87 A	652	656	640	652	631	606	603	285	296
C	720	740	729	717	689	666	580	308	315
E	696	706	689	682	672	662	578	307	286
98 A	628	652	644	638	638	628	560	285	286
C	628	669	662	632	632	600	464	274	274
E	679	702	699	679	659	620	412	292	292

The residual stress profiles of the unhardened races were uniform (S = 0) and the stress was nearly zero beyond 0.25 mm (0.010 in.) depth below the surface. Rapid heating and cooling of the surface significantly changed the state of residual stress. The residual stress at constant depth varied around the race, S ≠ 0. The condition of S ≠ 0 was caused by a variance in control of the surface hardening process. In an attempt to determine the source of the nonuniform stress, microhardness traverses were made, the amount of retained austenite was measured by x-ray diffraction, and the microstructure was characterized.

The measurements of microhardness for each of the three hardened races are shown in Table III. A good correlation was found between S and the range of microhardness, H, at each depth around the race (Fig. 6). H is least in race 98, the race with least distortion. In addition, it is clear that H approaches zero with increasing depth. This is evidence that the hardening process and not the state of residual stress prior to hardening caused S ≠ 0.

Another correlation was found between S and the difference between O.D. and I.D. hardness. The average of the three stress calculations at each depth is plotted in Fig. 7. A reference line of 548 Knoop, corresponding to 50 HRC, is indicated. The triangles indicate O.D. hardness and the circles indicate I.D. hardness. The difference between O.D. and I.D. hardness is greatest in race 98, the race having the lowest S and least OOR. The magnitude of the difference between O.D. and I.D. hardnesses of race 98 is attributable primarily to low O.D. hardness (< 548 Knoop). The difference between O.D. and I.D. hardnesses of race 87 is less than that of race 98. The difference between O.D. and I.D. hardnesses of race 82, the race with largest S and greatest OOR is opposite in sign to that of the races with lower S--the O.D. of race 82 is harder than the I.D.. S, the range of residual stress, is directly proportional to H, the range of microhardness, and inversely proportional to the difference between O.D. and I.D. hardnesses.

These differences in hardness profiles produced by surface hardening are associated with a difference in residual stress profiles. Clearly, the surfaces of the three races were not rapidly heated and cooled identically. Differences of hardness and residual stress resulting from surface hardening can occur in three ways: (1) Excessive austenitizing

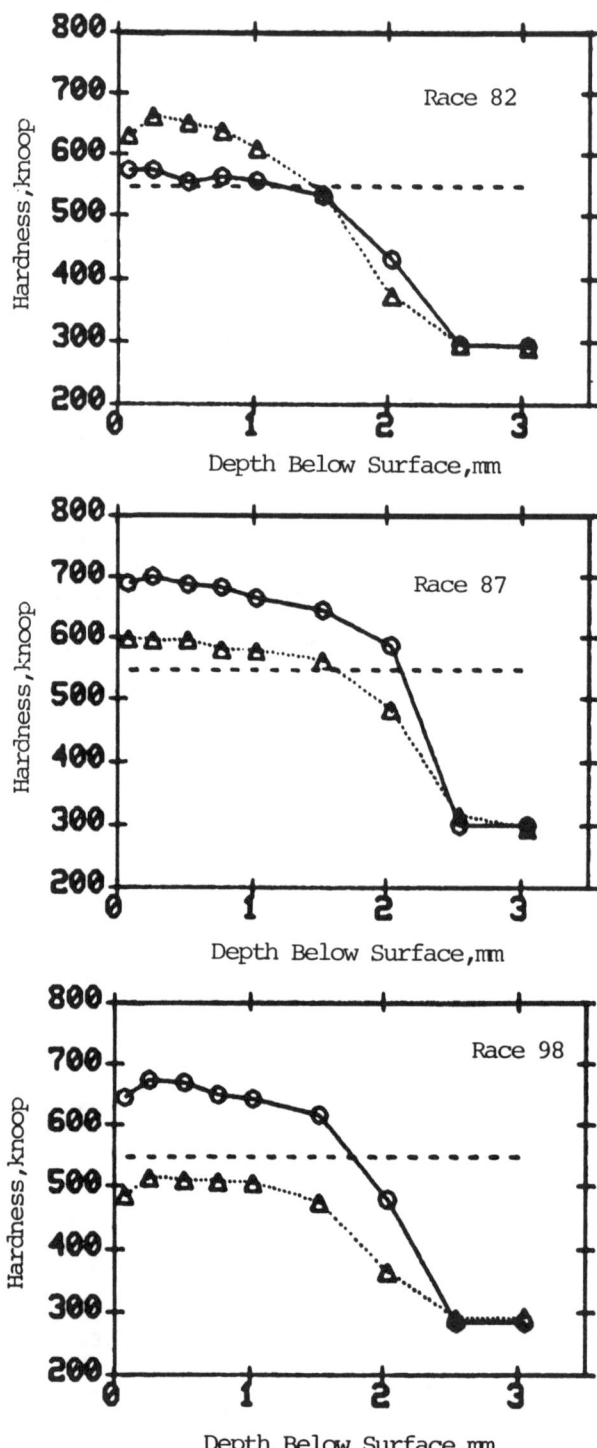

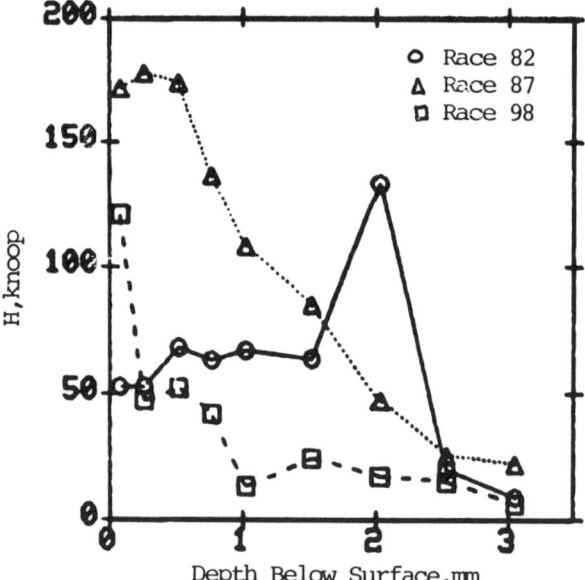

Fig. 6--H, the range of microhardness as a function of depth below the O.D. surface.

Fig. 7--Hardness profiles below I.D. (O) and O.D. (△) surfaces.

temperature resulting in an excessive amount of retained austenite, (2) Slow cooling rate resulting in soft non-martenistic transformation products, and (3) Insufficient austenitizing temperature resulting in insufficient dissolution of carbon in austenite. Inspection of the case microstructure revealed the presence of low, not high amounts of retained austenite; consequently, the austenitizing temperature was not excessive. No non-martensitic transformation products were present in the microstructure; therefore, the cooling rate was sufficient to prevent nucleation of soft phases. The source of the low hardness on the O.D. of race 98 is a lesser concentration of carbon in the martensite resulting from a low austenitizing temperature. The temperature of the O.D. of race 98 immediately prior to quenching must have been lower than that of the other two races. A difference in austenitizing temperature is caused by a difference in heat input to the O.D. surface.

The source of the variance of OOR was variation in heat input to the three races. Heat input to the O.D. and I.D. surfaces must be constant from race to race because it affects the extent of carbide dissolution, volume expansion during quenching, tempering of martensite and transformation of austenite of the unhardened microstructure, elastic modulus, and yield stress. All of these factors produce stress in the race; the stress may be sufficient to distort the race.

CONCLUSION

The source of the variance of OOR was variation in heat input to the outer surface of the races during surface hardening. Residual stress existing in the races prior to hardening was not a factor. Residual stress measurements of the hardened races correlate well with OOR. It was recommended that the control of heat input and initiation of quench be improved. In addition, it was recommended that the aim hardness of the O.D. be as low as possible. The result of implementation of both of these recommendations will be a reduction in the magnitude of stress and the variance of OOR.

ACKNOWLEDGMENT

The author thanks D. H. Gang for the measurement of residual stress, D. L. Pierce for the measurement of microhardness, and R. Fuquen for the critical review of the manuscript. The permission of the Timken Company to publish the results of this study is appreciated.

REFERENCES

1. G. E. Hollox and R. T. Von Bergen: Heat Treatment of Metals, $\underline{5}$ (1978.2) 27-31.
2. R. F. Kern: Heat Treating, $\underline{17}$ (Feb. 1985) 30-32.
3. D. T. Llewellyn and W. T. Cook: Metals Technology, $\underline{4}$ (1977) 265-278.
4. "Metals Handbook", 9th ed., vol.4, American Society For Metals, Metals Park, Ohio (1981), pp. 451-483.
5. ibid., pp. 507-517.
6. ibid., pp. 518-522.
7. ibid., pp. 484-506.
8. "Residual Stress Measurement by X-Ray Diffraction," SAE Handbook SAE-J784a, Society of Automotive Engineers, Warrendale, PA (1981).

THE INFLUENCE OF RESIDUAL STRESSES ON THE FATIGUE DESIGN OF WELDED STEEL STRUCTURES

J. Graham Wylde
Edison Welding Institute
Columbus, Ohio USA

ABSTRACT

High tensile residual stresses are formed in welded structures primarily as the result of differential contractions which occur as the weld metal cools. These residual stresses combine with stresses applied during service and can have a profound effect on the service performance of the structure. One of the major effects is that even applied compressive loads will become damaging and thus fatigue design must be based upon the entire stress range including both tensile and compressive loads. This paper describes the fatigue design approach for welded joints and discusses how this is influenced by the presence of residual stresses. The paper also discusses how the approach might be modified for stress relieved joints.

IT IS WELL KNOWN that welded joints may contain very high residual stresses (Refs. 1,2). These occur primarily because of differential contractions which occur as the weld metal solidifies and cools to ambient. Additional components of residual welding stress arise from thermal gradients during cooling and sometimes from a phase change in the solidifying metal.

The precise distribution of residual stresses in and surrounding a welded joint is complex and depends on a number of factors, including material composition, thickness, applied restraint, number and sequence of weld passes and level of preheat (Ref. 3). However, the contraction of the highly heated region in the immediate vicinity of the weld is considerably greater than the material yield strain and thus the values of tensile residual stresses in this region are generally of yield magnitude, with balancing areas of compression in the adjacent material.

In the direction parallel to the weld, the thermal contraction of the weld metal is restrained by the surrounding base metal. Thus yield magnitude tensile residual stresses occur throughout the weld region. These tensile residual stresses are balanced by regions of compression further from the weld as illustrated in Figure 1.

In the transverse direction the situation is more complex and the residual stress distribution will depend on the number and sequence of weld runs and the external restraint. If there is no applied restraint the transverse residual stress must be self-equilibriating across the thickness. Tensile residual stresses will be generated by shrinkage of the capping passes and these will produce compressive stresses in the center of the specimen and tension on the back face as illustrated in Figure 2. This distribution will be modified significantly by the presence of external restraint.

Although the precise distribution of residual stresses in any particular welded structure is complex, many experimental studies have confirmed the general distribution described above.

When welding relatively large or thick members, the restraint will generally be sufficient to allow the peak residual stresses to equal the yield strength of the material. For design purposes it is conventional to assume that yield magnitude residual stresses will be present in both the longitudinal and transverse directions. However, this assumption will generally not be true for laboratory specimens which are seldom of sufficient size to provide the restraint necessary to allow yield residual stresses to develop. This may be particularly significant when interpreting specimen test data.

THE COMBINATION OF APPLIED AND RESIDUAL STRESS - When load is applied to a weld joint that contains residual stresses, the applied and residual stresses will combine. As discussed above, the initial residual stress distribution in a weld joint will contain regions of high tensile stress equal to the yield strength of the material, balanced by areas of compression. As external load is applied local yielding occurs resulting in a redistribution of residual stresses.

Under cyclic loading the residual stresses shake down to a stable configuration over a period of about 20 cycles (Ref. 4). After shakedown is complete the maximum tensile residual stress will be reduced but will extend over a wider area than originally, as shown in Figure 3. Gurney (Ref. 5) has suggested a simplified model to account for this behavior which predicts that the peak level of residual stress remaining after shakedown will be given by:

$$\sigma_{res} = \sigma_y - K_t\sigma$$

where σ = applied stress
σ_y = yield stress
K_t = stress concentration factor.

Subsequent applied load cycles will now effectively pulsate downwards from tensile yield stress as shown in Figure 4.

The above simplified approach seems to apply very well for applied loads which are fully tensile, alternating tension-compression and zero-compression. For joints subjected to compression-compression loading the situation is more complex. In this case the maximum and minimum stresses in the cycle will be:

$$\sigma_{max} = \sigma_y - R K_t\sigma$$
$$\sigma_{min} = \sigma_y - K_t\sigma$$

where R = stress ratio (minimum stress/maximum stress). This is shown schematically in Figure 5.

Assuming that fatigue crack growth will only occur under effective tensile stress, Figure 5 suggests that at compressive stress ratios as the applied stress range is increased, the effective tensile stress range increases to a maximum value (when $\sigma = \sigma_y/K_t$) and then actually decreases with further increase in applied stress range. This model has been confirmed experimentally by Gurney (Ref. 5). Although this may have limited practical significance because few structures will be subjected to compression-compression loading, it does provide validation that the simple model is reasonable.

FATIGUE TEST DATA - From the above considerations on the interaction of applied and residual stresses, it is clear that the effect of high tensile residual stress produces an effective stress cycle which pulsates downwards from tensile yield by an amount equal to the applied stress range. This holds for fully tensile, alternating tension-compression and zero-compression loading. Thus, we would anticipate similar fatigue lives for welded joints tested at all positive stress ratios, R = 0, R = -1 and R = 0 (compression).

Figure 6 shows fatigue results obtained by Maddox (Ref. 6) from longitudinal fillet welded specimens in a structural C-Mn steel. The specimens were 150 mm wide, 13 mm thick and 800 mm long with an attachment 150 mm long and 13 mm thick fillet welded onto one side. Because the author was concerned that the specimens might not have sufficient restraint to allow full tensile yield residual stresses to develop, spot heating was applied to the weld ends where fatigue cracks would be most likely to initiate. The correct application of this technique produces yield tensile residual stresses in the center of the heated spot (Ref. 5).

The results show that there was no significant difference between tests carried out at R = 0.67 and 0.5 (high mean stress) R = 0 (zero to tension) R = -1 (alternating tension-compression) and R = 0 (compression). All data are plotted on the basis of the full stress range irrespective of the sense of the applied loading.

In the test specimens fatigue cracks initiated at the weld toe around one end of the attachment and propagated through the wall thickness of the main plate. In the test carried out at R = 0.67, 0.5, 0 and -1.0, the cracks propagated across the specimen width producing complete rupture. In the tests carried out at R = 0 (compression) the cracks arrested after they had grown to approximately 80 mm in length. At this point they had grown out of the tensile residual stress field. This provides a clear illustration of the importance of residual stresses.

In a real structure cracks in compression members may arrest after they have propagated away from areas of high tensile residual stress. It could be argued that this is sufficient reason to ignore the need to consider zero-compression loading. However, it is considered that this would be a dangerous approach since the residual stress distribution in a large structure may be very different from that in a laboratory specimen. Furthermore, once a crack propagates through the plate thickness the local stiffness of the member will be changed and secondary bending stresses may result. For example, fatigue cracks have been observed to propagate across the full compression flanges of beams even though this involved growth outside the influence of the initial tensile residual stress distribution.

Gurney (Ref. 5) carried out tests on similar specimens to those tested by Maddox (Ref. 6) under compression to compression loading. As discussed previously, the simple model for residual stress redistribution predicts that the effective tensile stress range should increase with applied stress up to a maximum value after which point further increase in applied stress would produce a decrease in the effective tensile stress range. Thus, the model predicts an abrupt slope change in the S-N curves for welded joints tested under fully compressive loading. Figure 7 shows the results of tests carried out at R = 0 (compression) and R = 0.1 (compression) to confirm this.

THE INFLUENCE OF STRESS RELIEF - From the previous section one might anticipate that stress relief would produce a significant improvement in the fatigue performance of welded structures in C-Mn steels. However, although stress relief may be beneficial under certain loading conditions, under tensile loading it only provides a small improvement in fatigue strength.

The primary reason for this is that the majority of the total fatigue life of welded joints in steel is occupied in crack propagation (Refs. 5,7); the initiation phase is negligible. Fatigue crack growth rates in structural steel are not strongly dependent upon the applied mean stress (Refs. 8,9). Figure 8 summarizes the results obtained by Booth, et al. (Ref. 9) and shows very little effect of mean stress for positive R values.

However, stress ratio does have a small influence on the fatigue crack threshold (Ref. 10). As stress ratio increases, threshold stress intensities are reduced. Thus, one might anticipate a higher fatigue limit in stress relieved joints than in as-welded. This is confirmed in fatigue test data for welded joints in the as-welded and stress-relieved conditions which generally show a small increase in the fatigue limit in the stress-relieved specimens. Typical results obtained by Gurney (Ref. 11) are shown in Figure 9. These data were also obtained for longitudinal fillet welded specimens.

In addition to the increased fatigue limit, the results indicate a slight change of slope with the results for stress-relieved specimens having a slightly shallower slope than the as-welded joints. This is typical and can be explained by the slight change in crack propagation rate between high and low R values. The curves intersect at an applied stress range of yield where one would clearly expect the results for stress-relieved and as-welded specimens to be identical.

A comprehensive analysis of data by Gurney (Ref. 12) indicated that on average this rotation of the S-N curve represented a factor of 0.815 at 2×10^6 cycles. This slope change is relatively small but does have significant consequences for those involved in laboratory fatigue testing of specimens to generate design data for structural applications.

There is a danger that if small specimens are used which are not large enough to hold full tensile residual stresses, the results obtained will lie on a shallow slope and will thus be unsafe for the design of large structures. In this connection it is important to appreciate that size effects will also tend to produce higher fatigue lives from thin specimens than thick specimens tested at the same applied stress range (Ref. 13). Consequently, extreme caution should always be exercised in the interpretation of small scale laboratory test data and its applicability to the design of welded structures.

For joints tested under full or partial compressive loading, stress relief can produce a major benefit. Tests carried out under alternating loading by Maddox (Ref. 6) generally show a significant improvement in fatigue strength as a result of stress relief. Test results are shown in Figure 10. It is interesting to note that some stress-relieved specimens showed very little improvement in fatigue life over the as-welded data. This presumably represents inconsistencies in the effectiveness of the postweld heat treatment (PWHT).

Thus, there is a major difficulty in trying to make allowances for stress relief in design. The uncertainty in the effectiveness of PWHT and other stress relief treatments makes it very difficult for the designer to guarantee the potential improvement. In this connection it is also worth noting that the presence of local weld repairs made after stress relief would be extremely detrimental.

For the above reasons code writing bodies (Ref. 14) have been reluctant to allow any relaxation in design stresses for stress relieved structures.

DESIGN APPROACH - On the basis of the fatigue test data available for welded joints a design approach has been proposed by Gurney (Ref. 15) and others which are now the basis for a range of international fatigue design codes for welded structures.

For as-welded structures fatigue design is based on the full applied stress range including all compressive components of applied stress. No stress ratio or mean stress corrections are relevant and so the same set of design S-N curves can be used for all loading situations.

This approach is clearly justified for as-welded structures and for stress-relieved structures subjected to full tensile loading. For stress-relieved structures subjected to partial compressive loading they will generally be overconservative. Here, the problem of relaxing the requirements for stress-relieved structures is not absence of test data, but doubts over the effectiveness of stress relief treatments. Sufficient experimental data exist to make design recommendations for stress-relieved structures. However, doubts linger over just how applicable these would be for large stress-relieved structures which may contain uncertain levels of residual stress and the problems associated with weld repairs.

In connection with weld repairs, major problems would occur if local repairs were required on a large structure which had been designed for higher stresses on the basis of stress-relief. Local stress-relief treatments are unlikely to provide a significant reduction in peak residual stress levels and, consequently, the fatigue strength of the repaired region could be significantly lower than the designer anticipated.

CONCLUSIONS

Although further work is required to fully investigate the influence of residual stresses on the fatigue performance of welded steel structures, it appears that the following conclusions can be drawn:

(1) Provided there is adequate restraint welded joints will contain high tensile residual stresses.

(2) For as-welded structures it is necessary to design on the basis of stress range including compressive components of applied stress.

(3) Stress relief will generally be beneficial if the applied stress is partially or fully compressive but has only a small effect if the applied stress is tensile.

(4) It is dangerous to assume a higher fatigue strength on the basis of stress relief because of uncertainties in the effectiveness of stress relief treatments and other sources of residual stress in large structures.

(5) Small scale laboratory test specimens may not have sufficient restraint to allow full tensile yield residual stresses to develop. Thus, caution is necessary when applying these data to the design of large structures.

REFERENCES

1. Masubuchi, K. 1980. Analysis of welded structures, Pergamon Press, Oxford, England.

2. Van Vlack, L.H. Elements of materials science, Addison-Wesley Publishing Co., Inc., Massachusetts, U.S.A.

3. Masubushi, K. December, 1980. Models of stresses and deformations due to welding - A review. Journal of Metals.

4. Wylde, J.G. December, 1982. Fatigue tests on 457 mm diameter welded tubular T-joints under out-of-plane bending. UKOSRP report.

5. Gurney, T.R. 1979. Fatigue of welded structures, Cambridge, Univ. Press, 2nd Edition.

6. Maddox, S.J. 1982. Influence of tensile residual stresses on the fatigue behavior of welded joints in steel. ASTM STP 776.

7. Signes, E.A., Baker, R.A., Harrison, J.D. and Burdekin, F.M. March, 1967. Factors affecting the fatigue strength of welded high strength steels. British Welding Journal.

8. Richards, C.E. and Lindley, T.C. 1973. The relevance of crack closure to fatigue crack propagation. Proc. Conf. Mechanics and Mechanisms of Crack Growth, BSC, Cambridge.

9. Maddox, S.J., Gurney, T.R., Mummery, A.M. and Booth, G.S. 1978. An investigation of the influence of applied stress ratio on fatigue crack propagation in structural steels. W.I. Members Report 72/1978/E.

10. Pook, L.P. July, 1971. Fatigue crack growth data for various materials deduced from the fatigue lives of precracked plates, NEL Report 484.

11. Gurney, T.R. November, 1977. Some recent work relating to the influence of residual stresses on fatigue strength. Proc. Conf. Residual Stresses in Welded Construction and their Effects, The Welding Institute.

12. Gurney, T.R. and Maddox, S.J. 1973. A re-analysis of fatigue data for welded joints in steel. Weld. Res. Int., vol. 3, no. 4.

13. Wylde, J.G. and McDonald, A. 1980. The influence of joint dimensions on the fatigue strength of welded tubular joints. Int. J. Fatigue.

14. Gurney, T.R. April, 1987. From past to future: A contemporary review relating to fatigue design rules for welded structures. Int. Conf. Fatigue of Welded Constructions, The Welding Institute, Brighton.

15. Gurney, T.R. May, 1976. Fatigue design rules for welded steel joints. W. I. Res. Bull., vol. 17.

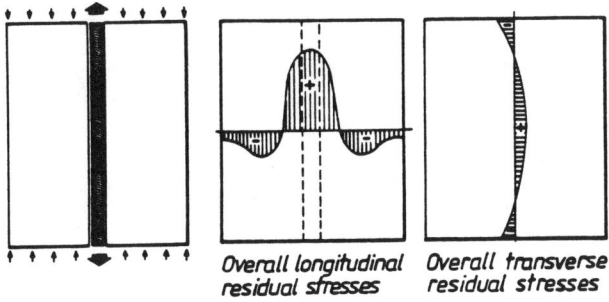

Figure 1. Basic residual stress distribution in groove weld

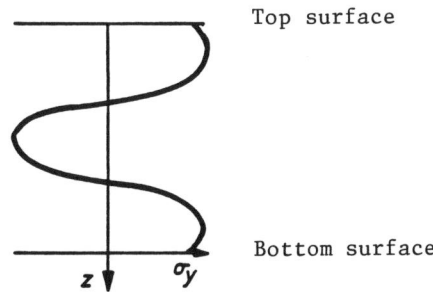

Figure 2. Typical through-thickness distribution on centerline of weld made without restraint

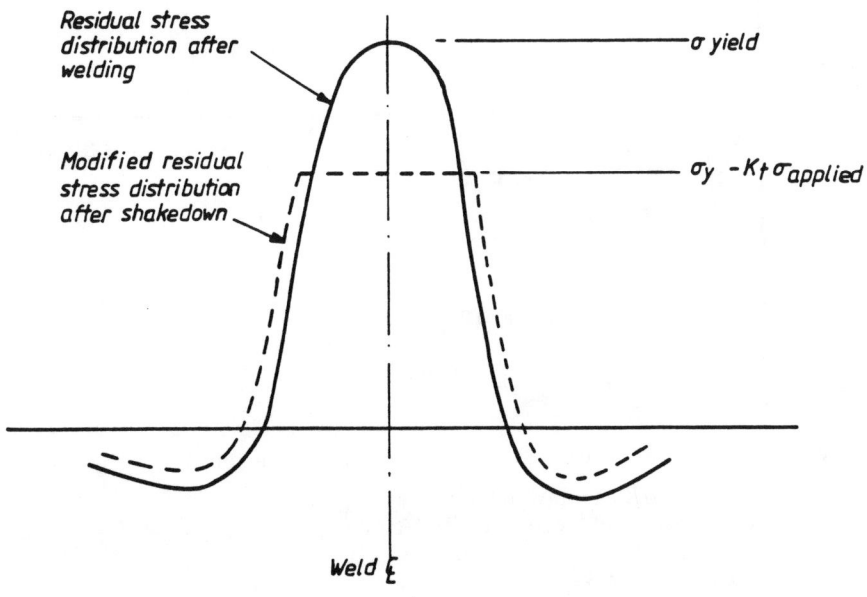

Figure 3. Relaxation of residual stresses due to applied loading

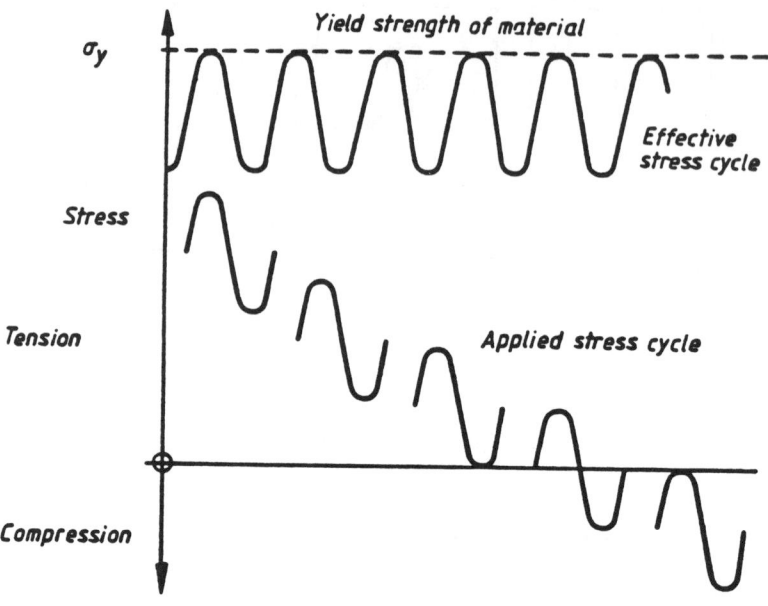

Figure 4. Effective stress cycle under fatigue loading

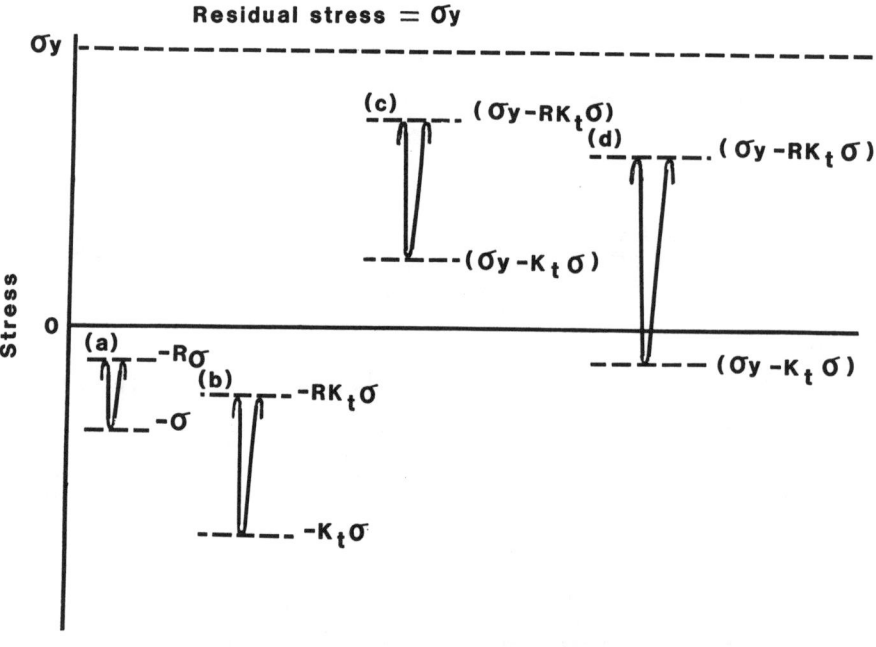

Figure 5. Effect of residual stresses under compression-compression loading
    (a) Nominal applied stress
    (b) Applied stress at weld toe
    (c) Effective stress at weld toe
    (d) Effective stress for larger applied stress

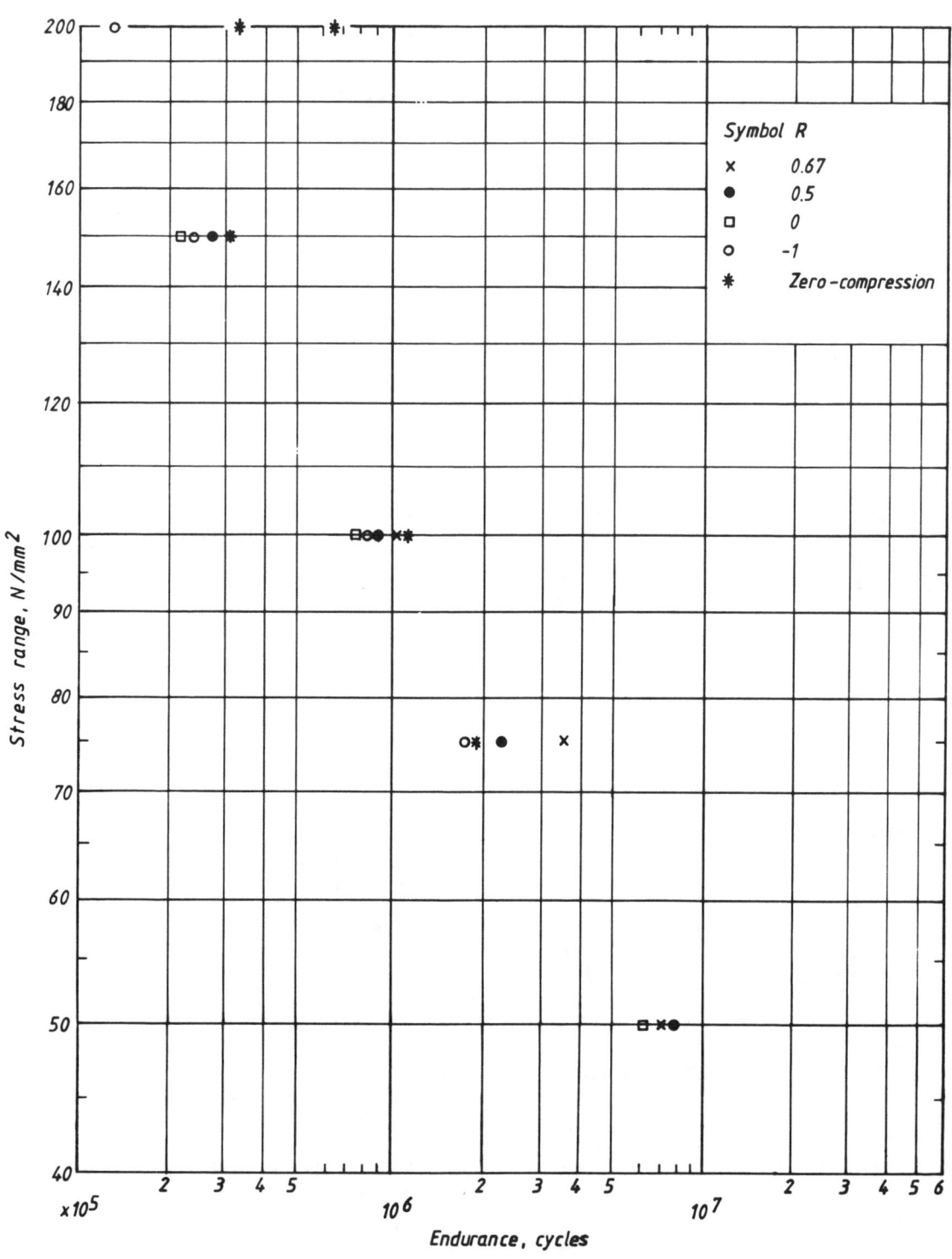

Figure 6. Fatigue results for longitudinal fillet welds tested at various applied stress ratios, (as-welded)

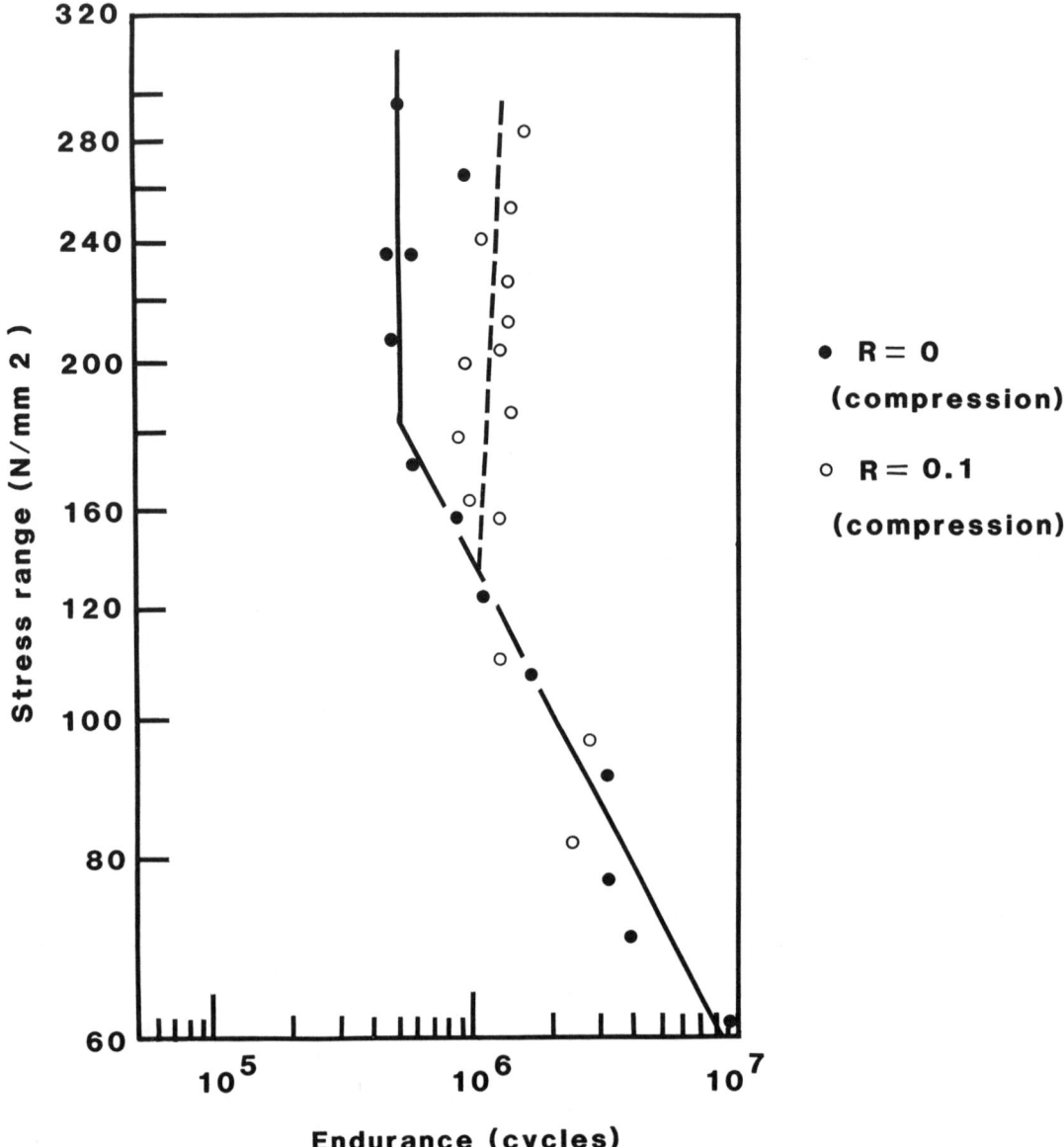

Figure 7. Fatigue test results for longitudinal fillet welded joints tested under fully compressive loading

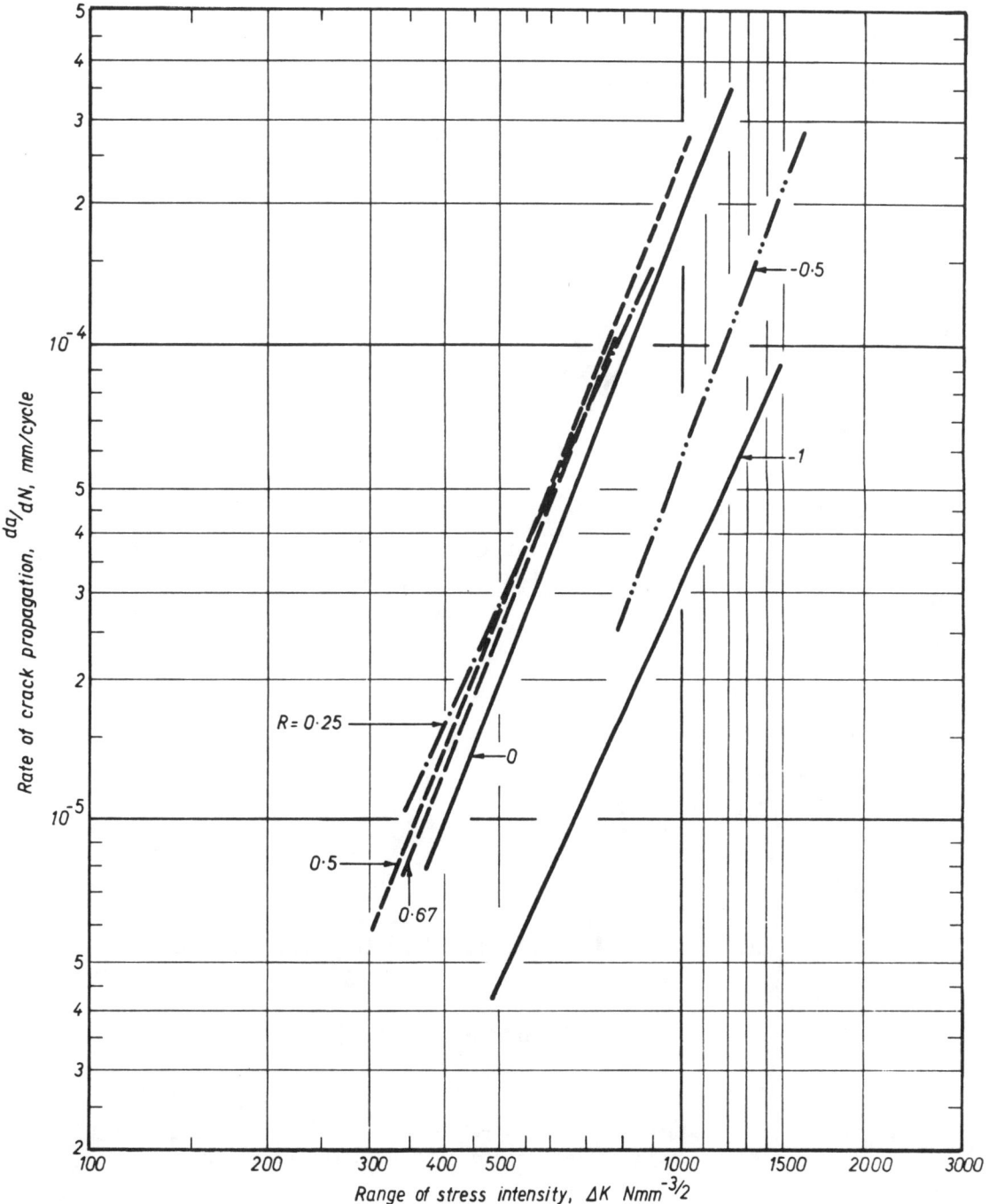

Figure 8. The influence of stress ratio on fatigue crack growth rate

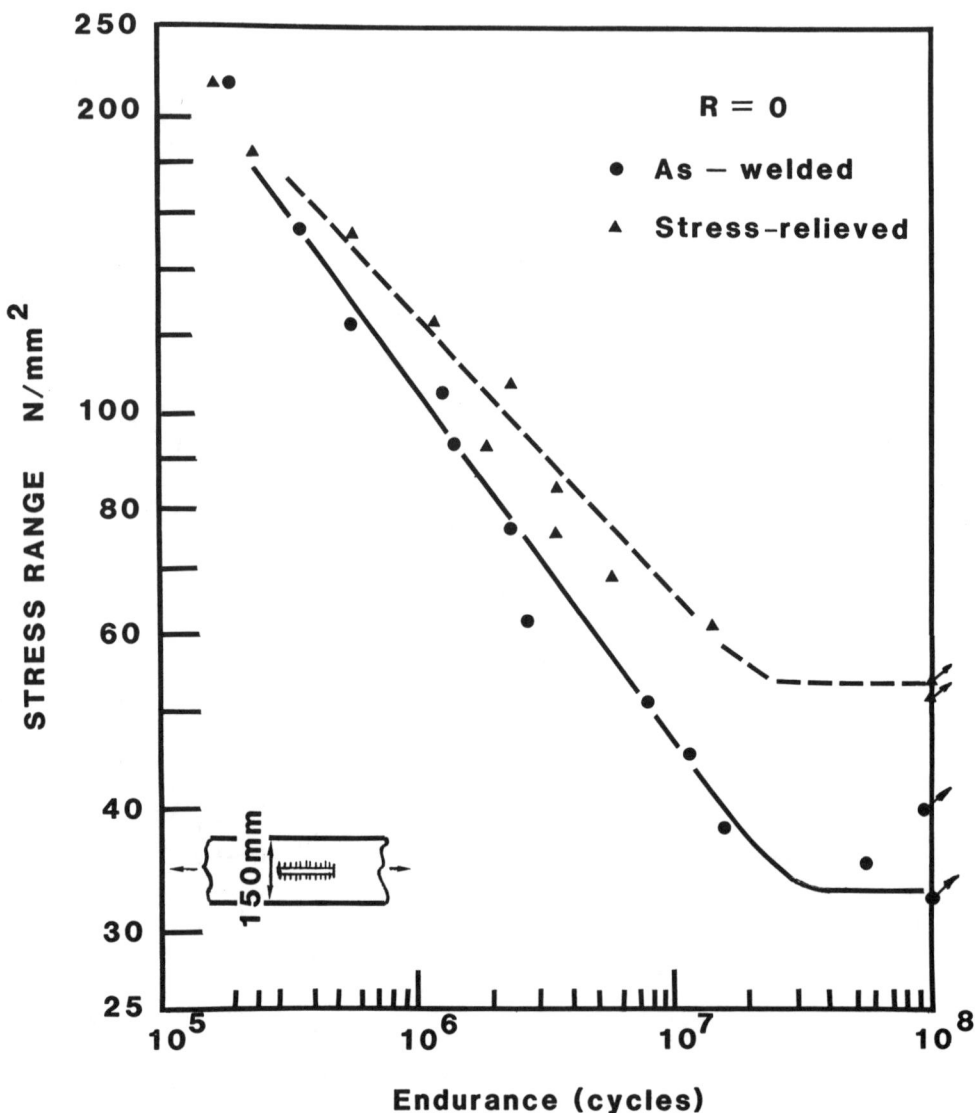

Figure 9. The influence of stress relief on the fatigue strength of longitudinal fillet welded joints tested at R = 0

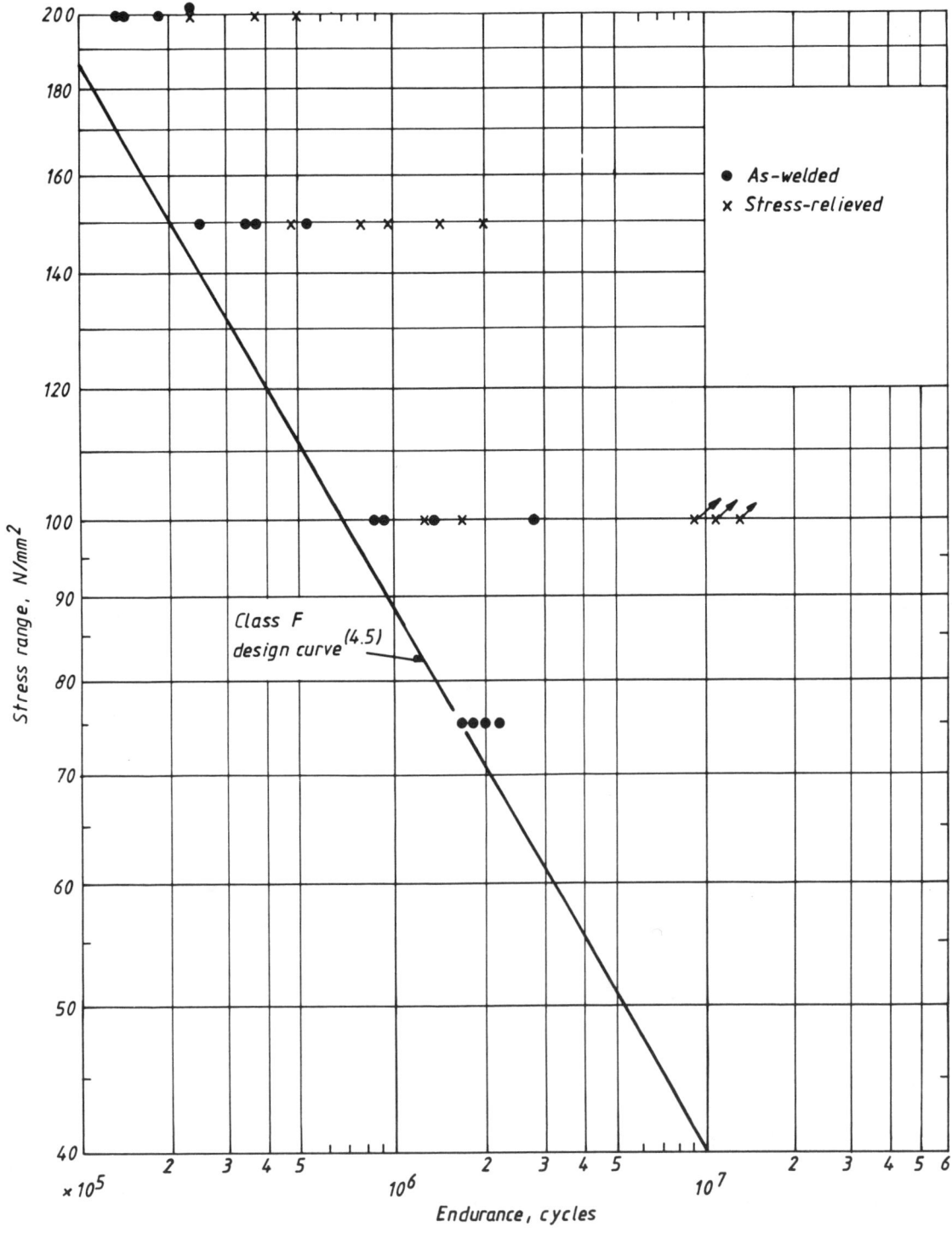

Figure 10. Fatigue test results obtained from as-welded and stress-relieved specimens under alternating loading (R = -1)

RESIDENTIAL STRESSES AT PINCH WELDS IN SMALL STAINLESS STEEL TUBES

W.C. Mosley
E.I. du Pont de Nemours and Co.
Aiken, South Carolina, USA

ABSTRACT

Stress corrosion cracking of pinch-welded Type 304L stainless steel tubes has been used to determine the locations and magnitudes of tensile residual stresses. Cracking severity, a measure of the extent of tensile stresses above a threshold level, has been correlated with welding parameters and annealing to define conditions that minimize the possibility of environment-assisted slow crack growth on the tube interior. Welding with low current, low electrode force and no restraint, and laser annealing after welding were found to minimize interior stress corrosion cracking at welds in cold drawn tubes. Preliminary work with annealed tubes indicates that pinch welding parameters influence tensile stresses differently.

DIFFUSION WELDS made by pinch welding are used to seal small stainless steel tubular fill stems on hydrogen gas bottles. Tensile residual stresses on the tube interior near the weld can contribute to through-wall failures by environment-assisted slow crack growth. Efforts are being made to identify welding parameters and annealing methods that minimize tensile stresses without compromising the strength or quality of the weld. This paper describes the application of a stress corrosion cracking technique to determine the locations and magnitudes of tensile stresses at pinch welds in stainless steel tubes. Tube size and weld geometry prohibit measurements of interior stresses using conventional methods such as strain gages and x-ray diffraction.

DISCUSSION

MATERIALS - This work was performed on 3.2-mm-O.D. x 1.6-mm-I.D. Type 304L stainless steel tubing. Pinch welds were made in cold-drawn and annealed tubes with hardness values of R_B95 and R_B66, respectively.

WELDING - Confined and unconfined pinch welds were made in the stainless steel tubes. Unconfined welds were made by compressing the tubing between cylindrical electrodes and applying electrical current to cause a solid state bond to form between the inner sides of the collapsed tube wall. Side anvils were used during confined pinch welding to restrict the width of the weld. Welding variables included electrode force, welding current, welding time, and axial restraint.

TESTING - Pinch welded tubes were exposed to boiling magnesium chloride solutions in an apparatus like that used for ASTM Standard Practice G36-73. Three different methods of exposure were employed.

Exposure of bare specimens produced stress corrosion cracking on both interior and exterior surfaces of pinch welded tubes. Exterior cracking was evaluated using optical and scanning electron microscopy.

Some welded tubes were sheathed in heat-shrinkable plastic tubing so that only the interiors were exposed to the boiling magnesium chloride solutions. This second method was used to simulate environment-assisted slow crack growth from the inside out. Interior cracking was evaluated by serial metallography.

A third exposure method allowed individual weld specimens to be exposed sequentially to magnesium chloride solutions boiling at progressively higher temperatures. The specimens were cut axially perpendicular to the weld depression. One half of each specimen was ground to remove flashing from cutting and produce a smooth surface for examination. Optical microscope examination after each exposure revealed the locations and sizes of both exterior and interior cracks. Specimens were examined by serial metallography following the final exposure.

RESULTS

TYPES OF CRACKING - Three types of internal cracking have been observed. Type 1 cracking occurs in the region of the closure bond. Type 2 cracking occurs outside the bond region where the tube wall is deformed by the pinch welding process. These two types of cracking are similar to the environment-assisted slow crack growth that can occur at pinch welds in fill stems. A third type of internal cracking that occurs at fairly large distances from the weld is thought to be caused by residual stress from the tube fabrication process.

Four types of external cracking have been observed. Type A cracking occurs just outside of the weld depression in the region where the weld causes the tube outer surface to be in tension. Type B cracking occurs within the weld depressions. Type C cracking occurs at the side of the weld depressions in confined welds. These types of external cracking indicate tensile stresses that can contribute to through cracking from environment-initiated cracks on the tube interior. A fourth type of external cracking occurs away from the weld at indentations caused by clamping the tube in restraining fixtures.

EFFECTS OF MAGNESIUM CHLORIDE SOLUTION BOILING TEMPERATURE - The boiling temperature of magnesium chloride solution increases with the magnesium chloride concentration. The boiling temperature at which external cracking occurs at confined pinch welds in cold-drawn Type 304L stainless steel tubes has been correlated with the tensile stress at the crack location (Figure 1). Residual stresses at ten locations at welds made with high, medium, and low welding currents were determined by the material removal method using miniature electrical strain gages. Companion welds were exposed bare to magnesium chloride solutions boiling at 105, 115, 120, 130, and 155°C. No external cracking was detected after 26 days at 105°C and 10 days at 115°C. Cracking was detected after three days at 120°C and after one day at 130°C and 155°C. Figure 1 shows the correlations between tensile stress and exterior cracking after ten days at 120°, 130°, and 155°C corresponding to cracking thresholds of 160 to 200, 90 to 110, and less than 45 MPA, respectively.

CRACKING SEVERITY - The severity of each type of cracking was judged by optical microscopy or metallography to be none, slight, moderate, or severe with ratings of 0, 1, 2, and 3, respectively. Slight cracking caused superficial cracks with very little depth or length. Severe cracking corresponded to deep or long cracks that frequently penetrated the wall of the tubing. Moderate cracking was intermediate between slight and severe. Cracking severity ratings were used to determine how variations in welding parameters, metallurgical conditions, and annealing influence residual stresses.

CRACKING ANALYSES

EFFECTS OF WELDING PARAMETERS - Confined pinch welds were made in cold-drawn Type 304L stainless steel tubes using welding currents, electrode forces, and welding times above and below procedural values and with full and no restraint. These samples were exposed for ten-day periods to magnesium chloride solutions boiling at 120°, 130°, 140°, and 155°C using the third exposure method (Figure 2). Sums of the severity ratings for the various types of cracking at both ends or sides of the pinch welds

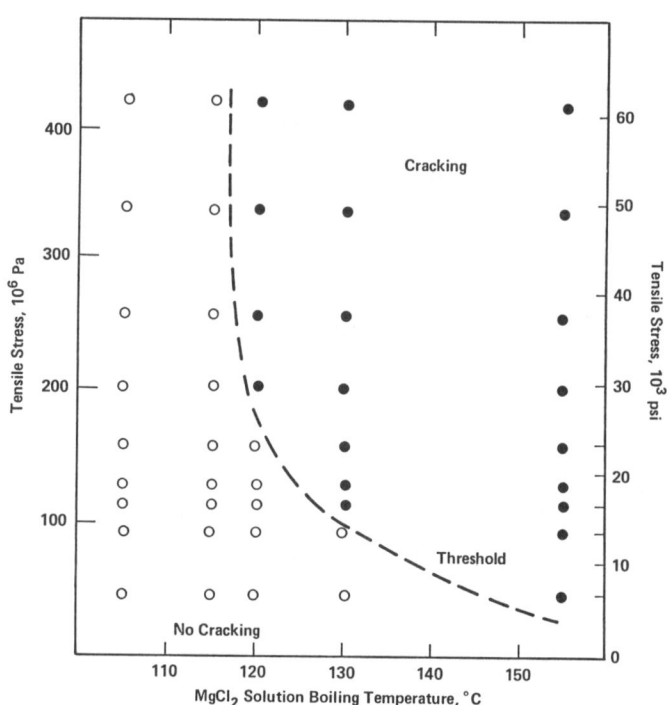

FIGURE 1. STRESS CORROSION CRACKING ON THE EXTERIOR OF CONFINED PINCH WELDS IN COLD DRAWN 304L STAINLESS STEEL TUBES AS FUNCTIONS OF TENSILE STRESS AND MAGNESIUM CHLORIDE BOILING TEMPERATURE.

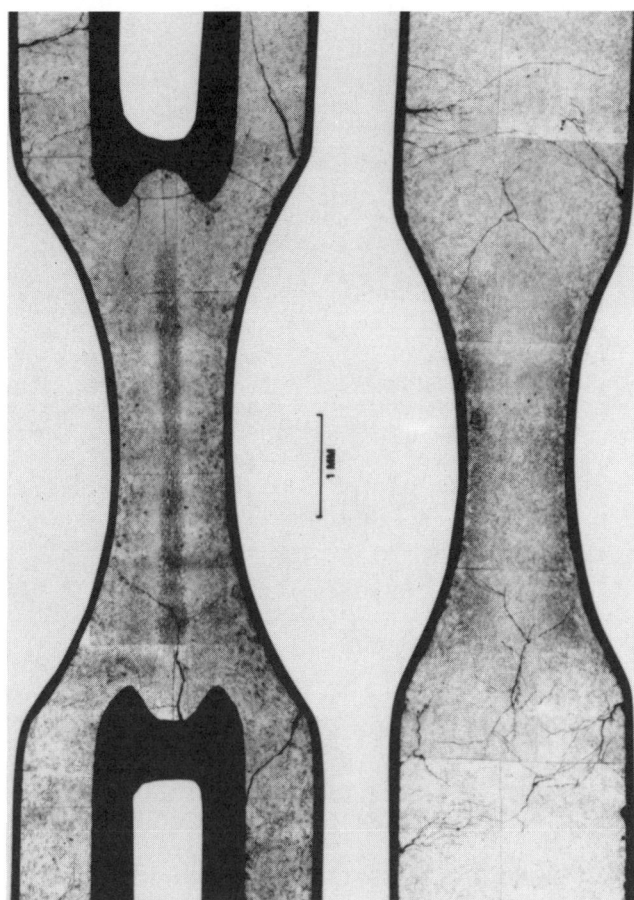

FIGURE 2. SERIAL METALLOGRAPHY OF A CONFINED PINCH WELD IN A COLD DRAWN 304L STAINLESS STEEL TUBE MADE AT HIGH CURRENT, HIGH FORCE, LONG WELDING TIME AND NO RESTRAINT REVEALS TYPES 1 AND 2 INTERNAL CRACKING AND TYPE A EXTERNAL CRACKING CAUSED BY EXPOSURE FOR TEN DAYS TO MAGNESIUM CHLORIDE BOILING AT 155°C.

Table 1. Severity Ratings for Stress Corrosion Cracking at Confined Pinch Welds in Cold-Drawn 304L Stainless Steel Tubes (Maximum = 6)

Hardness	Welding Current	Electrode Force	Welding Time	Restraint	Internal Cracking Type 1	Internal Cracking Type 2	External Cracking Type A	External Cracking Type B	External Cracking Type C
$R_B = 95$	High	High	Long	Full	1	4	4	0	0
				None	5	5	5	0	0
			Short	Full	6	6	5	0	0
				None	2	2	6	0	0
		Low	Long	Full	2	4	4	0	0
				None	3	2	5	0	0
			Short	Full	4	4	5	3	0
				None	0	2	5	2	0
	Low	High	Long	Full	0	4	5	0	2
				None	3	2	6	0	1
			Short	Full	2	5	6	0	1
				None	0	4	5	0	2
		Low	Long	Full	0	5	6	3	0
				None	0	0	5	0	0
			Short	Full	0	5	5	0	1
				None	3	2	5	0	1

Table 2. Totals of Cracking Severity Ratings for Confined Pinch Welds in Cold-Drawn 304L Stainless Steel Tubes

	Internal Cracking		External Cracking		
	Type 1	Type 2	Type A	Type B	Type C
Current					
High	23	29	39	5	0
Low	8	27	43	3	8
Force					
High	19	32	42	0	6
Low	12	24	40	8	2
Time					
Long	14	26	40	3	3
Short	17	30	42	5	5
Restraint					
Full	15	37	40	6	4
None	16	19	42	2	4

after the 155°C exposure are presented in Table 1. Totals of the severity ratings for high and low values of each welding parameter as given in Table 2 reveal the influences on cracking and, hence, the extent of tensile stresses above the threshold level. Thus, Type 1 internal cracking (within the bond region) is strongly relate to high current and moderately related to high electrode force. Type 2 internal cracking is related strongly to full restraint. In contrast, Type A external cracking does not appear to be related to welding parameter variations. Type B and Type C external cracking, while not as severe as the other types of cracking, seem to be related to low electrode force and high welding current, respectively.

EFFECTS OF METALLURGICAL CONDITION - Although the data are incomplete, Table 3 gives cracking severity ratings for confined pinch welds in annealed Type 304L stainless steel tubes. These welds were exposed and examined along with the welds in cold-drawn tubes (Figure 3). However, some correlations between the various types of cracking and welding parameters differ from those observed for welds in cold-drawn tubes. Type 1 internal cracking at welds in annealed tubes is strongly related to high current (same as for cold-drawn tubes) and moderately related to low electrode force and long welding time (different from cold-drawn tubes). No restraint appears to cause more severe Type 2 internal cracking at welds in annealed tubes whereas full restraint causes more severe Type 2 cracking at welds in cold-drawn tubes. Only one of fourteen welds in annealed tubes compared to all sixteen of the welds in cold-drawn tubes had Type A external cracking. Type B external cracking is slightly more severe at welds in annealed tubes than in cold-drawn tubes and is related to high rather

Table 3. Severity Ratings for Stress Corrosion Cracking at Confined Pinch Welds in Annealed 304L Stainless Steel Tubes (Maximum = 6)

Hardness	Welding Current	Electrode Force	Welding Time	Restraint	Internal Cracking Type 1	Internal Cracking Type 2	External Cracking Type A	External Cracking Type B	External Cracking Type C
$R_B = 66$	High	High	Long	Full	4	4	0	0	0
	High	High	Long	None	3	4	0	3	0
	High	High	Short	Full	0	0	0	3	0
	High	High	Short	None	1	2	0	0	0
	High	Low	Long	Full	4	0	0	0	0
	High	Low	Long	None	5	6	0	0	0
	High	Low	Short	Full	2	2	0	2	0
	High	Low	Short	None	4	3	0	0	0
	Low	High	Long	Full	-	-	-	-	-
	Low	High	Long	None	0	2	0	3	0
	Low	High	Short	Full	0	3	0	0	1
	Low	High	Short	None	0	5	3	0	2
	Low	Low	Long	Full	0	4	0	2	0
	Low	Low	Long	None	0	3	0	0	0
	Low	Low	Short	Full	-	-	-	-	-
	Low	Low	Short	None	0	4	0	0	0

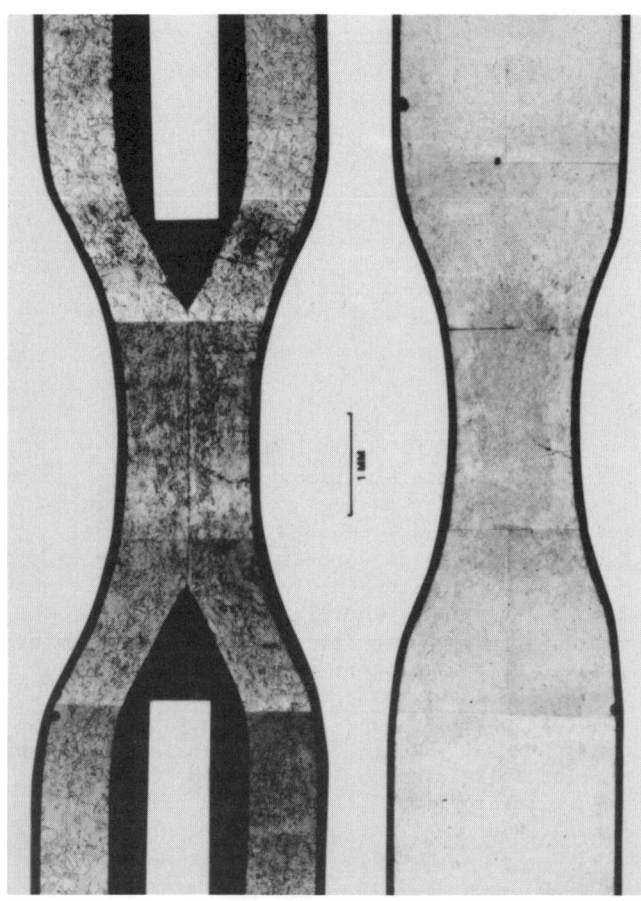

FIGURE 3. SERIAL METALLOGRAPHY OF A CONFINED PINCH WELD IN ANNEALED 304L STAINLESS STEEL TUBE MADE AT LOW CURRENT, HIGH FORCE, LONG WELDING TIME AND NO RESTRAINT REVEALS TYPE 2 INTERNAL CRACKING AND TYPE B EXTERNAL CRACKING CAUSED BY EXPOSURE FOR TEN DAYS TO MAGNESIUM CHLORIDE BOILING AT 155°C.

than low electrode force. Type C external cracking is less severe at welds in annealed tubes than in cold-drawn tubes but is related to low current for welds in both kinds of tubes.

EFFECTS OF WELD TYPE - Unconfined pinch welds have "V"-shaped closures in contrast to variable shaped closures in confined pinch welds. Type 1 internal cracking in unconfined pinch welds produces a characteristic chevron pattern as shown in Figure 4. This pattern shows that residual stresses in the bond region of unconfined welds have different orientations than those in confined welds.

EFFECTS OF POSTWELDING ANNEALING - Confined pinch welds in cold-drawn Type 304L stainless steel tubes that were annealed at 1050°C for ten minutes exhibited no cracking after three days exposure to magnesium chloride solution boiling at 155°C. This results shows that tensile stresses were removed by this annealing process. A laser welder was used to anneal a narrow band (~3 mm wide) on the exterior near one end of the weld depression of confined pinch welds in cold-drawn type 304L stainless steel tubes. Laser power and focus were adjusted to heat the steel to just below melting. Ten days of exposure to magnesium chloride solution boiling at 155°C caused no internal cracking at the annealed end compared to severe cracking at the other end (Figure 5). These results suggest that localized external annealing using a laser may be beneficial in relieving residual stresses that contribute to environment-assisted slow crack growth without changing the metallurgical conditions of the major portion of a fill stem.

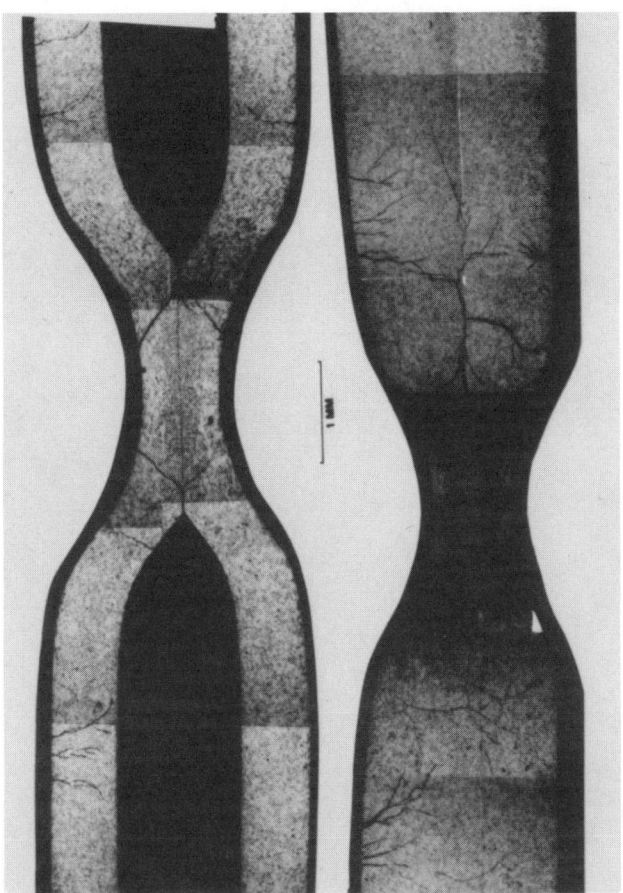

FIGURE 4. SERIAL METALLOGRAPHY OF AN UNCONFINED PINCH WELD IN A COLD DRAWN 304L STAINLESS STEEL TUBE EXPOSED BARE FOR THREE DAYS TO MAGNESIUM CHLORIDE SOLUTION BOILING AT 130°C REVEALS TYPE 1 INTERNAL CRACKING WITH A CHARACTERISTIC CHEVRON PATTERN, TYPE 2 INTERNAL CRACKING AND TYPE A EXTERNAL CRACKING.

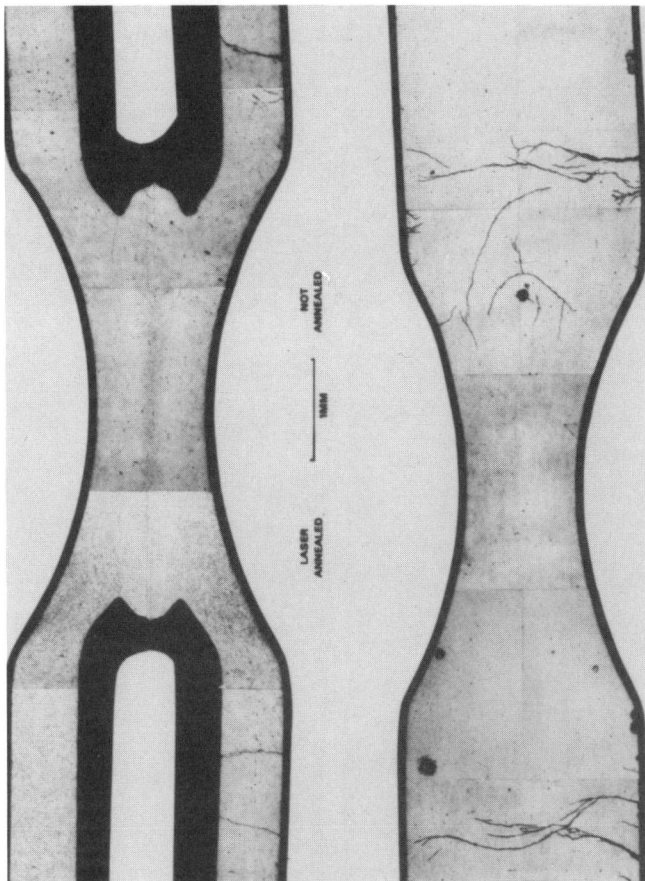

FIGURE 5. SERIAL METALLOGRAPHY OF A CONFINED PINCH WELD IN A COLD DRAWN 304L STAINLESS STEEL TUBE EXPOSED FOR TEN DAYS TO MAGNESIUM CHLORIDE SOLUTION BOILING AT 155°C REVEALS NO INTERNAL CRACKING AT THE LASER ANNEALED END AND SEVERE INTERNAL CRACKING AT THE UNANNEALED END.

FUTURE WORK - The continuation of this work will include more tests of pinch welds in Type 304L stainless steel tubes and new tests of pinch welded tubes of Types 21-6-9 and 316 stainless steels. Effects of metallurgical conditions produced by high energy rate-forging will be investigated. The goal is to identify welding and material conditions that minimized environment-assisted slow crack growth without compromising strength and quality of the pinch weld.

ACKNOWLEDGMENTS

The author is indebted to T. A. Jones and L. E. Johnson of the Savannah River Plant for performing the pinch welding and laser annealing. Measurements of exterior residual stresses by the material removal method performed by D. G. Berghaus of the Georgia Institute of Technology are gratefully acknowledged.

The information contained in this article was developed during the course of work under Contract No. DE-AC09-76SR00001 with the U.S. Department of Energy.

RESIDUAL STRESSES IN CERAMICS

C.O. Ruud, D.J. Snoha
Materials Research Laboratory
The Pennsylvania State University
University Park, Pennsylvania USA

C.P. Gazzara, P. Wong
U.S. Army Materials Technology Laboratory
Watertown, Massachusetts USA

ABSTRACT

The Pennsylvania State University has developed an advanced XRD stress measuring instrument which provides for unprecedented ease of stress measurement as well as speed consistent with accuracy. This advanced instrument has been applied to residual stress measurements in several types of ceramic materials to determine the relationship between processing parameters, breaking strength, and residual stress.

Application to alpha silicon carbide was studied by mapping the residual stresses in the radial and tangential directions along radial traverses on a cross-section of a ceramic 50 calibre gun tube liner. The measured stresses in the gun barrel liner compared well with predicted stresses. Two types of alpha-alumina specimens were studied. The first was represented by 40 broken halves of rods, 20 of which has been thermally treated to provide compressive surfaces residual stresses. A correlation between breaking strength and axial residual stresses was found to be 87 percent. This was in spite of the fact that stresses varied markedly along the axial dimension and that the stress was not measured at or even near the axial region where the breaking strength data had been obtained. For the second type of alpha alumina specimen studied, i.e., a 1/2 x 5.6 x 5.6 inch alumina armor tile, residual stresses were measured in four locations. The stress pattern found could be rationalized to result from air-cooling from sintering or some other type of high temperature thermal treatment.

Residual stress measurements have been performed on reaction bonded and hot pressed silicon nitride. Several silicon nitride specimens fabricated by various means have been subjected to thermo-mechanical loading, to simulate high temperature turbine application, and then recharacterized for residual stress. Results show some stress variation within and between supposedly homogeneous, identical specimens. The residual stress pattern in chemical vapor deposited beta silicon carbide has also been studied, and this work is described.

BACKGROUND

The inherent strength, high-temperature performance, and corrosion and erosion resistance of structural ceramics have promised marked improvement over certain metallic components in service performance. Unfortunately, ceramics not only are inherently much stronger than metals, they are also much less ductile, i.e., more brittle. The fact that metals are ductile, i.e. will plastically yield without fracturing, provides for relief of unexpected high-stress concentration in components prior to their failure. However, the virtual absence of plastic deformation that might relieve stress concentrations, means that the stresses must be accurately known everywhere in ceramic structural parts. Information on residual stresses becomes even more vital as the density of critical components is increased through elimination of pores, since pores have been a major cause of low fracture strengths. It is therefore important that the residual stresses resulting from a ceramic component's fabrication be known, especially in regions of high service stresses.

A few attempts have been made to measure or estimate residual stresses in ceramic bodies. Kirchner [1] in studying the beneficial effects of compressive residual stresses attempted to estimate them through tests relating the flexural strength and fracture mirror radius. Most researchers however have used rather crude mechanical techniques where the deflection or strain change in a component is measured as the result of material removal. The mechanical techniques are by nature destructive and provide only a coarse measurement of stress so that in situations where high-stress gradients exist, the true magnitude of the existing stresses cannot be determined.

Semple [2] proposed that x-ray diffraction (XRD) techniques be used to measure residual stress in alumina bodies. The XRD techniques are nondestructive for surface stress mapping and when used in conjunction with material removal techniques can provide excellent three-dimensional stress pattern determination. Unfortunately, until now the XRD techniques have been time consuming, ten to thirty minutes per reading, and in many cases awkward to apply. This is in spite of the fact that several companies have developed portable x-ray diffractometers for easier laboratory application and for field measurements.

The Materials Research Laboratory of The Pennsylvania State University has developed an advanced XRD stress measuring instrument which provides for unprecedented stress measurement speed consistent with excellent accuracy. The instrument was developed in order to facilitate a more thorough understanding of residual stress patterns in crystalline materials. Since its installation studies of nickel and copper base alloys as well as ferritic and stainless steels have been conducted providing an unprecedentedly detailed understanding of the stress patterns in the subject components. This unique XRD instrument is based upon the use of a position sensitive scintillation detector (PSSD), the principle of which is based upon the coherent conversion of the diffracted x-ray pattern into an optical signal, i.e., light; the conduction of this light signal, over several liner centimeters of coherent, flexible fiber optic bundles; the amplification of this signal by electro optical image intensification; the electronic conversion of the signal; and the transfer of the electronic signal to a computer for refinement and interpretation [3]. Another paper in these proceedings describes the PSSD device with more information [4].

One advantage of this PSSD-based stress analyzer is that its divergent XRD optics can be applied to give a very small irradiated area on the specimen so as to provide for excellent spatial resolution of stress readings. Also, the geometry of its x-ray optics allows for stress readings in very confined areas such as gear teeth, holes, and turbine vane bases. It should be noted that this PSSD instrument is not restricted to one residual stress measurement technique, e.g., the single exposure technique, but can be applied to the double-exposure (DET) and the sin-square-psi ($\sin^2\psi$) techniques [5, 6]. The rapidity of stress measurement provides for data collection over a period of a few seconds which is often a two order of magnitude improvement over conventional XRD instrumentation. Furthermore, this speed is consistent with excellent measurement accuracy; for example, recent tests in stainless steel have produced unequaled precision and accuracy of stress measurement on this material [7].

PROCEDURE AND RESULTS

Six types of ceramic specimens were selected for study: two were alpha alumina, one was alpha silicon carbide, one beta silicon carbide, and the last two were silicon nitride bars. The first type of alumina specimen was represented by forty broken halves of rods, twenty of which had been thermally treated to provide compressive surface residual stresses. The second type of alumina specimen was a ceramic tile for composite armor. The third type of specimen was an alpha silicon carbide gun tube insert, placed in a compressive stress state by a steel jacket. The beta silicon carbide material was a chemical vapor deposited (CVD) layer on a carbon substrate, and the silicon nitride specimens were rectangular bars, one set hot pressed and the other reaction bonded.

X-RAY RESIDUAL STRESS MEASUREMENTS - All residual stress measurement reported herein were performed by the single exposure technique using copper k-alpha x-radiation. Table I shows the Miller indices (hkl) of the interatomic planar spacing and the two theta angles used for each material as well as the time of data collection for a complete stress measurement and the estimated precision of the measurements. Accuracy of measurements could not be determined because the applicable elastic constants for the ceramics were not derived experimentally as is the recommended procedure [6, 8]. This derivation is recommended because the constants usually found in the literature are bulk values representing the average elastic properties for all

TABLE I

X-RAY STRESS MEASUREMENT AND ELASTIC PARAMETERS APPLICABLE TO CERAMIC SAMPLES STUDIED

	Alpha Alumina	Alpha Silicon Carbide	Beta Silicon Carbide	Silicon Nitride Hot Pressed	Silicon Nitride Reaction Bonded
(hkl)	(1 0 16)	(2 0 38)	(333, 511)	(3 2 6)	(3 2 6)
2θ	150.4	150.7	133.6	148.1	148.1
Data Collection Time (Sec)	15	60	15	7	7
Precision KSI (MPa)	4 (28)	4 (28)	11 (77)	6 (42)	6 (42)
E psi (MPa)	46×10^6 (324×10^3)	59.4×10^6 (418×10^3)	59.4×10^6 (418×10^3)	64×10^6 (451×10^3)	56×10^6 (394×10^3)
ν	0.21	0.142	0.142	0.17	0.17

crystallographic directions. However, in the XRD stress measurement technique, strain measurements in a single selected crystallographic direction, related to the Miller indices of the diffracting planes, are used. The difference between the bulk elastic values and the empirical XRD values can be substantial however, for the ceramic stress measurements in the study, bulk constants were used because specimens suitable for experimental determination of E and ν were not available.

Calibration of the PSSD instrument was performed using -400 mesh powders and setting the x-ray beam so that it was normal to the powder specimen surface, i.e., $\beta = 0°$. The incident x-ray beam in each case was collimated to provide the same shape and size of irradiated area that was used for stress measurement on the various specimens. Confirmation as to the validity of the calibration was performed on each powder with the x-ray beam incident at thirty degrees, i.e., $\beta = 30°$. Several readings were taken at various R_o distances on the powders between 38 and 42 mm and the mean and standard deviation of this value was obtained. The mean stress should be nearly zero if the calibration is valid, since a ceramic powder will not sustain a macro residual stress between the powder particles. Such a zero stress standard has been accepted by ASTM.

RESIDUAL STRESSES IN THERMALLY TREATED Al_2O_3 RODS - The forty alumina rods in which residual stresses were measured were actually the broken halves of twenty Al-300 alumina rods, 0.20 inches (5 mm) in diameter and are described by Tree et al. [9]. Ten of these rods had been annealed at 1000°C for 8 hours and slow cooled so as to provide as stress-free a material as possible, and ten had been annealed at 1000°C for 8 hours, equilibrated at 1450°C, then force air-quenched to provide a compressive surface residual stress.

Axial residual stress measurements were performed upon the broken halves of the twenty specimens by locating a rectangular-shaped irradiated area, 0.12 x 0.10 inches (3 x 2.5 mm) of copper K-alpha x-radiation on the specimen. Measurements were performed with the rods in a horizontal position and the incident beam at a sixty-degree angle to the axis of the convex surface at the irradiated area so as to provide a surface normal to incident x-ray beam angle (i.e., β angle) of 30 degrees [5, 6]. The rods were rotated at 0.15 rps during stress measurement so as to obtain the average of the stress around the circumference. All of the measurements were performed in less than six hours, including three repeats of data accumulation and stress calculation, and the placement of the specimens in the rotation device for x-ray data collection.

The XRD measured axial residual stresses are plotted versus the breaking strength data in Fig. 1. A linear regression fit to all the data from both types of specimens shows a correlation of 87%. This would indicate that the XRD residual stress measurements performed herein could be used to predict the breaking strength within a standard deviation of ± 9.7 KSI (68 MPa).

The XRD data indicates that the annealed and the quenched rods possessed an average axial stress of about -35 ± 3 KSI (-241 ± -21 MPa) and -18 ± 4 KSI (-124 ± 31 MPa), respectively. However, the XRD stresses were calculated using bulk elastic constants, a practice that can produce errors greater than ten percent in reasonably isotropic metals, and possibly much larger errors in an anisotropic material, such as alumina. However, such an error would affect the calculated residual stresses, i.e., -35 and -18 KSI, proportionally. The difference in the XRD results from the annealed versus the quenched rods indicates that the latter possess a surface stress about 17 KSI (120 MPa) more compressive than the former.

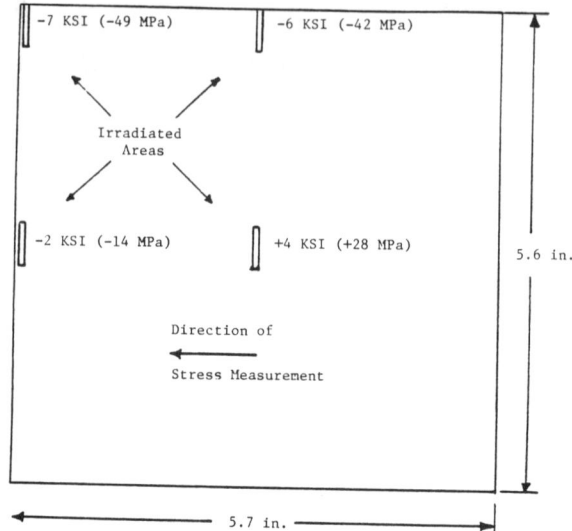

Fig. 2. Residual stress measurement results from a 0.5 inch (13 mm) thick Coors AD-94 sintered alumina armor tile.

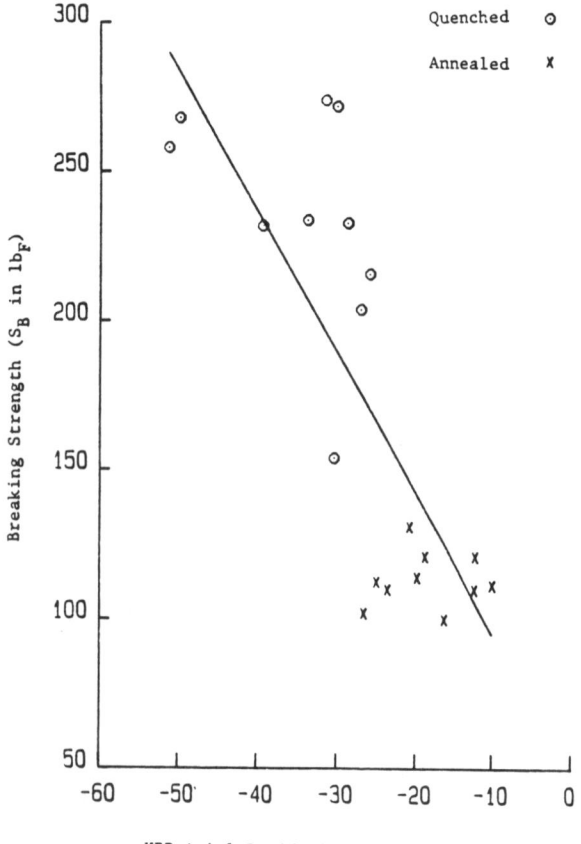

Fig. 1. Alpha-Alumina Rods Breaking Strength vs. Axial XRD Residual Stress

RESIDUAL STRESSES IN ALUMINA TILE - The second type of alpha alumina specimen in which residual stresses were measured in this study was a 0.5 x 5.6 x 5.7 inch (13 x 14.2 x 145 mm) Coors AD-94 sintered alumina armor tile. Measurement was performed in only four areas of the tile using the same size of irradiated area as used for the alumina rods, i.e., 0.04 x 0.51 inches (1 x 13 mm). Figure 2 shows a sketch of the tile and indicates the position, direction, and value of the stress readings. These measurements were made holding the specimen-to-detector distance within ±0.004 inches (±0.1 mm) of 1.79 inches (40 mm); therefore, the precision was better than that indicted in Table I and was less than ±3 KSI (21 MPa). The absolute accuracy of the residual stress measurements on the tile is not known; however, comparison of the readings is valid and indicates a trend toward more tensile residual stresses in the center of the tile. Further, the residual stresses perpendicular to the edges seem to be more tensile than those parallel to the edges. The aforementioned trends in residual stresses on the tile could be rationalized as due to cooling since the corner would cooled the fastest, thereby tending to develop compressive residual stress, and the center the slowest, thereby tending toward more tensive stress. However, caution must be observed in drawing conclusions from this cursory examination of an armor tile.

RESIDUAL STRESSES IN A SILICON CARBIDE LINER - The third type of specimen in which residual stresses were measured in this study was an alpha silicon carbide 50 calibre gun tube liner. The liner was the inner tube of a composite gun tube with two outer steel tubes. The two steel tubes were designated the sleeve and jacket; with the sleeve shrink fitted into the jacket with a 0.002 inch (0.05 mm) interference. The silicon carbide liner was shrink fitted into the sleeve-jacket assembly with a 0.003 inch (0.08 mm) interface. The result of these interferences was to produce compressive radial and tangential stresses on the ceramic liner. A paper by Bunning et al. [10] discusses the procedure in some detail and the calculated stress patterns shown in Fig. 3 were obtained from that reference.

The specimen examined was an approximately 2.1 inch (53 mm) long segment of the composite gun tube that had been sectioned through a plane normal to the axis from the longer original tube. The sawed surface had been subsequently mechanically polished and stresses in the ceramic were measured in that condition.

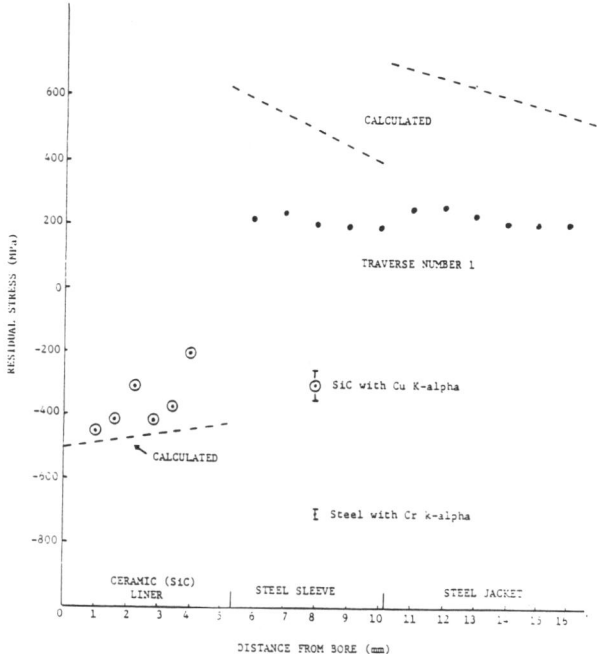

Fig. 3. Measured and calculated (predicted) tangential stresses induced by shrink-fit assembly of a composite 50 calibre gun tube.

However, subsequently, the steel cross section of the specimen was electropolished so as to remove the stresses induced by mechanical polishing prior to stress measurement. The tangential and radial stresses induced by the shrink fits in the ceramic liner were measured using both copper and iron K-alpha radiation. These same stress directions were measured in the steel cross-section using Cr K-alpha radiation. The stresses were measured along three radial traverses approximately one hundred and twenty degrees apart.

Figure 3 shows plots of the induced tangential patterns measured on the cross-section of the composite gun barrel from one of the radial traverses, i.e., traverse number one. The stresses in the steel are less tensile than the calculated (predicted) stresses as might be expected with the smaller shrink fit, i.e., 0.003 for the experimental, instead of 0.004 as used for the calculated stresses. The estimated precision and accuracy of the stress measurement on the steel is 2 KSI (14 MPa). The measured stresses on the ceramic liner are less compressive than predicted, which is logical since the steel sleeve and jacket stresses were lower.

It should be noted that Bunning et al. [10] predicted tensile stresses in the axial direction in the steel portion of the composite gun tube and compressive in the ceramic. Since the specimen studied was a cross-section of the original tube, it is reasonable to expect that the cutting of the specimen produced some stress relief in the axial direction due to the creation of the free surface studied. In other words, the axial stresses would have to be reduced to zero at the surface upon which the tangential and radial stress measurements were made. This stress relief would cause the stresses in the plane of the cross-section to become more tensile in the steel and more compressive in the ceramic. No correction for this free surface stress relief was made to the data reported.

RESIDUAL STRESSES IN BETA SILICON CARBIDE CVD COATING - Two samples of chemical vapor deposited (CVD) beta silicon nitride on carbon were mapped for residual stresses using CuKα radiation reflected from the (333, 511) crystallographic planes, see table I. The samples were squares about 1 x 1 x 0.3 inches (25 x 25 x 8 mm) and stresses were measured in a grid of 25 locations evenly spread over the surface of the samples. The size of the irradiated area was 0.08 x 0.08 inches (2 x 2 mm) and stresses were measured in two orthogonal directions. All residual stresses in the CVD coating were highly compressive and about the same magnitude in the two orthogonal directions indicating a uniform stress over the surface both in direction and magnitude. The stress on one of the samples average about -150 KSI (-1056 MPa) and on the other about -125 KSI (880 MPa). The results also indicated what seemed to be a strong preferred orientation of the (111) crystallographic planes parallel to the surface. The high residual compressive stress indicated in these results may well be a major factor in allowing CVD beta silicon carbide coatings to provide an oxidation barrier for carbon substrates.

RESIDUAL STRESSES IN SILICON NITRIDE BARS - Two sets of four samples each of hot pressed and reaction bonded silicon nitride rectangular bars 0.5 x 3.25 x 0.15 inches (13 x 83 x 4 mm) in size were measured for magnitude and distribution of residual stresses. Stresses were measured in the longitudinal direction of the bars at fifteen places equidistant along each bar length. The irradiated area was 0.04 x 0.24 inches (1 x 6 mm) with the short dimension parallel to the long dimension of the bars. Table II lists the mean longitudinal residual stresses on the eight silicon nitride bars. The mean residual stress in the hot pressed bars were within one standard deviation of zero and were essentially uniform along the bars. The mean residual stress tended to be near 20 KSI (140 MPa) tensile on three of the

TABLE II

MEAN RESIDUAL STRESSES (ksi) MEASURED ON THE TOP AND BOTTOM OF SILICON NITRIDE (Si_3N_4) RECTANGULAR BARS

			$2\theta \simeq 148.1°$ (326), $\beta = 30°$	
SPECIMEN	FACE	STRESS DIRECTION	HOT PRESSED	REACTION BONDED
1	Top	Longitudinal	-2.7	20.1
	Bottom	Longitudinal	-1.6	24.9
2	Top	Longitudinal	0.4	6.3
	Bottom	Longitudinal	9.4	5.3
3	Top	Longitudinal	-0.2	21.6
	Bottom	Longitudinal	-0.6	12.4
4	Top	Longitudinal	4.7	21.9
	Bottom	Longitudinal	1.7	25.9

reaction bonded bars and about 5 KSI (35 MPa) tension on one of the bars. In contrast to the hot rolled bars the stresses varied sufficiently on the reaction bonded bars so that they could not be considered uniform. The possible reason for the non-uniformity of stresses on the reaction bonded bars could be due to varying amounts of microporosity.

SUMMARY AND RECOMMENDATIONS

Residual stress measurements in the several structural ceramics reported herein showed that the XRD instrument applied provided improved speed, precision, resolution, and ease of measurements over previously attempted measurements described in the literature. Further, the data described in this paper represents a significant contribution to the total number of XRD residual stress measurements on ceramics reported to date.

ACKNOWLEDGEMENT

The investigation described herein was in part made possible by funding from the U.S. Department of the Army through the Army Materials Technology Laboratory under contract number DAAG46-83-K-0036. The authors are grateful to Dr. D. M. Rose of U.S. Army TACOM, Warren, MI for providing the Beta Silicon Carbide samples and E. Lenoe for Si_3N_4 bars.

REFERENCES

1. H. P. Kirchner, *Strengthening of Ceramics, Treatments, Tests, and Design Applications*, Marcel Dekker Inc., N.Y. β Basel, 1 1979.

2. C. W. Semple, 'Residual Stress Determinations in Alumina Bodies,' Report No. AMMRL TR70-14, June 1970.

3. C. O. Ruud, 'A Unique Position Sensitive Detector for X-Ray Powder Diffraction,' Ind. Res. and Dev. Magazine, December 1982.

4. C. O. Ruud and D. J. Snoha, 'Residual Stress Mapping of Gears and Bearings,' Proceedings of ASM's 1987 Conference on Residual Stress--In Design Process and Materials Selection, April 1987, Cincinnatti, OH.

5. C. O. Ruud and D. J. Snoha, 'Displacement Errors in the Application of Portable X-Ray Diffraction Stress Measurement Instrumentation,' J. of Met., Vol. 36, No. 2, February 1984, pp. 32-38.

6. SAE, 'Residual Stress Measurement by X-Ray Diffraction-J78a,' Soc. of Auto. Eng., Warrendale, PA, 1971.

7. C. O. Ruud, P. S. DiMascio, and D. M. Melcher, 'Application of a Position Sensitive Scintillation Detector for Nondestructive Residual Stress Measurement Inside Stainless Steel Piping,' Adv. in X-Ray Anal., Plenum Press, Vol. 29, 1983.

8. P.S. Prevey, 'A Method of Determining the Elastic Properties of Alloys in Selected Crystallographic Directions for X-Ray Diffraction Residual Stress Measurement,' Adv. in X-Ray Anal., Vol. 20, Plenum Press, 1977, pp. 345-354.

9. Y. Tree, A. Venkateswaran, and D. P. H. Hasselman, 'Observations on the Fracture and Deformation Behavior During Annealing of Residually Stressed Polycrystalline Alumina Oxides,' J. Mat. Sci., Vol. 18, 1983, pp. 2135-2148.

10. E. J. Bunning, D. R. Claxton, and R. A. Giles, 'Ceramic Liners for Gun Tubes: A Feasibility Study,' Presented at the Ceramic-Metal Systems Division, Fifth Annual Conf., Merrit Island, FL, January 20, 1981.

DEVELOPMENT OF A MODIFIED BLIND HOLE TEST FOR THE MEASUREMENT OF RESIDUAL STRESS GRADIENTS IN THIN WEAR RESISTANT COATINGS

L.C. Cox
Union Carbide Corporation
Indianapolis, Indiana USA

Thin, wear resistant tungsten carbide coatings tend to be very hard and lack toughness. Efforts to develop tougher coatings have led to the need to measure and understand the residual stress state of the coatings. The capability of measuring stress gradients was also desired, since the local stress of the coating surface could be very different from the average stress through the thickness of the coating. This paper describes the use of the blind hole test to measure residual stress gradients in detonation gun (D-Gun) coatings.

The D-Gun is a proprietary process in which a mixture of acetylene, oxygen and powder is placed in a gun barrel and the charge is ignited. The detonation wave accelerates the powder to about 2400 ft/sec while heating it to temperatures close to or above its melting temperature, depending on the chemistry of the initial gas mix. The powder is deposited on the target in a circle approximately 1" in diameter and anywhere from 0.2-0.8 mils thick.[1] The coating is composed of these slugs of powder overlying each other, and is usually 5-20 mils thick.

Coating residual stress is a seemingly unavoidable by-product of the D-Gun operation. A first order approximation of the residual stress can be achieved by assuming that at impact, the coating material is at its melting temperature and the substrate is at room temperature. Assuming no plasticity of either material, and a substrate thickness much greater than the coating thickness, the coating stress is composed entirely of the thermal stress resulting from the constrained contraction of the coating as it cools. Assuming a plate substrate at room temperature, T_R, and a deposition temperature, T_D, the coating stress is given by

$$\sigma = \frac{E}{1-\nu} \alpha \Delta T \qquad \text{Equation 1}$$

where E = Elastic Modulus of the Coating
ν = Poisson's Ratio
α = Coefficient of Thermal Expansion of the Coating
$\Delta T = T_D - T_R$

For a ΔT of 1000°C. Equation 1 gives a value for the stress in a WC-Co coating (E = 25 x 10^6 psi. ν = 0.3, α = 8.5 x 10^{-6}/°C) greater than 300,000 psi tension. This value is much larger than the measured strength of the coating. Clearly, a reliable method of stress measurement is needed.

Several methods have been used at Union Carbide to measure coating residual stress. Among them are

1. Removing the substrate from a cylindrical coating and measuring the deflection of the coating when longitudinal and

circumferential cuts are made (comonly called "parting out")(2,3),

2. Coating a 2" wide x 0.032" thick x 6" ID ring to incremental coating thicknesses and measuring the change in diameter of the ring and,

3. Coating the face of a 1" diameter disc, thickness on the order of 3/16", and measuring the out-of-plane displacement in units of He lightbands.

The first test gives only the average stress through the thickness of the coating. The stress at the outer layer may be very different from the average stress, and is more important in the initiation of coating cracks.

The second and third tests allow the stress to be calculated in incremental layers. However, they both depend on an initial data point taken prior to coating, and assume that the reference is not affected by the coating process. Particularly in the D-Gun process where very large kinetic energies are carried by the powder, this assumption is not easy to justify. Much more likely is that the stress in either the grit blasted substrate or the already deposited coating will be changed by the coating arriving later in the process, altering the initial reference point.

Thus, a more general method of measuring stress is needed. Required attributes of the test are that it be capable of giving stress gradient information, and it does not depend on a reference point taken before the part is coated. The blind hole test satisfies these two requirements.

The blind hole test has gained acceptance among stress analysts as a semi-destructive residual stress test. An ASTM standard[5] has been written to guide experimenters in the use of this test, but the standard is limited to constant stress profiles. Building on work published in the literature, the basic procedure has been modified to make it applicable to coatings.

Kelsey[8] was the first to try to apply the blind hole test to measure stress gradients. He postulated that the extension of the test was valid so long as the measured surface strain is a function only of the residual stress in the incremental layer being drilled, assuming the strain gage rosette, hole geometry and material properties are constant. Kelsey did experiments on samples in pure tension and pure bending and concluded that at hole depths less than 1/2 the hole diameter, the relieved surface strain was indeed independent of the stress gradient in the part. He concluded that the blind hole test could be used to measure stress gradients.

Procedure

A strain gage rosette with strain gages oriented at $0°$, $90°$ and $-135°$ was used to measure the relieved strain around the blind hole. The particular gage used is marketed by Texas Measurements (FRS 2). The centerline of the gages lies on a circle with a diameter of 0.202". The gage length is 0.059".

The hole was drilled using a jet of 27 μm Al_2O_3 carried in an argon stream. One limitation of using an abrasive jet to drill a hole is that the hole is typically not flat bottomed, and not always round. This is a serious liability, particularly when the depth of the hole is increased by small increments. Proctor and Beaney[7] addressed this problem and found that very satisfactory results are obtained when the hole is trepanned, i.e., when the nozzle is rotated off-center. This is the technique that was adopted by Coatings Service.

The sensitivity of the test is controlled by the diameter of the blind hole, assuming the strain gage rosette geometry remains fixed. Larger diameter holes result in increased sensitivity at the surface of the part, at the expense of information about stresses at deeper layers. For thin coatings, with thickness on the order of 0.005"-

0.020", a great deal of sensitivity is needed. Thus, the largest hole size obtainable without damaging the strain gages was required.

The accuracy of the blind hole test depends on the ability to consistently position the hole so that it is concentric with the strain gage rosette. An off-center hole will give erroneous strain readings, and may result in a strain gage being eroded and destroyed.

Thus, the key to making the blind hole test sensitive and accurate enough to measure stresses in thin coatings involved optimization of three techniques: masking the gages for protection from the grit blast medium, accurately positioning the hole in the center of the strain gage rosette, and reproducibly creating the largest trepanned hole possible. The key to achieving all three tasks lay with the masking procedure. A mask accurately positioned on the rosette protected the gage, and at the same time served as a reference for the abrasive jet nozzle. Initially, the nozzle was rotated by hand, and the sides of the mask were used to guide the nozzle. Later, a machine was built to rotate the nozzle. The outer diameter of the rotating nozzle was only slightly less than the inner diameter of the mask. The part was positioned so that the nozzle did not touch the mask at any point during a revolution. This fixed the center of the hole within a few mils of the center of the mask. The task, then, became one of adequately positioning the mask. This was done by choosing a mask with outer diameter matching that of the outer diameter of the gage. A microscope was used to position the mask so that none of the gage could be seen around the circumference of the mask. This located the mask concentricly to the strain gage rosette to within a couple mills tolerance. A masked, tested sample is shown in Figure 1.

When analyzing the data, use was made of the fact that the coating lands on the substrate and begins to contract in a radially symmetric manner. All stresses, therefore, lie in the plane of the coating, and are independent of direction. Consequently, all three strain gages should give identical readings when the blind hole is introduced. Figure 2 confirms that all three gages do tend to move together. Unfortunately, the reading at each gage is controlled not only by how much residual strain is in the incremental layer, but also by where the gage is positioned relative to the hole. The three gages are usually close to being the same, but have some error in their readings, most likely due to the hole being located slightly off-center with respect to the strain gage rosette. This situation was handled by averaging the readings of all three gages. In doing so, it was assumed that the error is reduced and the average value is what would be obtained if the hole were perfectly centered in the rosette.

Three hole diameters have been used with this test, depending on the technique used to introduce the hole. When the hole was drilled by hand, the diameter was 0.075". Later, a machine was built to rotate the nozzle, and the diameter was enlarged to 0.110". This diameter caused problems with the Al_2O_3, damaging the strain gage, so the diameter was dropped to 0.098". With the first two hole sizes, the test was calibrated to relate the relieved strains to the residual stress present prior to the introduction of the blind hole. The calibration consisted of applying a strain gage to a 1/4" thick tensile bar, drilling a hole, subjecting the bar to known stresses and comparing the measured strain around the hole to the theoretical strain without the hole.

The equation required to relate the residual stresses to the relieved stresses is given in the ASTM standard.

$$\sigma_x = \frac{\epsilon_1 + \epsilon_3}{4\bar{A}} + \frac{\sqrt{2}}{4\bar{B}}\sqrt{(\epsilon_1 - \epsilon_2)^2 + (\epsilon_2 - \epsilon_3)^2} \quad \text{Equation 2a}$$

where $\bar{A}$ & $\bar{B}$ are constants, determined by calibration experiments.

Taking into account the assumption that all three strain gages should read the same, equation 2 reduces to

$$\sigma = \varepsilon_A / 2\bar{A} \qquad \text{Equation 2b}$$

where ε_A is the average of the readings for all gages.

ASTM E-837 further defines the calibration constant $\bar{A}$ as

$$\bar{A} = \frac{(\varepsilon_3)_{cal} + (\varepsilon_1)_{cal}}{2\sigma} \qquad \text{Equation 3}$$

where $(\varepsilon)_{cal}$ = difference in the recorded strain with and without the blind hole

σ = stress applied to the calibration bar

and the subscripts 1 and 3 refer to the direction parallel and perpendicular to the applied stress.

Equation 3 can be rewritten using equations from theoretical elasticity.

$$\sigma = E \, \varepsilon_{APP} \qquad \text{Equation 4a}$$

and

$$(\varepsilon_3)_{cal} = -\nu \, (\varepsilon_1)_{cal} \qquad \text{Equation 4b}$$

to yield

$$\bar{A} = \left[(1 - \nu)/2E\right] \times \left[(\varepsilon_1)_{cal}/\varepsilon_{APP}\right] \qquad \text{Equation 5}$$

where E and ν are the elastic modulus and Poisson's ratio of the material, and ε_{APP} is the strain applied to the calibration bar.

Equation 4b requires the assumption that the effect of the hole is independent of orientation. That this is so has been reported in the literature [7], and proven with the use of the finite element technique.

Defining

$$\bar{a} = (\varepsilon_1)_{cal}/\varepsilon_{APP}$$

equation 2b can be written

$$\sigma = \left[E/(1-\nu)\right] (\varepsilon_A/\bar{a}) \qquad \text{Equation 6}$$

The constant $\bar{a}$ now becomes the subject of the calibration experiments. The form of equation 6 is particularly attractive, since $\bar{a}$ is a geometrical term, independent of the elastic properties of the material being tested.

The results of the calibration experiments are shown in Figure 3 for two hole diameters. Some difficulty was encountered trying to measure hole depths of 0.002"-0.010", when the strain gage itself was 0.003". For the smaller hole diameter, the depth was measured with a dial indicator. For the larger diameter, a focusing microscope was used which was considerably more accurate. In both cases, once the hole went through the gage, the relative hole depths could be determined within $\pm$ 0.0002".

The mechanics of the blind hole test indicate that the amount of stress relief should be proportional to $1/R^2$, where R is the radius of the hole. Thus, for any given hole radius, the slope of the calibration curve divided by the square of the radius should be constant. Table I shows that this is so for the parameters used here.

Figure 4 shows the complete calibration curve for the larger hole diameter. This curve has the same shape as those shown in Reference 9 for other hole diameters. In that paper, the axes were normalized so that the parameter plotted on the axis was hole depth divided by hole diameter, and the parameter plotted on the y axis was strain released divided by the maximum strain release recorded at deep hole depths. The hole diameter was non-dimensionalized by dividing by the diameter of the centerline of the strain gage rosette. The data presented in Reference 9, along with the data from Figure 4 are shown in Figure 5. Since the calibration experiments presented in Figure 4 were not carried out to deeper hole depths, it was assumed that the data point at Z = 0.019 in

Figure 4 represented 95% of the maximum strain release.

To make a comparison of the four curves shown in Figure 5, the hole depth required to achieve 40% of the maximum strain release was plotted in Figure 6. The points all lie on a straight line. The hole diameter used for this work lies outside the limits recommended by ASTM. However, no technical reason is given by the ASTM Committee for setting the limits at their current values. Figure 6 implies that the limits could be extended without sacrificing accuracy.

The method suggested by Micromeasurements publication TN-503 was used to relate the relieved strains to the stress in the coating. That is, it was assumed that in any depth increment, the residual stress was constant, and all movement of the strain gage was due to the stress in the incremental layer.

The stress in a layer was thus given by

$$\sigma_{\Delta Z} = \frac{E}{1-\nu} \frac{\Delta \varepsilon_{AVE}}{a \, \Delta Z} \quad \text{Equation 7}$$

where $\sigma(\Delta Z)$ = residual stress layer
 E = elastic modulus of the coating
 ν = Poisson's ratio of the coating
 $\Delta \varepsilon_{AVE}$ = avg. of strain gage readings in increment
 a = slope of calibration curve
 ΔZ = incremental layer

The results of the blind hole test were verified with the layer removal method of residual stress measurement. A large plate was coated, then a beam (1/2" x 10" x 1/16") was cut from the sample. Layer removal was accomplished with a diamond grinding wheel, removing 0.00025"/pass. After each 0.001" increment was removed, the deflection of the sample was recorded. The stress in the layer removed was given by

$$\sigma(Z) = \frac{E}{6}(t-Z)^2 \frac{dC}{dZ} - \frac{2E}{3}(t-Z)C$$

$$+ \frac{E}{3} \int_0^Z C \, dZ \quad \text{Equation 8}$$

where $\sigma(Z)$ = stress in layer removed
 E = elastic modulus of the substrate
 t = thickness of the beam before grinding
 Z = thickness of the removed layer
 C = radius of curvature of the beam

Results and Discussion

Figures 7 and 8 show a comparison of the blind hole and the layer removal measurements made on three different coatings. In both cases, the sample used for the measurements by layer removal was cut from the sample on which the blind hole measurement was performed. Clearly, the agreement is excellent, even at blind hole increments of 0.002". The plots do appear to be displaced relative to each other on the x-axis (location of depth increment). This is not cause for concern. D-Gun coatings have some roughness associated with them, making the initial reference point difficult to define. In two cases, approximately 0.003" was removed before the surface roughness was totally removed. This uncertainty created the discrepancy in locating the data along the x-axis. However, by using as a reference the point at which the stress gradient changes direction (presumably the point at which the blind hole moves into the substrate), the curves are nearly identical.

Another feature present in Figures 7 and 8 is the fact that the blind hole test loses accuracy at depths greater than 0.015". The depth is considerably less than one-half the hole diameter -- the value Kelsey cited as the limiting hole depth. In this case, accuracy is lost at depths greater than a value of z/D = 0.20. As discussed earlier, this is an expected result of having increased the hole diameter in order to raise the sensitivity at shallow hole depths.

Similar data are shown in Figure 9 for coatings made from the same powder, but under three different coating conditions. Again, agreement between the two stress measurement techniques is very good. Thus, the blind hole test is capable of giving reliable stress data over depth increments as small as 0.002". To the author's knowledge, this is the smallest depth increment used with the blond hole test.

Figure 10 shows data for two samples where the coating is the same but the surface preparation of the substrate is different. In one case, the substrate was grit blasted prior to coating, resulting in compressive stresses in the substrate. In the second case, the substrate was not grit blasted, and the substrate had tensile residual stresses. In both cases, the coating has relatively low stress, unaffected by the surface preparation of the substrate. However the difference in the residual stress state of the substrate is immediately apparent, and discernible with the blind hole test. It is concluded that the blind hole test is capable of discerning very large differences in magnitude and direction of the stress tensor, as well as the very small gradients shown earlier.

Finally, it is interesting to note that many of the coatings measured have compressive residual stress, not the tensile stress predicted by the first order approximation. Compressive residual stress is very beneficial from the user's viewpoint since coatings usually crack in tension.

Also note that in many cases the residual stress is not constant through the coating thickness, but is tensile at the surface and compressive in layers closer to the substrate. This fact is important since coating failures usually start either at the surface, in the case of cracking, or at the substrate, in the case of spalling. In predicting coating failure, the stress where failure starts is important. Thus, an analysis of a coating which uses the average value of the residual stress in the coating could be misleading.

Conclusions

1. The blind hole test is capable of measuring residual stress gradients in coatings where the total coating thickness is on the order of 0.015", and incremental layer thicknesses are as small as 0.002". Constant stress through a coating and stress that changes sign and gradient direction have been measured, as well as situations that lie between these two extremes.

2. Contrary to a simple mathematical model for residual stress which assumes that the coating stress results from constrained contraction during cooling, compressive stresses have been measured in many coatings.

3. Although the hole diameter used in this work is larger than that recommended by ASTM, the results of the calibration experiments fall on a straight line when plotted with similar results reported in the literature.

Acknowledgments

This work could not have been done without the help of Lori Claudy and Jim Price. These two helped develop the experimental technique, served as sounding boards, and participated in many useful discussions during the development and fine tuning of the test. Their help is greatly appreciated.

References

1. Tucker, R. C., "Plasma and Detonation Gun Deposition Techniques", Deposition Technologies for Films and Coatings, Noyes Publications, 1982.

2. R. C. Tucker, Jr., Internal UCAR report.

3. SAE Information Report, "Methods of Residual Stress Measurement - SAE J936", December 1965.

4. Perakh, M., "Calculation of Spontaneous Macrostress in Deposits from Deformation of Substrates and from Storing (or Restraining) Force

Factors", Surface Technology (1979) 265-309.

5. ASTM Standard E-387.

6. "Measurement of Residual Stresses by the Blind Hole Drilling Method", Measurements Group Technical Note 503.

7. Proctor, E., and Beaney, E.M., "Recent Developments in Centre-Hole Techniques for Residual Stress Measurement", Proceedings, 1981 Fall Meeting, SEM.

8. Kelsey, R. A., "Measuring Non-Uniform Residual Stresses by the Hole Drilling Method", Annual Meeting of the Society of Experimental Stress Analysis, November 1955.

9. Micromeasurement Group, TN-503

TABLE I

Relationship Between the Radius of the Blind Hole
and the Slope of the Calibration Curve

R	Slope of (a) Calibration Curve	a/R^2
0.0375	0.7	504.9
0.055	1.59	525.0

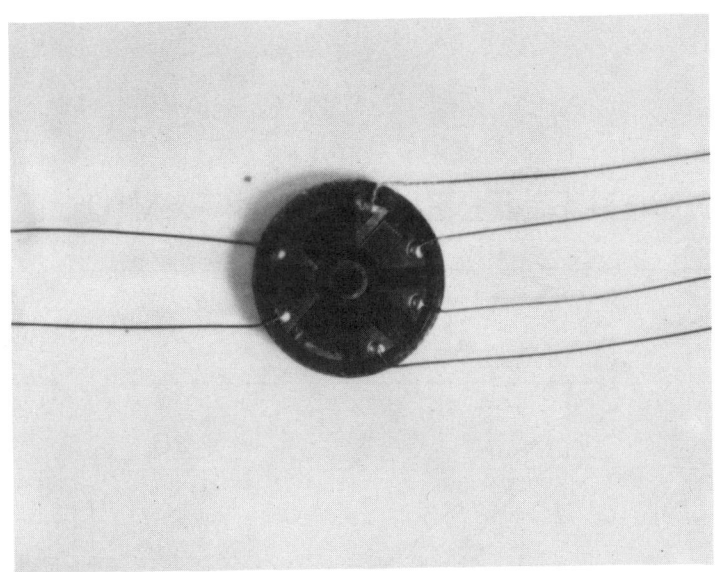

Figure 1a) A strain gage rosette is shown. Underneath the strain gage is the mask used to locate the hole. Note that the mask does not extend outside the outer rim of the strain gage rosette.

Figure 1b) A tested sample is shown. The strain gage and the mask are the same size, allowing more accurate location of the mask. The hole was made by a mechanical device designed to rotate the nozzle. Note that the rotation is so accurate that the island in the middle still has the strain gage attached.

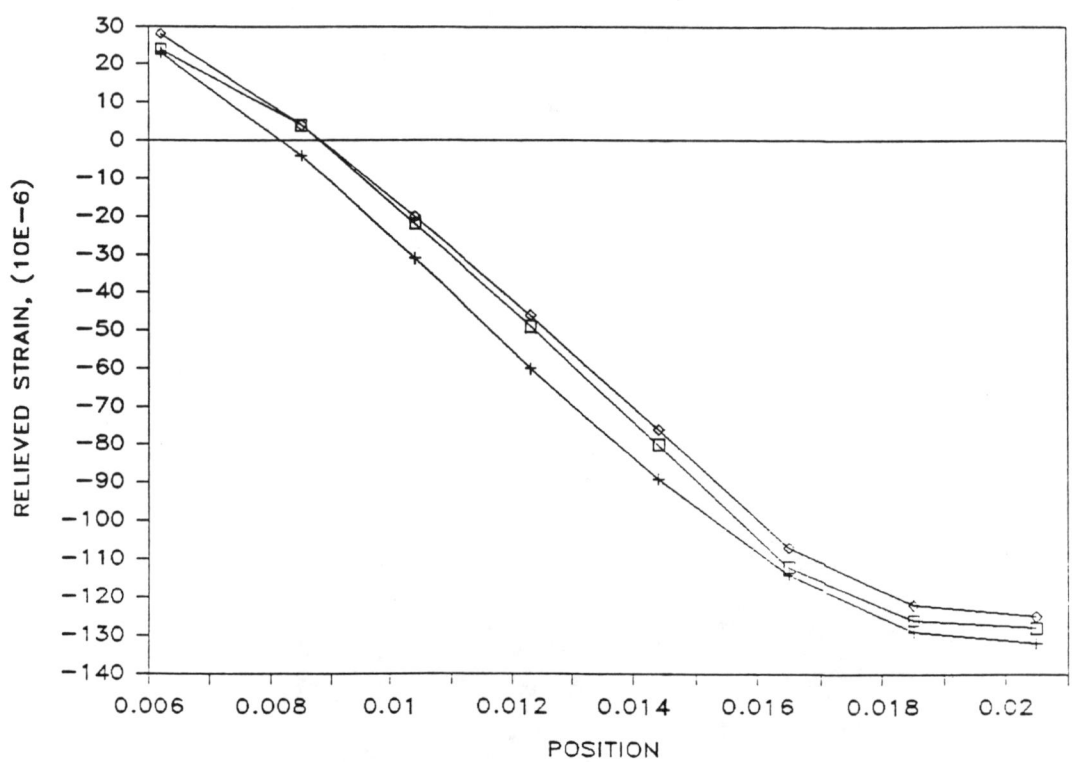

Figure 2. The readings of three individual gages are shown as a function of hole depth. The readings are nearly identical and move together. This is an expected result for a material in a state of uniform, biaxial stress. The readings of the three gages are averaged to give a single value of strain.

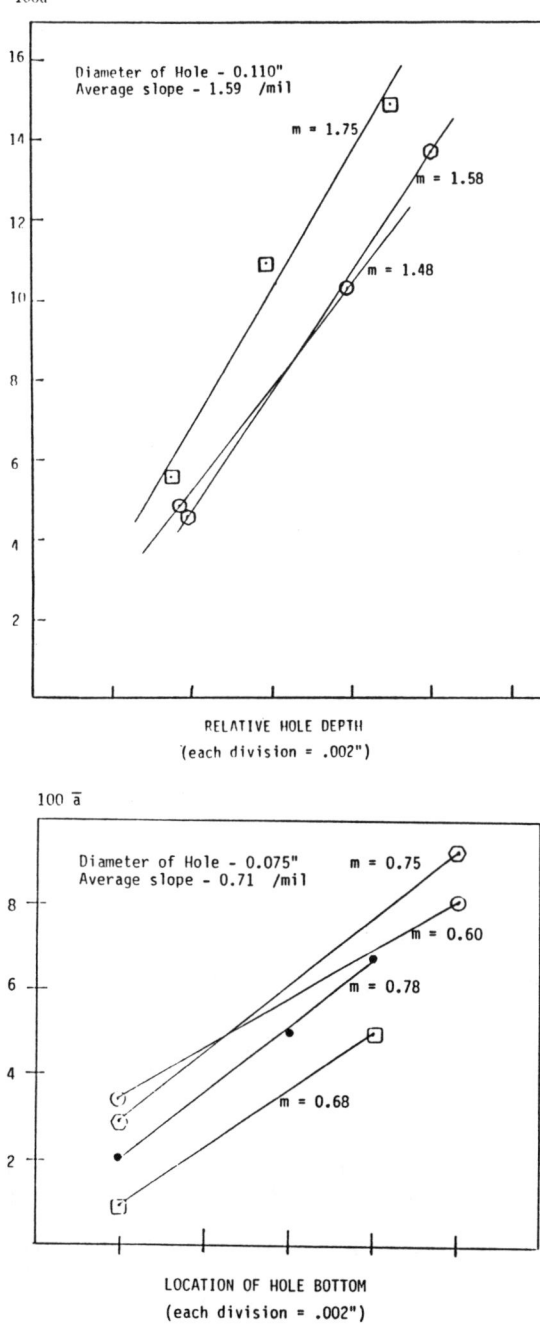

Figure 3. The results of several calibration experiments are shown. An absolute value of hole depth was difficult to obtain, due to the presence of the strain gage. However, once the gage was penetrated, the relative depth of a hole for two increments could be accurately determined. Assuming that within the region of interest the points should fall on a straight line, the slopes of the line were calculated and averaged.

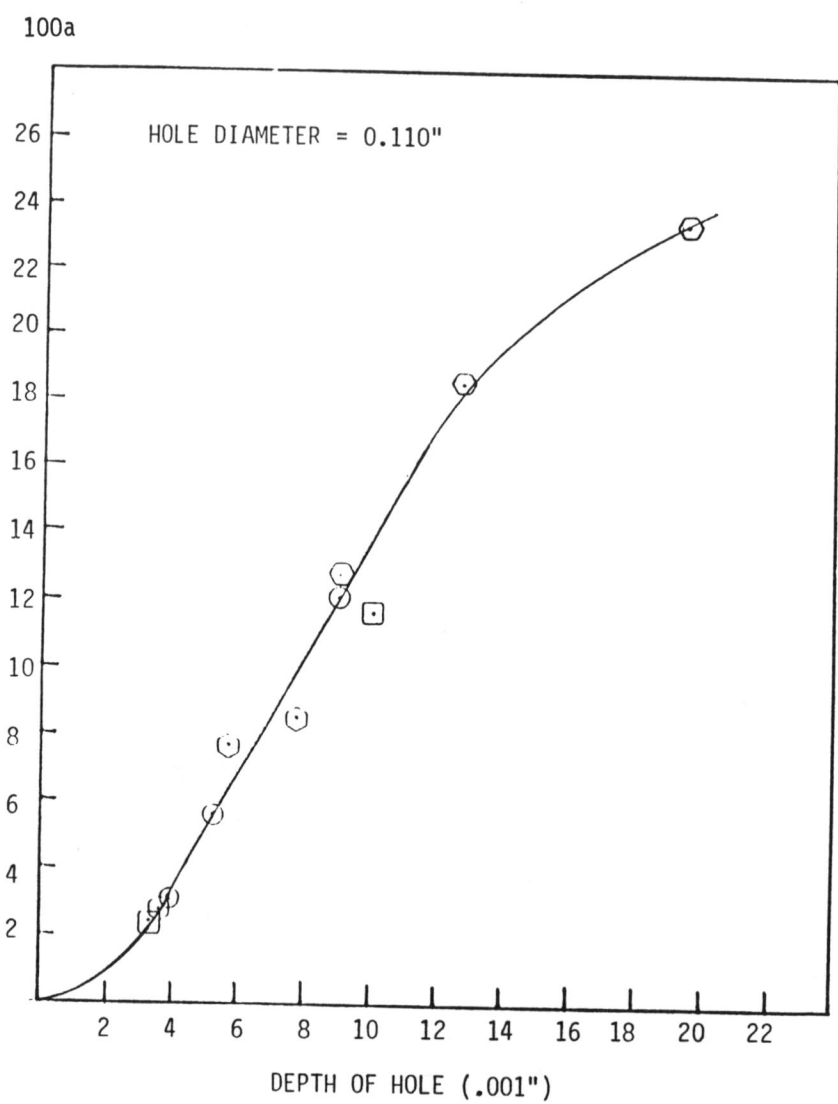

Figure 4. The complete calibration curve for a hole diameter of 0.110" is shown. The form of this curve is similar to that reported in the literature for other hole diameters.

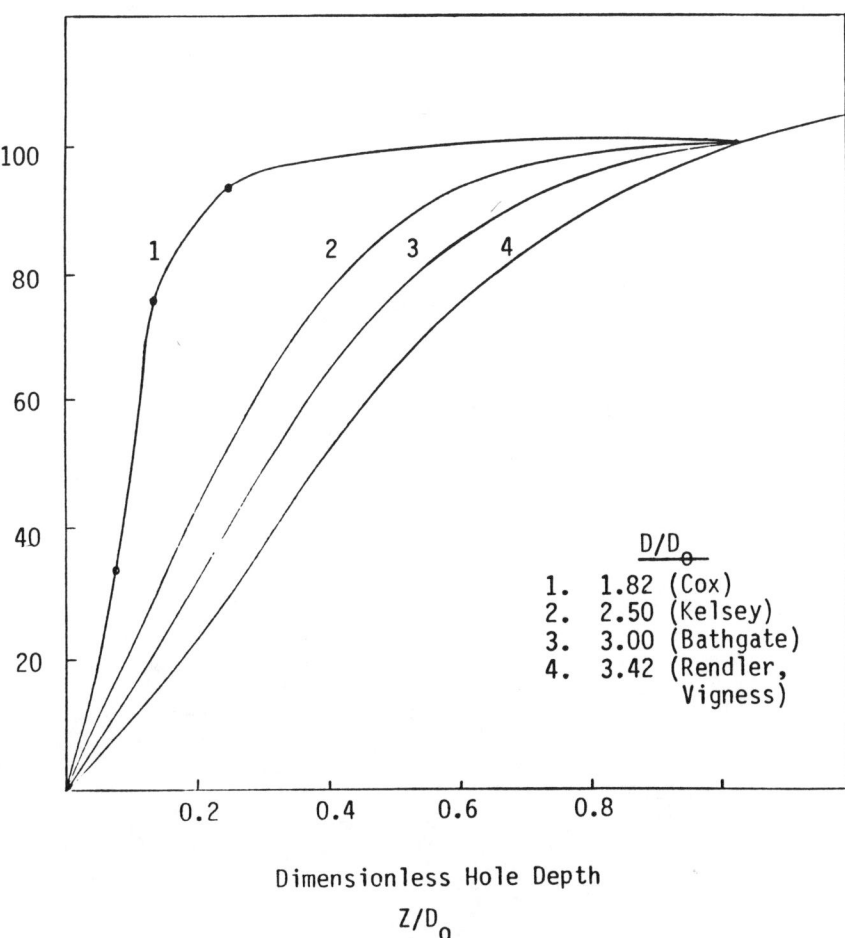

Figure 5. The data from Figure 4 are plotted in non-dimensional form, along with other data from the literature (see Reference 9). To get the data, it was assumed that the data point taken at a hole depth of 0.019" in Figure 4 represented 95% of the maximum strain relief, which would be recorded if the hole were drilled further.

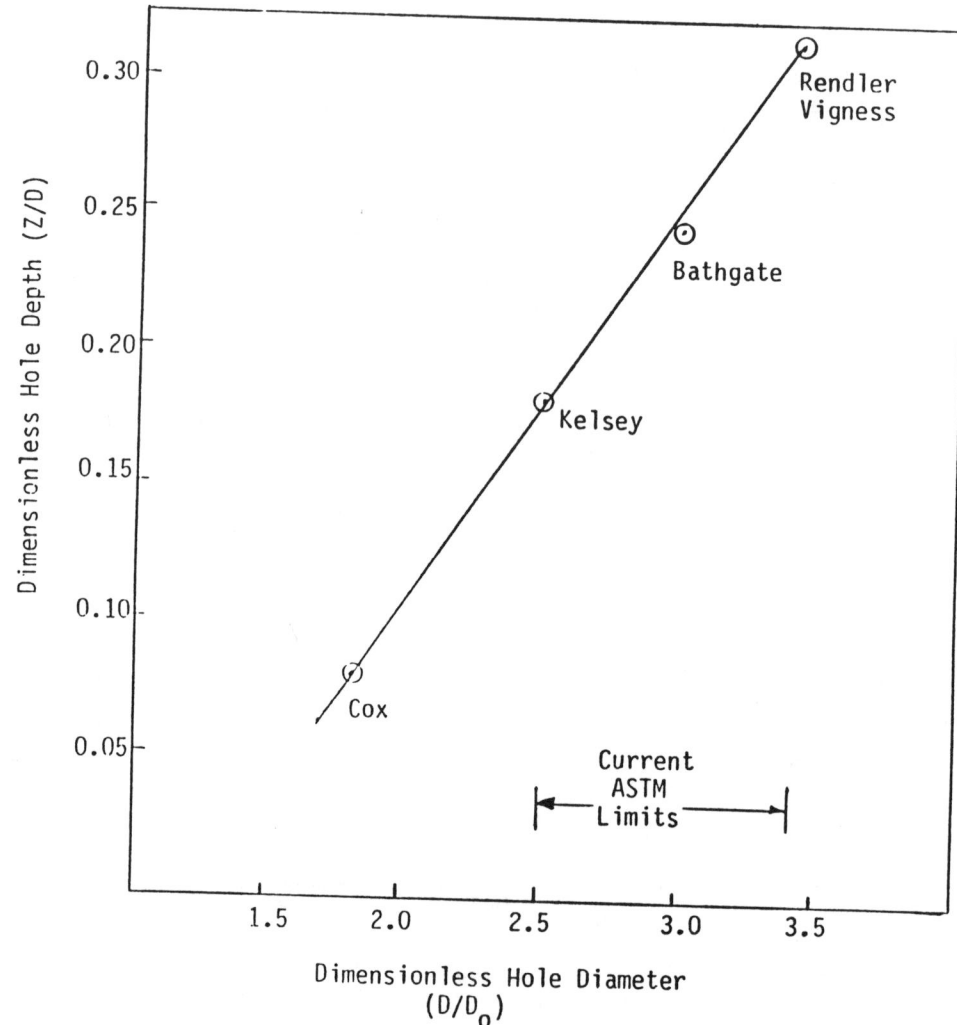

Figure 6. The hole depths required to achieve 40% of the strain relief were taken from Figure 5, and plotted as a function of hole diameter. All the points fall on a straight line. Adjusting the hole diameter to give maximum information at a particular depth range does not change the mechanics of stress relief around the hole.

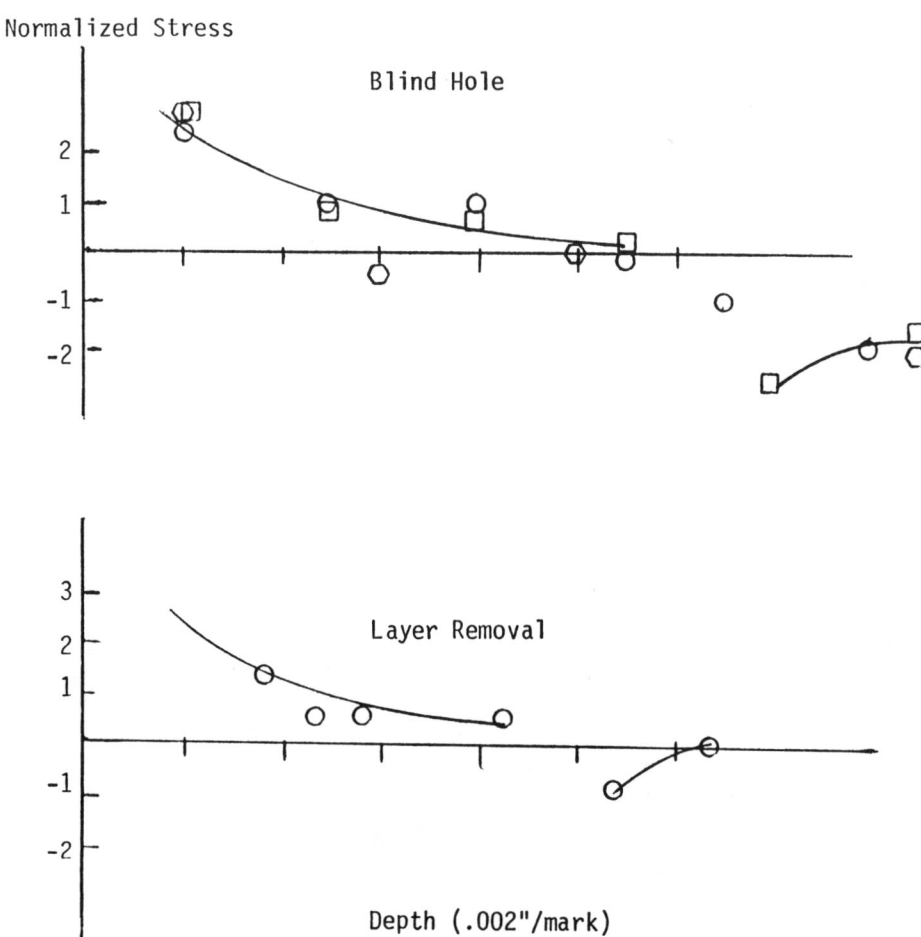

Figure 7. Comparison of data from blind hole and layer removal tests. Agreement between the two tests is very good.

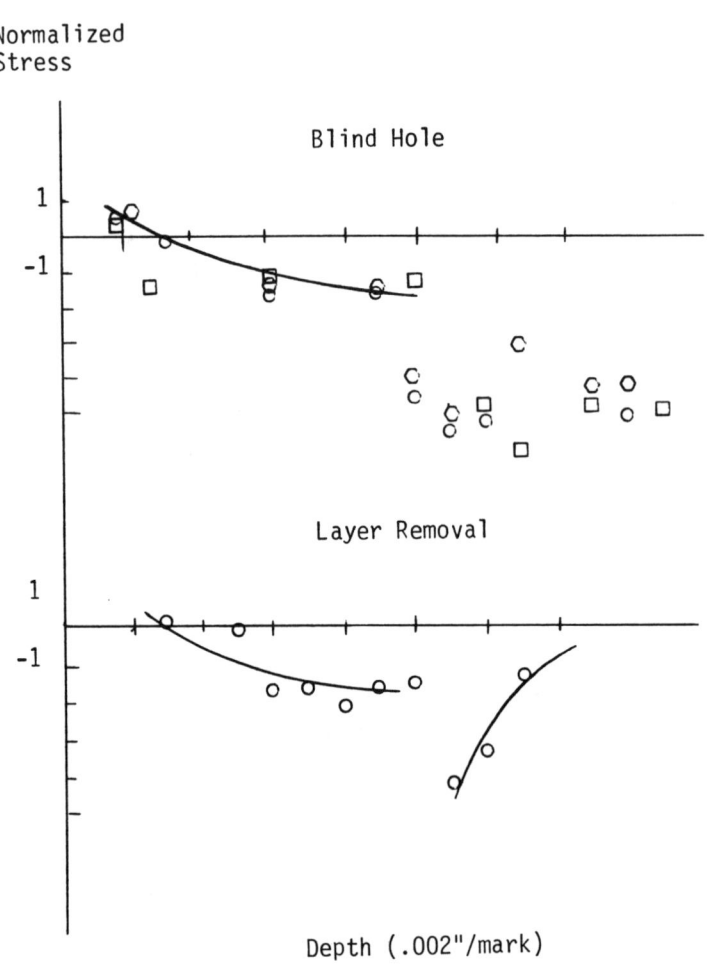

Figure 8. Comparison of the blind hole test and the layer removal test.

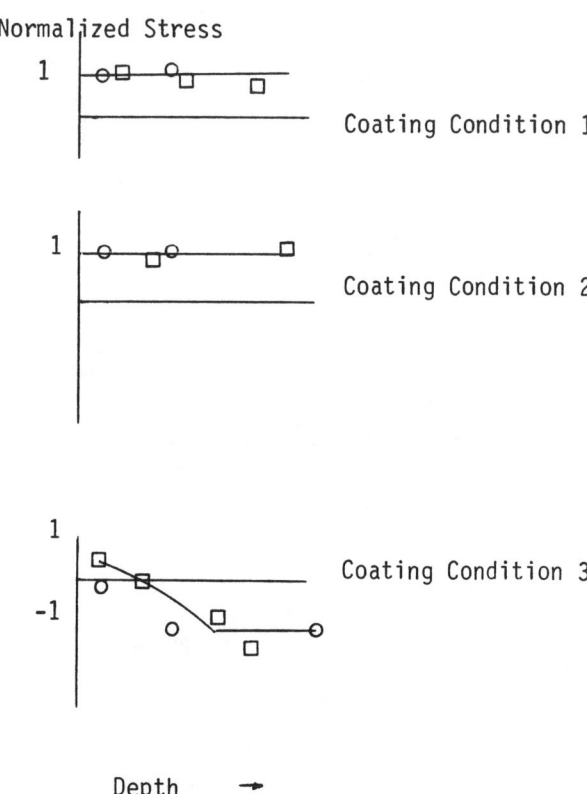

Figure 9. Comparison of data for blind hole and layer removal tests. In all cases, agreement between the two tests was excellent. Squares represent data obtained from the layer removal method.

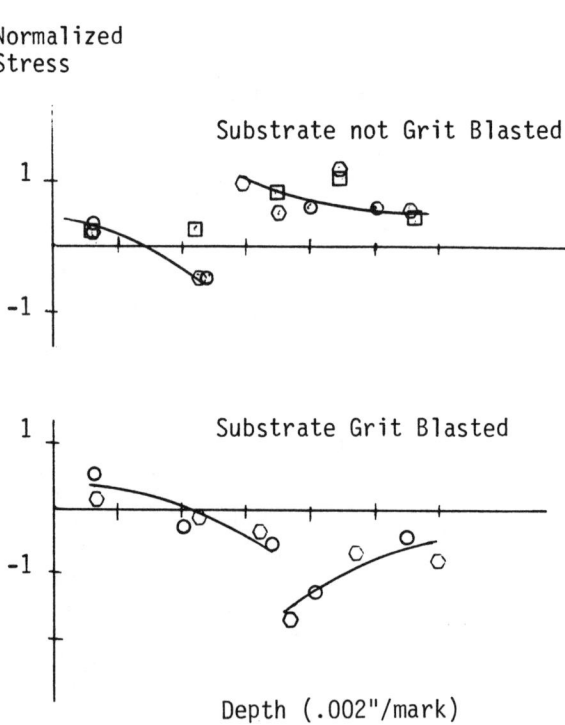

Figure 10. Comparison of residual stress profiles for a coating applied to substrates with different preparations. Note that the blind hole test gave the same result for the coating, and distinguished the difference in substrate stress due to the different surface preparation.

RESIDUAL STRESS ANALYSIS AND LCF TEST RESULTS FOR PEENED BOLT HOLE AND DOVETAIL CONFIGURATIONS

Marvin B. Happ, David P. Mourer, Richard L. Schmidt
General Electric Company
Lynn, Massachusetts, USA

ABSTRACT

Gas turbine superalloy compressor and turbine disks are shot peened to improve fatigue performance. The usual practice of peening with external nozzles must rely on uncontrolled ricocheted shot to cover areas such as bolt holes and dovetail pressure faces which are not directly in the shot stream. This practice is sensitive to peening conditions, especially when the depth of the bolt hole is considerably greater than the hole diameter or the length of the dovetail slot is considerably greater than its width.

Controlled deflector peening has been used effectively for holes in engine components where there is a large thickness-to-diameter ratio.[1] This paper reviews the peening of bolt holes and describes the adaptation of that process to single and double tang dovetails in compressor and turbine disks respectively. The intensity calibration procedure is discussed. Residual stress depth profiles obtained using X-ray diffraction are presented for various locations within the bolt hole and dovetail, and compared with similar measurements taken on a flat surface peened in the conventional manner. Differences in the residual stress profiles are discussed. Low cycle fatigue data illustrating the improvements achieved by deflector peening are presented for bolt hole specimens.

INTRODUCTION

Controlled deflector peening has been used effectively for holes in engine components where there is a large thickness-to-diameter ratio.[1] More recently this technique has been applied to single and double tang dovetails in compressor and turbine disks.

The usual practice of peening with external nozzles must rely on uncontrolled ricocheted shot to cover several key areas of high service stress, such as bolt hole surfaces and dovetail pressure faces, which are not directly in the shot stream. This practice is extremely sensitive to peening technique, especially when the length of the dovetail slot or a bolt hole is considerably greater than its width or diameter. Deflector peening represents a reproducible, controlled method of peening such features.

The results of four separate deflector peening programs are presented herein. It will be shown that these results verify the effectiveness of deflector peening. Residual stress as a function of depth is presented for bolt holes in Rene 95™ and IN718 material and for single tang and double tang dovetail slots in IN718. Percent cold work as a function of depth is also presented for the double tang dovetail configuration. Low cycle fatigue test results are reviewed for the Rene 95™ bolt hole specimens. Both Rene 95™, a nickel base superalloy, and IN718, an iron-nickel base superalloy, are extensively used by GE for fatigue sensitive aircraft gas turbine engine disks.

PROCEDURE AND DISCUSSION OF RESULTS

The procedure for deflector peening a bolt hole has been previously described.[1] To establish criteria for peening the dovetail configurations, shown in Figures 1 and 2, the first step was to locate a position in the dovetail tang equidistant between the design location of maximum loading on each pressure face (Plane Z) and the dovetail root. This was accomplished by scribing a circle on a layout of the dovetail cross-section, as shown in Figure 3. A similar procedure was followed for the outer tang dovetail in the turbine disk, except that the circle scribed there nested on the inner tang protrusions. The center of each circle defined the location of the tip of the reciprocating deflector during the peening cycle. Next, a calibration fixture was constructed having a groove in the shape of a keyhole with a diameter (width) equal to that of the scribed circles, as shown in Figure 4. An Almen strip to be peened in a band across its width equal to the keyhole width was located in the calibration fixture at the base of the keyhole. Intensity calibration was then performed in the manner previously described for holes.[1] Calibration was based on a peening intensity of 6A using S110 cast steel shot. Peening time was 125% of the time required to achieve saturation intensity. Shot hardness conformed to MIL-S-851 (Rockwell C 42-52).

Once peening parameters were established for a particular fixture keyhole width, the number of required deflector oscillations was determined proportionally via the relative circle diameter for a particular dovetail configuration. For example, the number of deflector oscillations used for a dovetail with a 6.10 mm (.240 inch) diameter scribed circle would be .240/.254 or 94% of that used to achieve peening intensity for a fixture with a keyhole 6.45 mm (.254 inch) wide. Shot deflector pins were machined from hardened steel to a diameter approximately half that of each circle.

Disks were positioned in a holding fixture so that the deflector pin would oscillate along the axis of the dovetail as determined by the center of each scribed circle. Shot entered through a nozzle positioned against the dovetail face. Using the parameters established during calibration, each dovetail slot was peened via shot ricocheted off the deflector pin and directed 90° from its incident direction against exposed dovetail surfaces. Double tang dovetails in the turbine disks were peened with two deflectors, one for each tang, to assure development of a uniform compression zone. Peening was done at the Metal Improvement Company, Windsor, CT.

RESIDUAL STRESS GRADIENTS - X-ray diffraction residual stress and percent cold work measurements were made by Lambda Research, Inc., Cincinnati, OH, with the two-inclined angle technique[2] utilizing the diffraction of Cu K-α radiation from the (420) planes of the FCC Rene 95 and the IN718 material. Details of specimen geometry and the X-ray diffraction parameters are presented in Table I. The elastic constant ($E/1 + \nu$) used in calculation of the residual stress was determined in a direction normal to the (420) planes by measuring the change in spacing of the (420) planes with stress on a beam loaded in four point bending on the diffractometer.[3] Irradiated areas were rectangular in shape with the long axis normal to the direction of stress measurement for the bolt holes (circ. dir.) and free surfaces, and, because of geometric restrictions, parallel to the direction of stress measurement for the dovetail slots (axial direction). Residual stress was determined as a function of depth starting at the surface. Initial readings were obtained at .0127 mm (.0005") to .025 mm (.001") increments. A total of approximately ten data points for each specimen location were obtained. Sectioning of the samples was necessary prior to X-ray diffraction in order to provide access for the incident and diffracted X-ray beams. Measured residual stresses were corrected for any stress relaxation occurring on sectioning. The relaxation stresses were obtained by placement of a strain gauge at the measurement location prior to sectioning. Material was removed electrolytically for subsurface measurements and further corrections applied as appropriate. Also corrections were applied for the effects of the penetration of radiation into the subsurface stress gradient. All residual stress values reported herein are fully corrected values. Experimental random error related to diffraction peak position determination, sample positioning and instrument alignment are very small being up to only several ksi for the larger magnitude compressive stresses.

In the case of the double tang dovetail program the percent cold work was also determined as a function of depth. This was determined by using the full-width-at-half-maximum height of the (420) diffraction peak obtained in the 0° rotation. Percent cold work was derived from an empirical relationship established by preparing tension and compression specimens to produce known true plastic strains up to a magnitude of approximately 30%.

Residual stress profile data is presented graphically in Figure 5 for the Rene 95 disk bolt hole program. Data is presented in Figure 6 for the IN718 disk bolt hole program, Figure 7 for the IN718 disk single tang dovetail program, Figures 8 and 9 for the IN718 double tang dovetail program. A summary of data for all programs is presented in Table II.

All residual stress profiles show a compressive zone at least .114 mm (.0045 inch) deep and a maximum compressive stress in the neighborhood of 760 to 1100 MPa (100-160 ksi) for IN718 and 1550 to 1725 MPa (225 to 250 ksi) for Rene 95 at .025-.050 mm (.001-.002 inch) below the surface. Compressive stress values at the surface vary from 1000 to 1200 MPa (146 to 176 ksi) for the Rene 95 and from 220 to 850 MPa (32 to 123 ksi) for the IN718. Uniform compression is exhibited along the entire length of the peened surface, as shown for the single tang dovetail (Locations A and B). Locations A and B represent stress profiles at 6.35 mm (.25") from each end of the dovetail slot. Lower residual surface stresses generally correlate with a greater depth for the depth of maximum compressive stress, e.g. the double tang dovetail slot pressure face locations (Locations 2 and 3) which show higher compressive stress at surface than root and free surface locations (Locations 1 and 4) show a shallower depth for achievement of maximum compressive stress. The pressure face residual stress gradient curves exhibit less of an inverted bell shape than that characteristically exhibited by the other data. The absence of a bell shape is believed to be associated with an overall balance of stresses that must be maintained in the peened part. This balance is effected by percent cold work, which in the case of the double tang dovetail is shown to correlate inversely with the magnitude of surface residual compressive stress. Surface residual stress is expected to increase with increasing cold work, but beyond some threshold will decrease, as was apparently the case here. This threshold is likely related to the onset of peening induced surface "overwork" resulting in a reduction in surface compressive stress.

LOW CYCLE FATIGUE TESTING - Experiments were performed to verify the effect of peening and the resultant residual stress state on fatigue behavior. Bolt hole test specimens, as shown in Figure 10, were manufactured from PM Rene 95 alloy intentionally contaminated with nominal .51 mm (.020") dia. alumina to represent naturally occurring flaws and to provide a range of defect locations along the bolt hole and radius.

For peening, the bolt hole specimens manufactured from this material were sandwiched between two rubber plugs with hole diameters equal to the specimen gauge hole to simulate the thickness of the actual disk bolt hole (Fig. 11). Peening and calibration was performed in the manner described previously. The exterior surface of the specimen was also peened to 6A intensity to preclude spurious fatigue initiations at defects removed from the bolt hole region.

After peening the specimens were fatigue tested at 750°F in a load controlled mode at twenty cycles per minute with a triangular waveform at an A ratio (ratio of alternating stress/mean stress) of 0.95. Testing was performed both on peened and unpeened specimens as no baseline was available for material produced in this fashion. The data obtained are depicted in Figure 12. As can be readily seen from the figure the peening improved fatigue life at all but the highest stress level with an improvement factor greater than an order of magnitude at the lower stress levels. All specimens were observed to fail from the bolt hole initiating at the seeded inclusions making a direct comparison totally valid. In addition the high density of the seeds (20K particles/lb of Rene 95) provides for a very severe test of peening uniformity as each specimen likely had numerous defect sites in the bolt hole high stress region.

COMMENTS - The residual stress profiles and LCF data obtained show that deflector peening is an effective technique for imparting favorable residual compressive stresses on both bolt hole and dovetail surfaces. The difference in the shape of the residual stress profiles for the double tang dovetail may be attributed to excessive peening of the root radius resulting from the second deflector. Other factors may also account for differences. For example, the dovetail, being non-circular, had its surface exposed to ricocheted shot impinging from different incident angles, which may result in less than ideal surface compression at the root. Note that the flat surface which also exhibited an inverted bell shape stress gradient profile, was peened with shot directed not at 90° to the surface, but at a 45° angle. This condition could also influence the stress gradient. The preceding explanation is speculative and does not explain why in comparison to the bolt hole and single tang stress profiles the double tang pressure face profiles are unique. However, it does point to the need for further work to gain an understanding of the peening process variables and their relationship to the shape of the residual stress profile. It is believed that the ability to control profile shape at the near surface is certainly key to achieving optimum fatigue performance, thereby maximizing the benefits obtainable from shot peening.

REFERENCES

1. M.B. Happ, "Shot Peening Bolt Holes in Aircraft Engine Hardware," <u>Proceedings Second International Conference on Shot Peening</u> (1984), pp 43-49.

2. Prevey, P.S., "X-Ray Diffraction Procedure for Residual Stress Measurement", Lambda Research Inc. publication, Jan., 1980.

3. Prevey, P.S., "A Method of Determining the Elastic Properties of Alloys in Selected Crystallographic Directions of X-Ray Diffraction Residual Stress Measurements," <u>Advances in X-Ray Analysis</u>, Vol. 20, Plenum Press, 1977, pp 345-354.

Table I - X-Ray Diffractometer Fixturing Parameters

	R95 Bolt Hole Program	IN718 Bolt Hole Program	IN718 Single Tang Dovetail Program	IN718 Double Tang Dovetail Program
Geometry				
Diameter	7.62mm(.300"), 6.45mm(.254")	6.35mm(.250")	See Figure 3	See Figure 3
Axial Length	20.83mm(.820")	19.05mm(.750")	6.35mm(2.5")	14.0mm(.55")
Measurement Location	Mid length	Mid length	See Figure 3; 6.35mm (.25") from each end	See Figure 3 (Mid length)
Direction of Measurement	Circ.	Circ.	Axial	Axial
Irradiated area	8.9mm x 2.0mm (.35 x .08"), Long axis $\perp$ dir. measurement	5.1mm x 2.0mm (.2 x .08"), Long axis $\perp$ dir. measurement	7.6mm x 1.8mm (.3 x .07"), Long axis // measurement dir.	6.4mm x 1.3mm (.25 x .05"), Long axis // measurement dir.
Incident beam divergence	1°	1°	3°	3°
Detector	Si (Li) set for 90° acceptance of copper K-alpha energy			
Counts per point	10,000	10,000	50,000	65,000
Rotation	0-35°	0-45°	0-45°	0-45°
$E/(1+\nu)$, MPa (ksi)	$16.9 \pm .07 \times 10^4$ ($2.45 \pm .01 \times 10^4$)	$14.0 \pm .21 \times 10^4$ ($2.03 \pm .03 \times 10^4$)	$13.3 \pm .14 \times 10^4$ ($1.93 \pm .02 \times 10^4$)	$13.3 \pm .14 \times 10^4$ ($1.93 \pm .02 \times 10^4$)

Table II - Summary of Residual Stress Profile Data

Program	Description	Surface Stress MPa	(ksi)	Surface % CW	Maximum Compressive Stress & Depth MPa	(ksi)	mm	(in)
Bolt Hole (R95)	Web (Flat Surface)	-1009	(-146)	-	-1549	(-224)	.048	(.0019)
	Small Bolt Hole	-1069	(-155)	-	-1696	(-246)	.013	(.0005)
	Large Bolt Hole	-1217	(-176)	-	-1553	(-225)	.038	(.0015)
Bolt Hole (IN718)	Flat Surface	-440	(-64)	-	-1091	(-158)	.046	(.0018)
	Bolt Hole	-663	(-96)	-	-1044	(-151)	.015	(.0006)
Single Tang Dovetail (IN718)	Dovetail Tang-Loc A	-312	(-45)	-	-875	(-127)	.025	(.001)
	Dovetail Tang-Loc B	-218	(-32)	-	-780	(-113)	.048	(.0019)
Double Tang Dovetail (IN718)	Flat Surface-Loc 4	-523	(-76)	17.3	-954	(-138)	.038	(.0015)
	Root - Loc 1	-303	(-44)	19.1	-853	(-124)	.051	(.002)
	Lower Tang-Loc 3	-848	(-123)	13.1	-902	(-131)	.015	(.0006)
	Upper Tang-Loc 2	-814	(-118)	13.4	-893	(-129)	.028	(.0011)

Figure 1 - Compressor Disk Single Tang Dovetails

Figure 2 - Turbine Disk Double Tang Dovetails

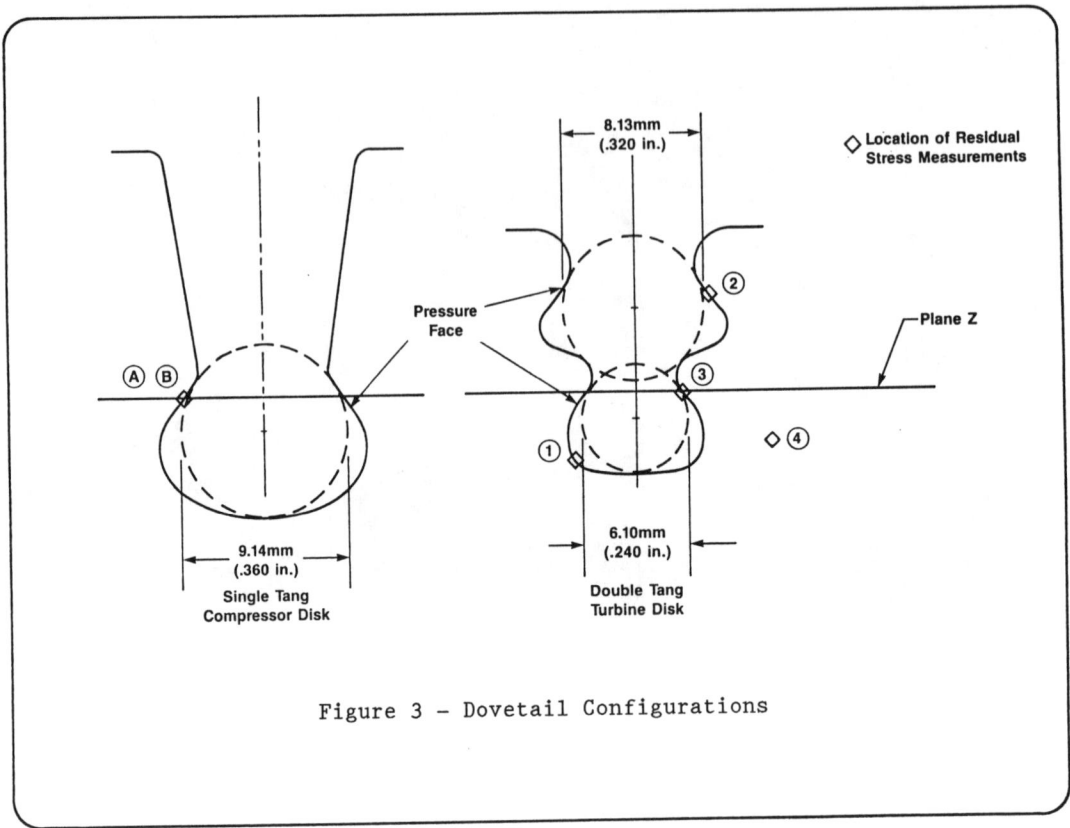

Figure 3 – Dovetail Configurations

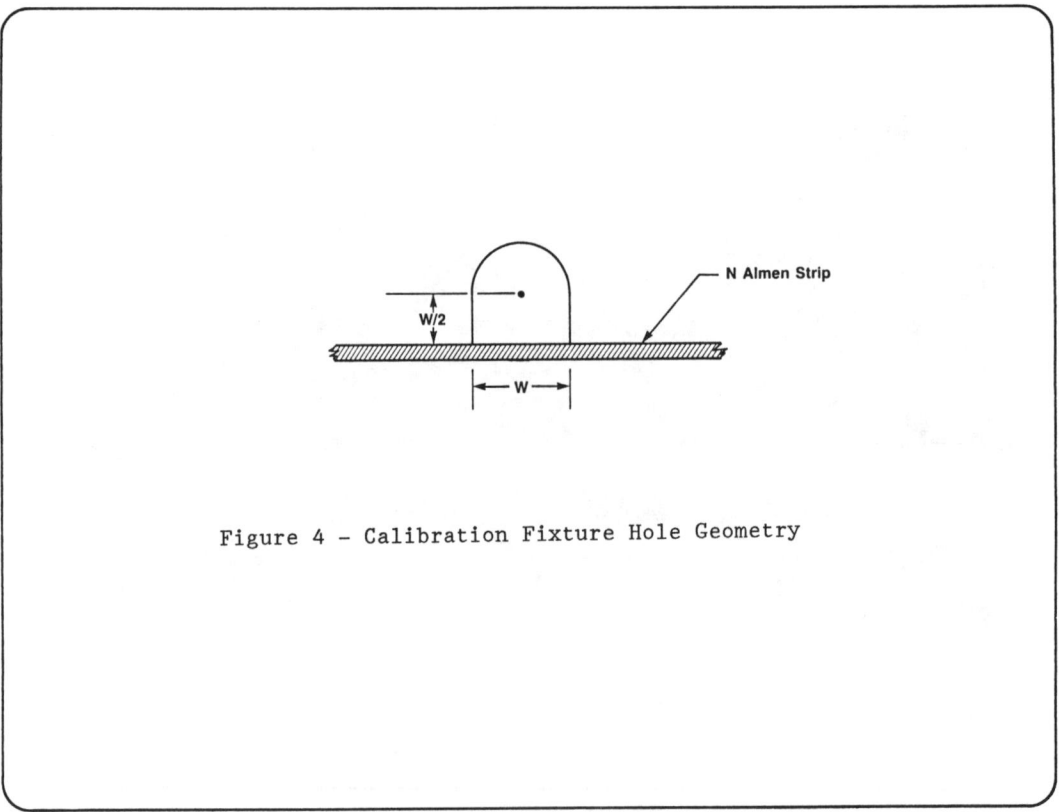

Figure 4 – Calibration Fixture Hole Geometry

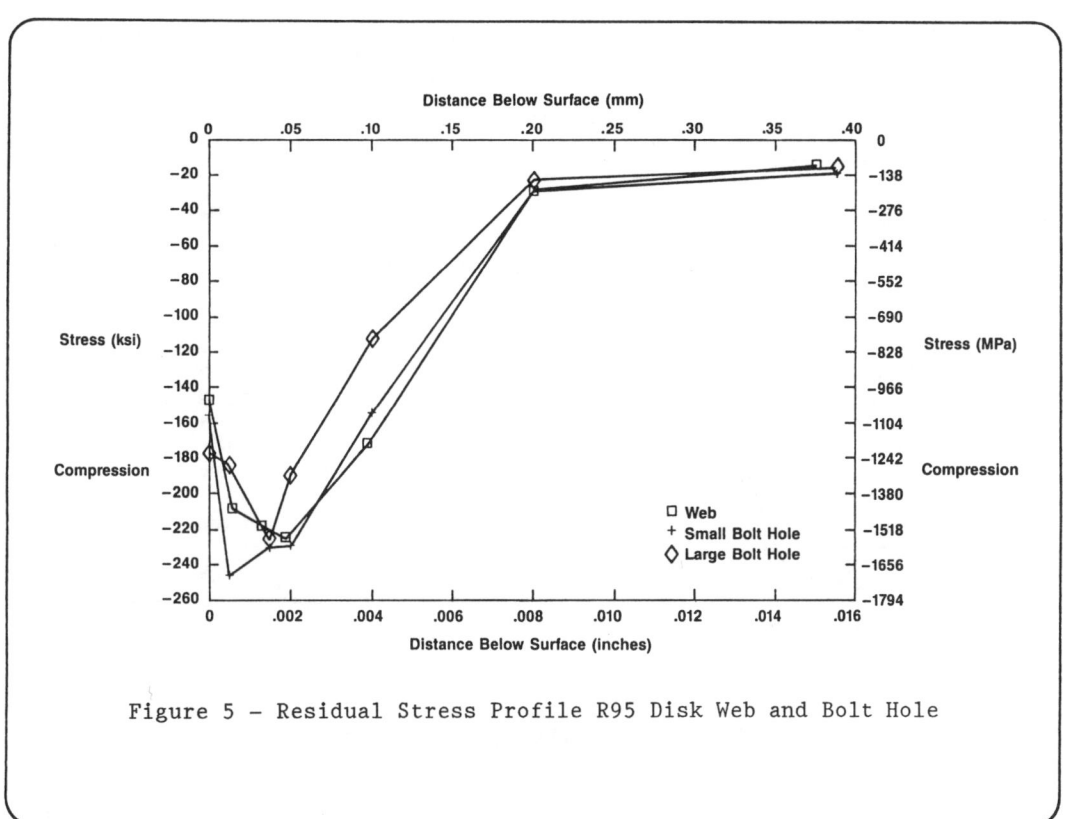

Figure 5 - Residual Stress Profile R95 Disk Web and Bolt Hole

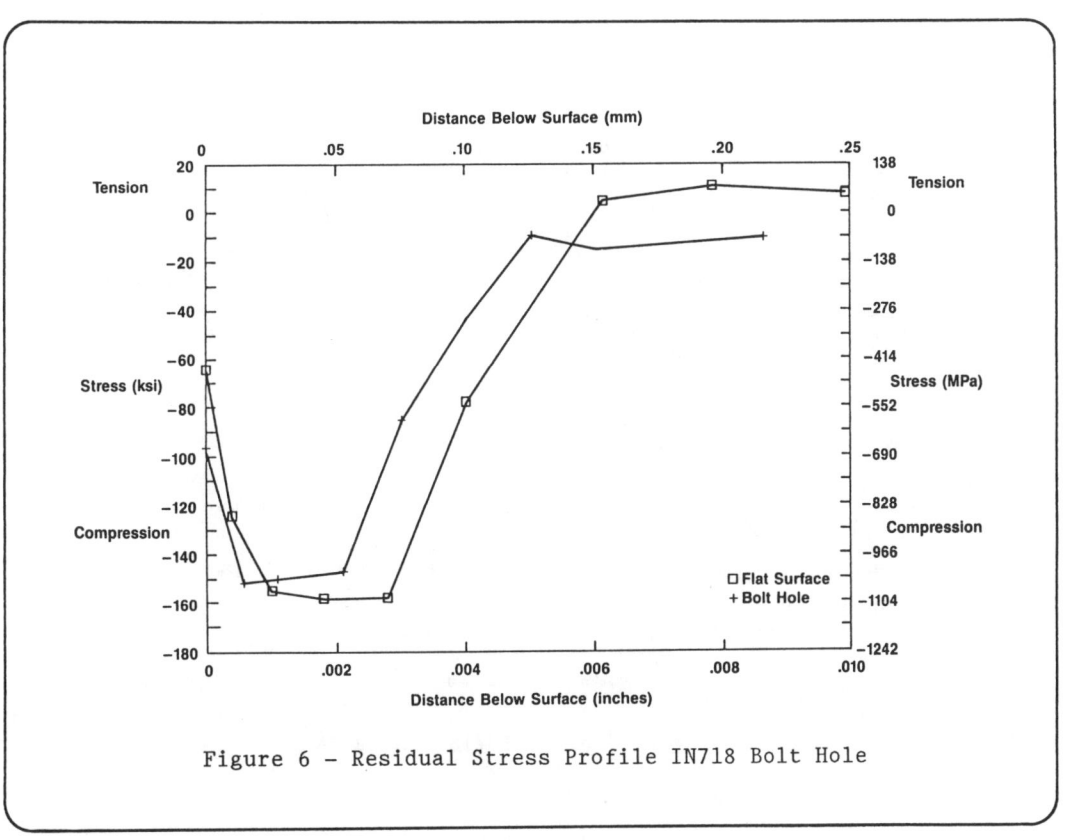

Figure 6 - Residual Stress Profile IN718 Bolt Hole

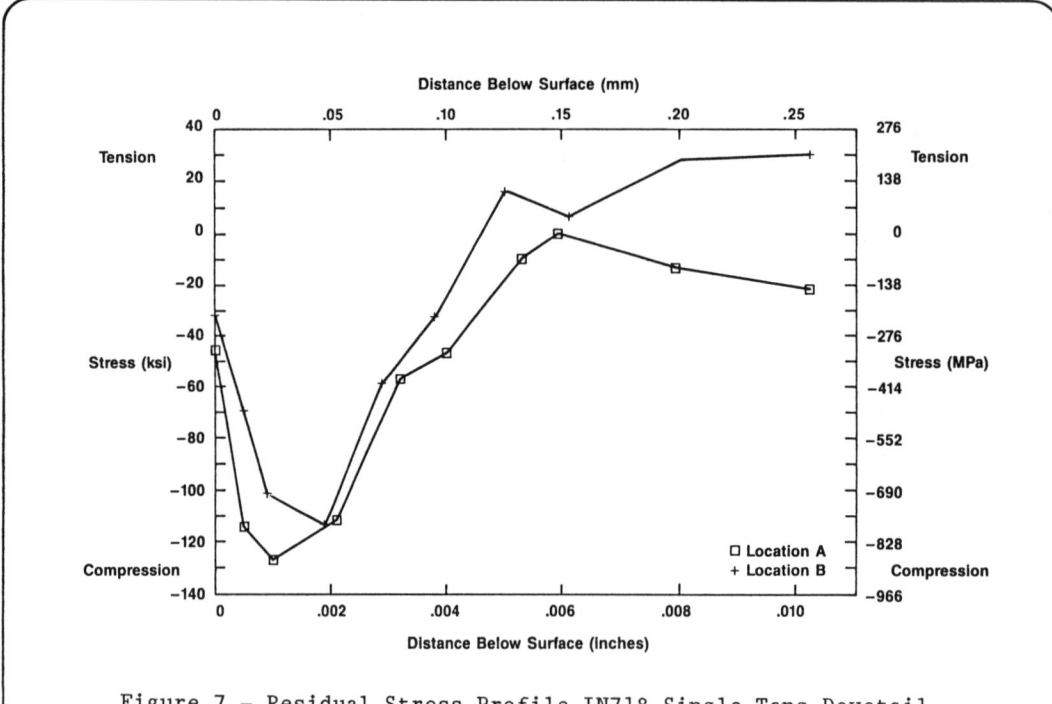

Figure 7 - Residual Stress Profile IN718 Single Tang Dovetail

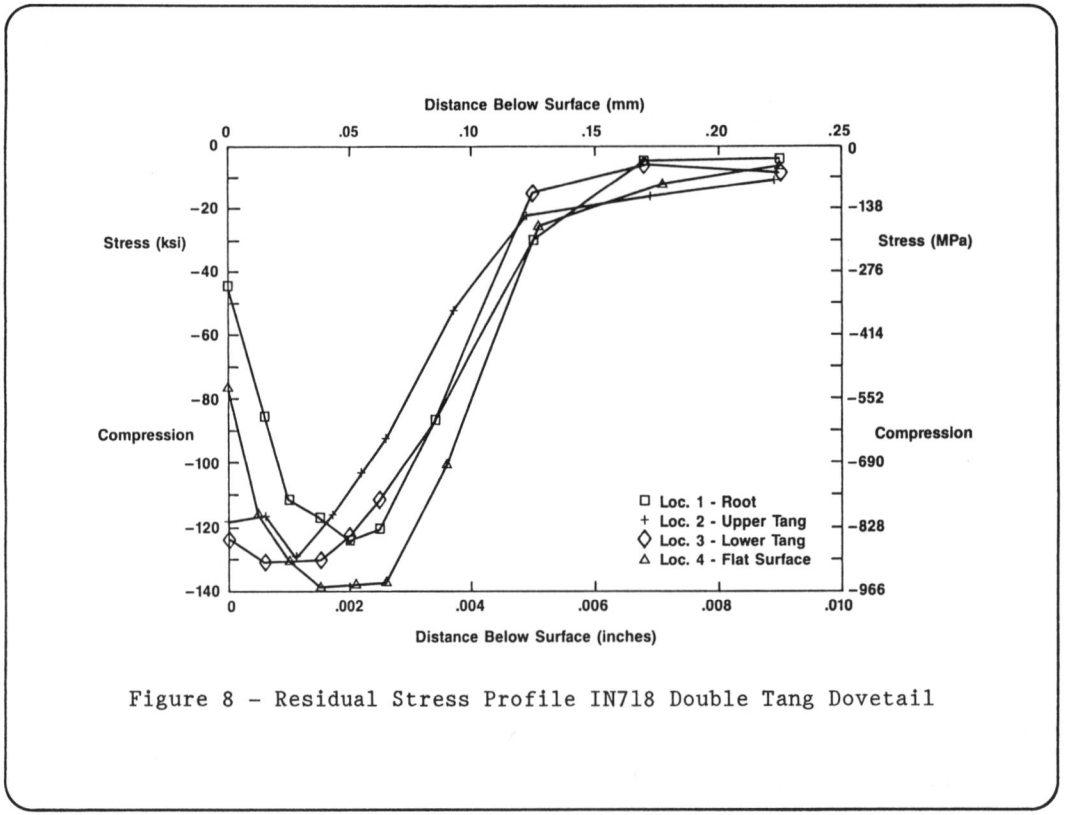

Figure 8 - Residual Stress Profile IN718 Double Tang Dovetail

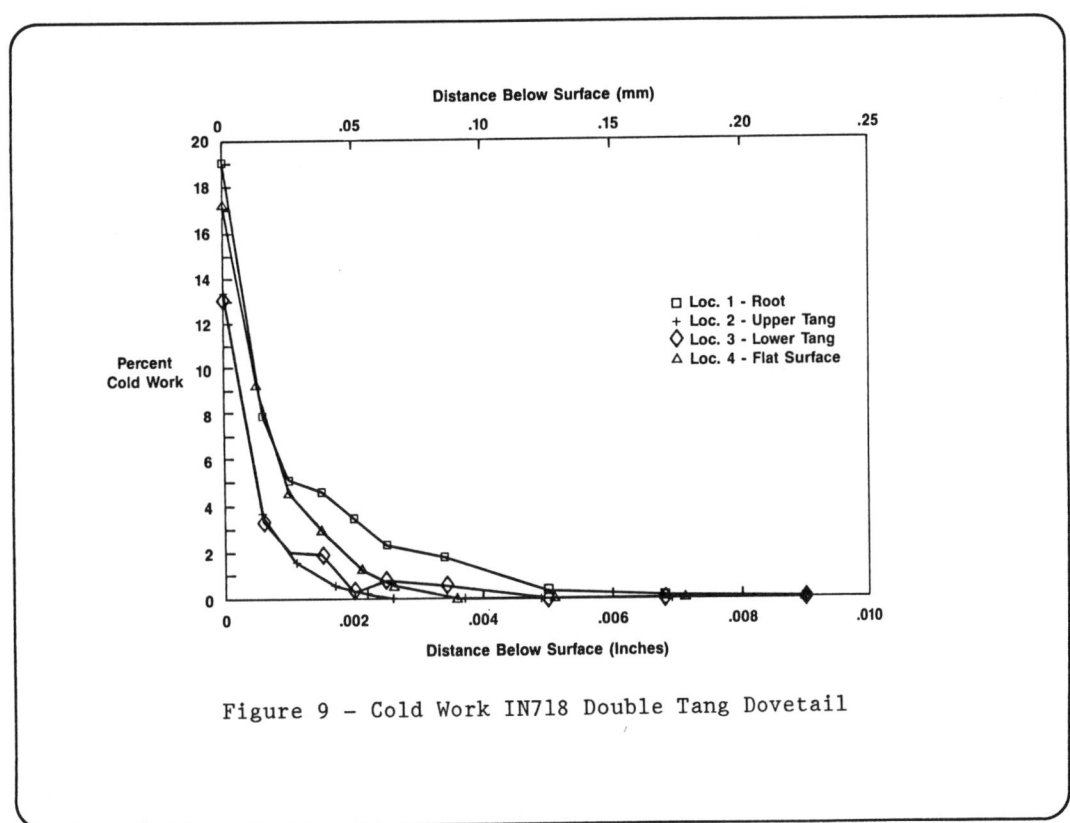

Figure 9 - Cold Work IN718 Double Tang Dovetail

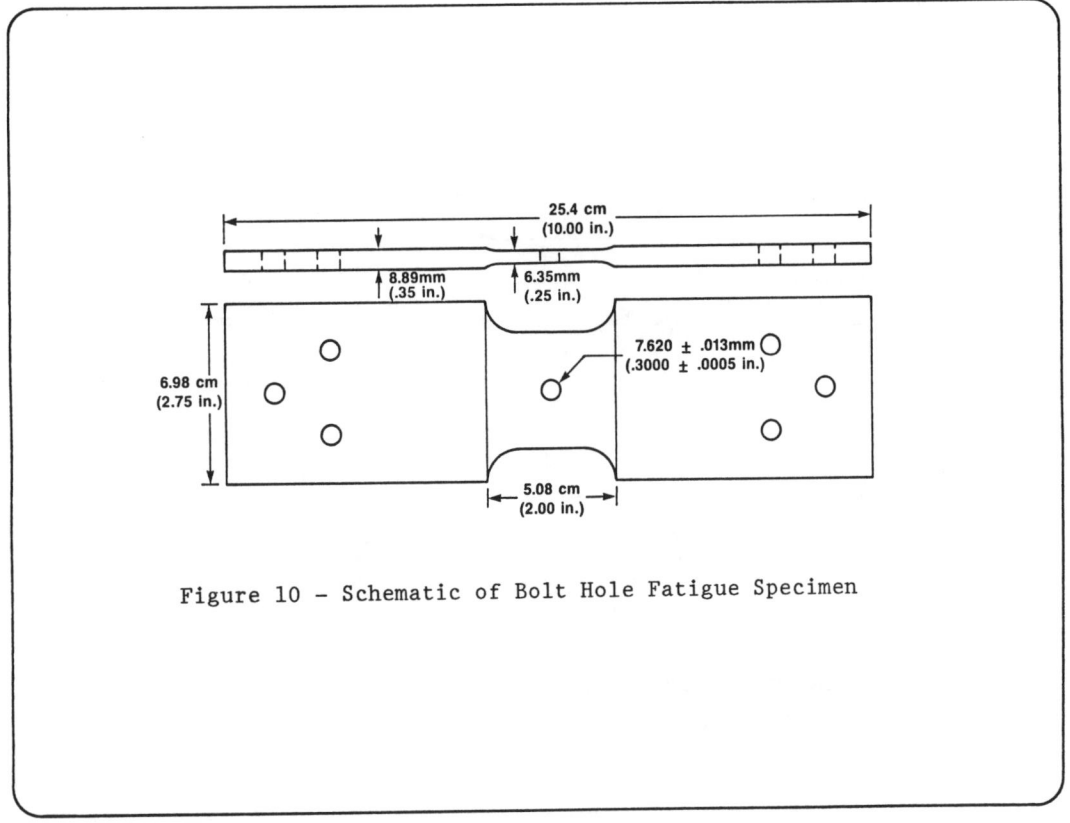

Figure 10 - Schematic of Bolt Hole Fatigue Specimen

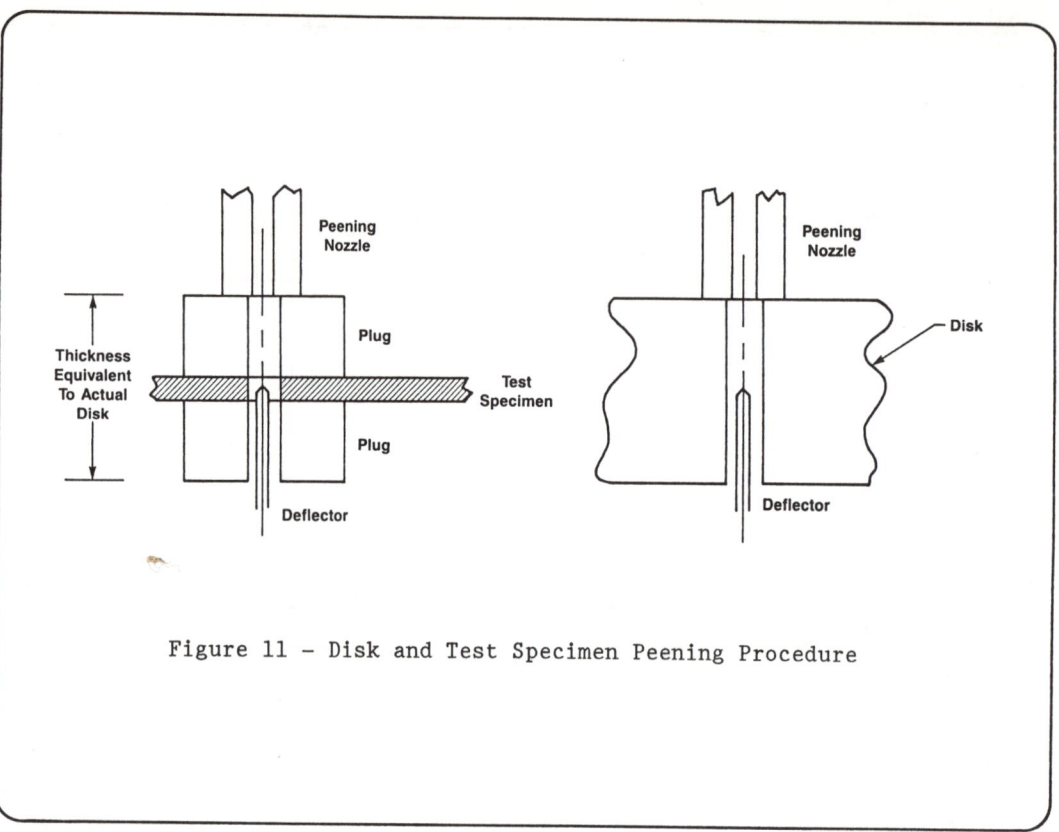

Figure 11 - Disk and Test Specimen Peening Procedure

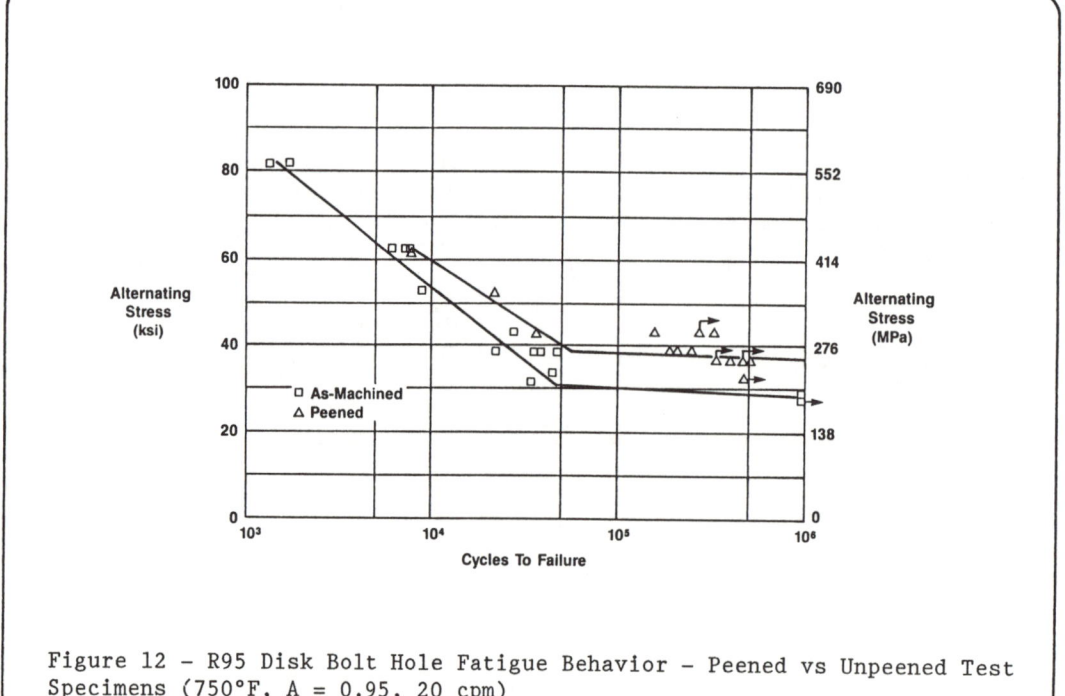

Figure 12 - R95 Disk Bolt Hole Fatigue Behavior - Peened vs Unpeened Test Specimens (750°F, A = 0.95, 20 cpm)

FATIGUE TESTING AND RESIDUAL STRESS MEASUREMENTS OF GRIT-BLASTED ALUMINUM ALLOYS

R. Myllymaki
Defence Research Establishment (Pacific)
Victoria, B.C. Canada

ABSTRACT

Concern that the fatigue strength of soft aluminum alloys might be impaired by grit blasting of the metal surfaces led to the fatigue testing of two aluminum alloys. The test results showed no detrimental effects on the fatigue life from the grit blasting studied in this report. Indeed, in one case, the fatigue life was significantly improved.

X-ray diffraction stress measurements confirmed that the grit blasting caused the formation of compressive residual stresses on the fatigue specimens' surfaces. Fine notches produced by the grit blasting may under some conditions offset the beneficial effects of the compressive residual stresses, e.g. when corrosion, wear, or other damage removes the compressive stress on the surface. However, no detrimental effect is envisaged for normal applications, especially when controlled procedures are used for grit blasting.

GRIT BLASTING OF ALUMINUM ALLOYS as a preparation for painting has been proposed as an alternative to the slower and costlier methods now in use, e.g. paint removers, scraping, heat, power sanders. Some concern was expressed, however, that grit blasting of soft aluminum alloys might impair the mechanical properties, particularly fatigue strength. Accordingly, a series of tests was carried out to determine the effects of grit blasting on three common structural aluminum alloys, namely 5083, 6061, and 7075.

The first part of the program examined the weight loss, thickness loss, surface roughness, and amount of warping caused by grit blasting. This work concluded that structural aluminum alloys can be grit blasted safely with standard dockyard type equipment and procedures, provided that the grit used is fine enough.

The second part of the program, the fatigue testing of grit blasted panels, is reported here. One concern was that the notches in the surface caused by the sharp grit might reduce the high cycle fatigue strength of the aluminum alloys. On the other hand, shot peening is commonly used to induce compressive surface stresses which improve the fatigue strength. It was suggested that grit blasting may have a similar effect. A non-heat-treatable alloy, 5083, and a heat-treatable alloy, 7075, were chosen for fatigue testing.

X-ray diffraction equipment designed for measuring residual stress was used to measure the residual surface stress on a number of fatigue specimens, both with and without grit blasting. The effects that grit size and blasting time had on residual stress were assessed on alloy 5083 to determine the most favourable grit size and blasting times relative to the surface residual stresses and fatigue strength.

MATERIALS EXAMINED

Two common structural aluminum alloys (7075-T6 and 5083-H32) in the form of 3 mm sheet were tested. Two different lots of alloy 7075 were used. Typical mechanical properties for these alloys were:

7075-T6 572 MPa UTS; 503 MPa YS; 11% elongation

5083-H32 290 MPa UTS; 207 MPa YS; 16% elongation

Table I gives the chemical composition of the alloys examined.

EXPERIMENTAL DETAILS

Figure 1 shows the type of fatigue test specimen that was tested on a Tatnall-Krause type fatigue test machine operating at 1200 cycles per second. An attempt was made to remove any small stress raisers by hand polishing the edges of the tapered section. The cantilever arrangement applies a uniform stress to the tapered section of the specimen. Thus, fracture anywhere in the tapered section gives a valid result. The specimens were tested to complete failure or to greater than 10^7 cycles. It is known that overstressing may improve the fatigue propagation properties of materials,[1,2], thus one set of tests was carried out with 2 cycles of 275 MPa stress, every 10^5 cycles to evaluate the effect of overstressing on fatigue crack initiation.

Two types of grit used for grit blasting were
(a) a medium industrial crushed slag grit (Black Beauty) used in the Dockyard, and
(b) a very fine #80 Alundum used in a laboratory grit blasting cabinet.

A jig was used to hold the test pieces and both sides were blasted in order to minimize the warpage of the specimens.

The grit blasted surfaces were examined with the binocular, metallurgical, and scanning electron microscopes. As a measure of surface profile, the fine focus on a metallurgical microscope was used to measure the distance between low and high points on the grit blasted surface. The transmission electron microscope was used to examine replicas from the fracture surface of one of the fatigued specimens.

One set of as-received specimens and one set of medium grit blasted specimens were heat treated. The specimens were solution treated at 480°C for 1-1/2 hours, water quenched, and aged at 120°C for 24 hours.

X-ray diffraction equipment specially designed for residual stress measurements, using V-filtered CrK_α radiation, was used to measure the residual surface stress on a number of fatigue specimens. The multiple exposure or $\sin^2 \psi$ method of X-ray diffraction residual stress measurement was used.[3] The ψ angles used were 0°, 15°, 30°, and 45°. The bulk values for the elastic modulus E (6.9×10^4 MPa) and Poisson's ratio (0.34) were used to calculate the residual stress. The aluminum (222) peak locations were established using the three-point parabola method after the diffraction intensities were corrected using the appropriate Lorentz-polarization-absorption factors.

Residual stress measurements were made on fatigue tested specimens representing a number of surface conditions. There was considerable scatter in the results so the residual stress was determined at eleven points on the surface of an as-received piece of 7075-T6, lot 2.

The effect that the grit size and time of grit blasting had on the residual stress was determined using another series of specimens. Alundum, medium, and fine industrial grade grits were used to grit blast 5083-H32 material on one side only. Four grit blasting times were used, namely 1T, 2T, 4T, and 8T. Time "T" for laboratory conditions was 25 seconds and for the industrial blast, 7 seconds. Residual stress was determined on three spots at 50 mm intervals along the centre line of the 70 mm x 200 mm specimens. In some cases, the transverse residual stress was also measured.

The residual stress was also determined at a fourth spot on the reverse side of the specimen, which had not been grit blasted.

The hardness of the as-received materials was measured on a "Rockwell" hardness tester. In addition, hardness measurements were made on some specimens before and after fatigue testing.

RESULTS

HARDNESS - The hardness measurements showed that the hardness of the second lot of 7075 material is lower ($4R_B$) than the first lot but it was still within commercial tolerances for 7075 in the T6 heat treated condition. The laboratory heat treated material was slightly softer and had a larger scatter in results. There is no evidence of a change in hardness due to fatigue testing.

SURFACE ROUGHNESS - The surface roughness for the 7075 alloy grit blasted with very fine Alundum and medium grit was 30 microns and 130 respectively. Not enough measurements were made to establish a reliable estimate of the scatter in the roughness results.

Figures 2 and 3 show cross-sections through samples grit blasted with very fine and medium grit respectively, while Figures 4 and 5 are scanning electron micrographs of the surfaces of the same specimens.

It is evident from these photographs that grit blasting produces a surface with a multitude of small notches, some of which are very sharp. The random nature of the grit impingement ensures that a substantial number of these notches will be positioned transversely to the applied load and act as stress raisers.

FATIGUE TEST RESULTS - Figure 6 shows a typical transmission electron micrograph of the fatigue and ductile fracture surfaces of a fatigue specimen. The multiple crack initiation sites of the grit blasted specimens are compared to the single crack initiation site of an unblasted specimen in Figure 11.

The fatigue test results are tabulated in Table 2 and plotted as S-N curves in Figures 7-10. There was considerable scatter in results which makes interpretation somewhat difficult and speculative. The grit-blasted specimens performed at least as well as, and in some cases, better than the as-received materials. Specimens which were heat-treated after grit-blasting had fatigue lives considerably lower than the as-received materials (Figure 9). Figure 8 shows that overstressing every 10^5 cycles increased the fatigue life of the as-received materials.

RESIDUAL STRESS MEASUREMENTS - Although the precision of the X-ray diffraction equipment and technique used for these tests can be ± 15 MPa under well-controlled conditions, the accuracy

of the present residual stress results must be treated with some caution because:
a. The bulk modulus of elasticity and the bulk Poissons ratio were used to calculate the stress.
b. No corrections were made for possible stress gradients near the surface of the specimens.
c. No corrections were made for non-linearity in the lattice spacing (d) versus $\sin^2 \psi$ relationship caused by texture in the aluminum alloys.[3] Whereas some of the results from the as-received material were very poor and had to be disregarded, the results from both the grit blasted and laboratory heat treated specimens were quite linear with high correlation coefficients.

Tables 3, 4, and 5 list the results of the residual stress measurements carried out on specimens in the as-received condition as well as on specimens which had been fatigue tested. Unless noted differently, all residual stress measurements were taken from the middle of the tapered section of the fatigue specimens (see Figure 1) in the longitudinal direction (parallel to the rolling direction). The signs + and − refer to tensile and compression stresses respectively.

As is evident in Tables 3 - 5, there was a lot of scatter in the residual stress results. This is very evident in Table 5, which gives results from an 11-point grid on a 75 mm x 200 mm specimen.

In all samples tested, grit-blasting produced compressive surface stresses. Figures 12 and 13 show that the finest grit produced the highest surface residual stress and that this value peaked at about 4T (T - time for white metal finish).

DISCUSSION

The fatigue test results have been compared to other test results found in the literature [1,4,5,6]. At $N=10^7$ cycles, the fatigue strength of Lot 1 material falls into the middle of the scatter band (125 - 210MPa) for 7075-T6 material given in Reference 5 and at the bottom edge of the scatter band (150 - 180MPa) for 550 MPa UTS aluminum alloys given in Reference 1. The results for 7075-T6 Lot 2 material (120 MPa at 10^7 cycles) fall below all the relevant results given in References 1,4,5, and 6. This suggests that 7075-T6 Lot 2 may have been a rogue material.

One expects rather more scatter in fatigue strength from grit blasted surfaces than from the smoother machined surfaces of material used for the test results given in References 1, 4, 5, and 6. The literature results refer to rolled plate, extrusions, and forgings tested for the most part with rotating beam specimens. Thus, the present results from smooth and grit blasted surfaces are not directly comparable to those presented in the literature for different forms and testing methods.

The chemical analysis and hardness of the two lots of 7075-T6 material differ somewhat, but not enough to explain the difference in fatigue strength. All of the Lot 1 material was used before the tensile properties of the two lots could be compared. However, the hardness of the two lots did not differ greatly and it follows that the tensile strength would also not differ greatly.

The residual stress measurements (Table 2) show that the stress on as-received Lot 1 material was compressive whereas the stress on the Lot 2 material ranged from slightly compressive to tensile.

Results given in Table 3 for Lot 2 material, show that the residual stresses in the transverse direction were tensile whereas in the longitudinal direction the stresses were both compressive and tensile at the various points tested. The difference in fatigue strength between the two material lots may be related to the difference in residual stress. However, the scatter in results is significant (see Table 3), not many samples were tested, and therefore, interpretation of the results must be treated with some care.

The fatigue strength for 5083-H32 material corresponds to results given in Reference 7, about 110 - 125MPa at $N = 10^7$ cycles. The present results also fall into the scatter band of results (90 - 170MPa) for 275MPa UTS aluminum alloy given in Reference 1. Table 2 shows that the longitudinal residual stress on the surface of 5083-H32 may be compressive or tensile.

The residual stress in the as-received materials is not high relative to the yield strength but is significant relative to the fatigue strength of these aluminum alloys. These results must be treated with some reservation however, because for aluminum irradiated with Cr-K_α X-rays, 95% of the diffraction comes from less than 40 microns below the surface. For the 7075-T6 alloy shown in Figure 3, which corresponds to only 3 - 4 grains from the surface. This very shallow surface layer may be affected by heat treatment (temperature and cooling rate), rolling, handling, straightening, and other effects. The scatter in the residual stress results for even these small samples was considerable and may be even greater if a large plate, e.g. 3 x 4 meters were systematically sampled.

The results given in Figure 7 suggest that the grit blasting had a somewhat beneficial effect on the fatigue properties of the 7075-T6 materials. Figure 10 shows that the very fine Alundum grit had a beneficial effect on the fatigue strength of the 5083-H32 material whereas the medium grit appeared to have little effect. Because all the results fall into the scatter band (90 - 170MPa) for fatigue results given in Reference 1, the apparent beneficial effect of the very fine grit may not be as significant if more tests were carried out.

The compressive stress in the specimen surface should act to improve fatigue strength and, conversely, the sharp notches left by the grit should act to reduce the fatigue strength. To clarify this latter effect, several specimens were heat treated after grit blasting to relieve the residual stresses. As seen in Figure 9, the fatigue strength after heat treatment was significantly below all the other test results for 7075-T6 material. The fatigue strength of 80MPa at $N = 10^7$ cycles for the heat streated specimens is the same as that given in Reference 4 and falls in the 55 - 100MPa range given in Reference 5 for notched specimens.

The heat treatment reduced the beneficial compressive stresses as illustrated by specimen H8 in Table 2 but the sharp notches were still present. Figure 11 compares the multiple fracture initiation of the heat treated specimen with the single point of initiation found on most of the smooth specimens. Without the benefit of the compressive stress, the fatigue initiation starts from any of the many available sites seen in Figures 2 and 3. Spherical shot, although not as efficient for paint removal as grit, might be considered as a means to prevent the formation of sharp notches.

The specimens used in the fatigue tests were grit blasted till a "white metal" finish was produced. This is a rather subjective control and it would have been preferable to use an Almen gauge.[8] In order to reproduce the condition of a grit blasted surface, a combination of variables must be controlled, e.g. grit size, shape, and hardness; time of blast; air pressure; nozzle size; distance of nozzle to work; angle of nozzle to surface. Precise control was not maintained in the tests reported and some of the results may reflect the inevitable variability of surface condition. However, grit blasting used industrially to remove paint and surface scale is not carefully controlled and this should be borne in mind when specifying grit blasting.

The effect that the blast time had on the residual stress is shown in Figure 12. Except for one result, very fine Alundum at time 2T, the compressive stress increased with blast time to about 4T and then levelled off or decreased somewhat. A similar curve (saturation intensity curve) is seen when shot peening, and is used as a quality control test.[8]

Figure 12 suggests that excessive exposure time may lead to "over blasting" of the surface as the compressive stress decreases between time 4T and time 8T.

As mentioned previously, the X-rays do not penetrate very deep and the residual stress results are from the first few grains on the surface. Because the maximum effect of shot peening is often somewhat below the surface,[9] the results for residual stress given in Tables 2 and 3 and Figures 12 and 13 may not be the maximum obtained. This is particularly relevant to the results for the medium grit, which are lower than for the very fine grit, but may be higher at some point below the immediate surface.

As seen in Table 2, specimen A9 had compressive stresses on the surface, and A15 had compressive stresses on one surface and tensile on the other (both specimens in the as-received condition. Table 4 shows that specimens grit blasted on one side had compressive stresses on both surfaces. In order to maintain equilibrium, the material must have tensile stresses in the core to balance the surface compressive stresses. If the surface compressive stresses are confined to a very thin layer, then the counter balancing tensile stresses will be close to the surface. With a shallow surface compressive stress, corrosion, wear, or other damage could quite easily expose the tensile stresses from below the surface.

CONCLUSIONS

1. The fatigue test results fall within the rather wide scatter band of results found in the literature, except for the 7075-T6 Lot 2 material which fell below published results.
2. The grit blasting did not have a significantly adverse effect on the fatigue strength of either the 7075-T6 or the 5083-H32 materials and, indeed, had a beneficial effect on 5083 grit blasted with the very fine grit. Thus, it would be useful to investigate finer grades of grit than the medium "Black Beauty" used for the fatigue tests in the present report.
3. The beneficial effect resulted from compressive residual stresses, which were confirmed by X-ray stress measurements.
4. The deleterious effect on fatigue of the notches produced by grit blasting may under some conditions offset the beneficial effects of the residual stress, e.g. when corrosion, wear, or other damage removes the compressive stress on the surface.
5. Periodic overstressing improved the resistance to fatigue crack initiation of 7075-T6.
6. Favourable results of fatigue testing support the use of grit blasting for paint removal on aluminum alloys. Although the damage caused by the small notches formed when grit blasting may in some cases be detrimental to the fatigue life, it will not be worse than the local damage caused by the paint removal methods currently used, e.g. scarping, heat (excess), and power sanding.
7. Well-prepared, grit blasted surfaces will require less frequent painting for corrosion protection. Paint removal and repainting for purely cosmetic purposes should be avoided because the long term effects of repeated grit blasting, painting, and corrosion cycles are not known. It is also likely that repeated

grit blasting will produce unsightly distortion similar to that seen on the hulls of some ships.
8. The need for a grit blasting procedure specification cannot be over emphasized. In order to reproduce an acceptable surface finish, a specification must be established which clearly defines all the variables which affect the grit blasting results.

REFERENCES

1. T.R. Gurney, "Fatigue of Welded Structures", Cambridge University Press, 1968, Chapter 7. p.133.
2. J. Schijve, "Observations on the Prediction of Fatigue Crack Growth Propagation Under Variable-Amplitude Loading", Fatigue Crack Growth Under Spectrum Loads ASTM STP 595, ASTM 1976, p. 3-32.
3. M.E. Hilley, (Ed.), "Residual Stress Measurement by X-Ray Diffraction - SAE J784a", Society of Automotive Engineers 1971.
4. ASM Metals Handbook, Volume 1, Eighth Edition, p. 881.
5. T.H. Sanders and J.T. Staley, "Review of Fatigue and Fracture Research on High Strength Aluminum Alloys", ASM Materials Science Seminar, 1978, ST. Louis.
6. Aluminum Standards Data 1979, The Aluminum Association.
7. F.V. Lawrence, W.H. Munse, and J.D. Burk, "Effect of Porosity on Fatigue Properties of 5083 Aluminum Alloy Weldments", Welding Research Council Bulletin 206, June 1975.
8. MIL-S-13165B - "Shot Peening of Metal Parts".
9. D.V. Nelson, R.E. Ricklefs, and W.P. Evans, "The Role of Residual Stresses in Increasing Long Life Fatigue Strength of Notched Machine Members", Achievement of High Resistance in Metals and Alloys, ASTM STP 467, ASTM 1970, p. 228-253.

TABLE 1

Chemical Analysis (Percent)

	Si	Fe	Cu	Mn	Mg	Cr	Ni	Zn	Be	Al
7075 - Lot 1	0.2	0.37	1.57	0.05	2.7	0.21	--	5.5	0.001	rem
7075 - Lot 2	0.13	0.33	1.64	0.03	2.3	0.19	--	5.5	0.001	rem
5083 - H32	0.18	0.25	--	0.67	4.4	0.07	--	--	<0.001	rem

TABLE 2

Residual Stress on Fatigue Cracked Specimens

Alloy	Surface Condition	Specimen Number	Fatigue Stress (MPa)	Fatigue Cycles →(No Failure)	Residual Stress (MPa)
7075-T6 (Lot 1)	As Received	37	170	587,800	-59
		38	170	386,400(g)	-17
		3	210	264,600	-29 (grip area)
		40	170	413,300	-13
		8	--	--	-32 (grip area)
	Medium Grit (BB)	3	210	264,600	-98/-104
		10	--	--	-34
		8	--	--	-51
	Very Fine Grit (Alundum)	40	170	413,300	-136
		10	--	--	-140
Heat Treated	Smooth Part	H8	85	10,383,100→	+9
	Medium Grit (BB)	H8	85	--	+10 (1st side) -10 (2nd side)
7075-T6 (Lot 2)	As Received	55	120	10,094,300→	+42
		53	140	1,023,900	+21 (1st side)
		53	140	1,023,900	+36 (2nd side)
		S2	170	3,058,500 (g)	-2 (1st side)
		S2	170	3,058,500 (g)	-3 (2nd side)
5083-H32	As Received	A9	155	10^7→	-21 (grip area)
		A23	105	11,260,100→	+14 (grip area)
		A15	155	20,462,000→	+21 (grip area 1st side)
		A15	155	20,462,000→	-10 (grip area 2nd side)
	Medium Grit (BB)	A23	105	11,260,100→	-19
		A20	130	4,090,300→	-20
		A15	155	20,462,000→	-54
	Very Fine	A9	155	10^7→	-47 (1st side) -41 (2nd side)

TABLE 3

Residual Stress for 7075 - Lot 2 - As Received

			Stress Direction	
Material	Condition	Spot	Transverse (MPa)	Longitudinal (MPa)
7075-T6	As Received	1	+6	+5
		2	+34	+5
		3	+106	+4
		4	+13	+4
		5	+41	-11
		6	+15	+18
		7	+67	-2
		8	+45	+32
		9	+48	+26
		10	+29	+15
		11	+36	-29
		Average	+40 $\pm$ 27	+6.0 $\pm$ 16

TABLE 4

Residual Stress for 5083-H32

Specimen No.	Grit	Time	Stress Direction	Residual Stress (MPa)				Spot 4 Reverse Side
				Spot 1	Spot 2	Spot 3	Average 1-3	
5.1	Alundum	1T	Long	38	47	38	40.8 $\pm$ 4.3	
5.2		2T	Long	34	29	31	31.4 $\pm$ 2.8	27
			Trans	49	57	31	45.9 $\pm$ 11.0	30
5.3		4T	Long	56	51	50	53.3 $\pm$ 4.3	33
			Trans	40	31	55	42.0 $\pm$ 9.9	38
5.4		8T	Long	41	48	40	43.1 $\pm$ 3.6	--
5.25	Fine	1T	Long	18	12	16	15.6 $\pm$ 4.3	--
5.28	Black	2T	Long	25	23	16	20.0 $\pm$ 6.1	--
5.31	Beauty	4T	Long	26	31	22	26.1 $\pm$ 4.7	--
5.34		8T	Long	24	18	18	19.8 $\pm$ 3.5	--
5.13	Medium	1T	Long	26	20	21	22.4 $\pm$ 4.5	--
5.16	Black	2T	Long	19	31	30	26.9 $\pm$ 5.4	--
	Beauty		Trans	--	26	--	--	--
5.19		4T	Long	40	30	32	33.7 $\pm$ 4.2	--
			Trans	43	30	--	--	--
5.23		8T	Long	38	22	28	29.1 $\pm$ 7.3	--

NOTE: All stresses are compressive.

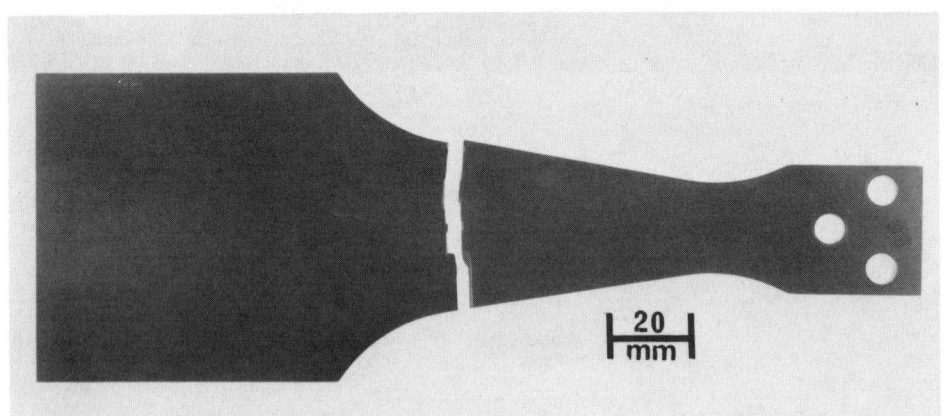

Figure 1. Tatnal-Krause Fatigue Test Specimen.

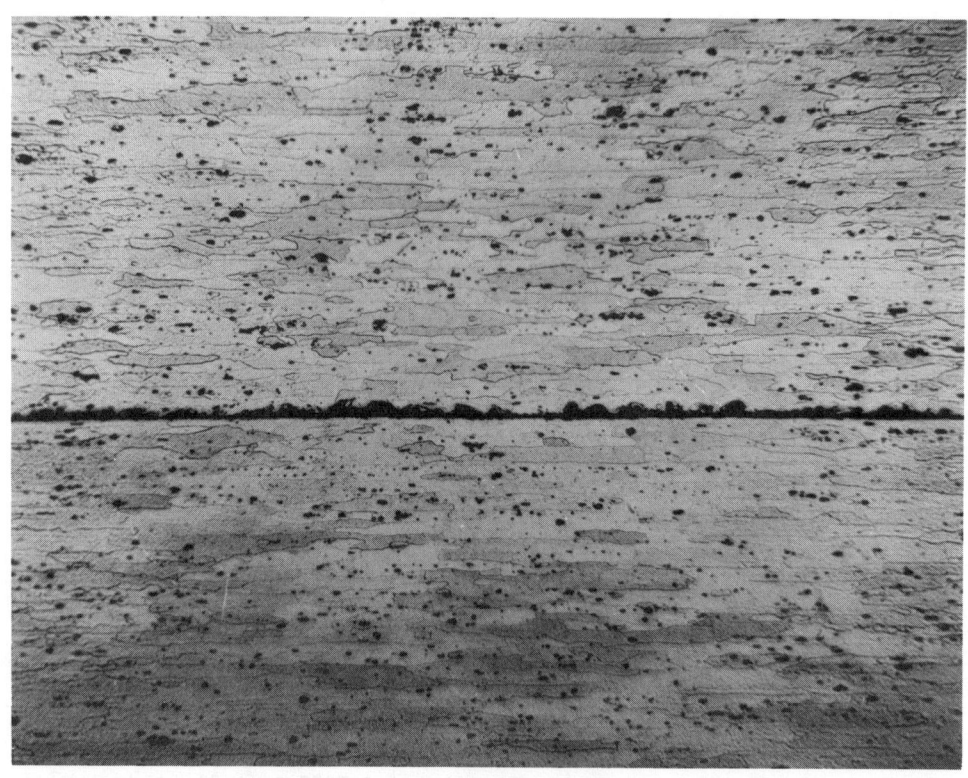

Figure 2. Cross Section Through Specimen (top) of 7075 Grit Blasted with Very Fine Grit, Alundum #80. (X170)

Figure 3. Cross Section Through Specimen (Top) of 7075 Grit Blasted with Medium Industrial Crushed Slag Grit (Black Beauty). (X170)

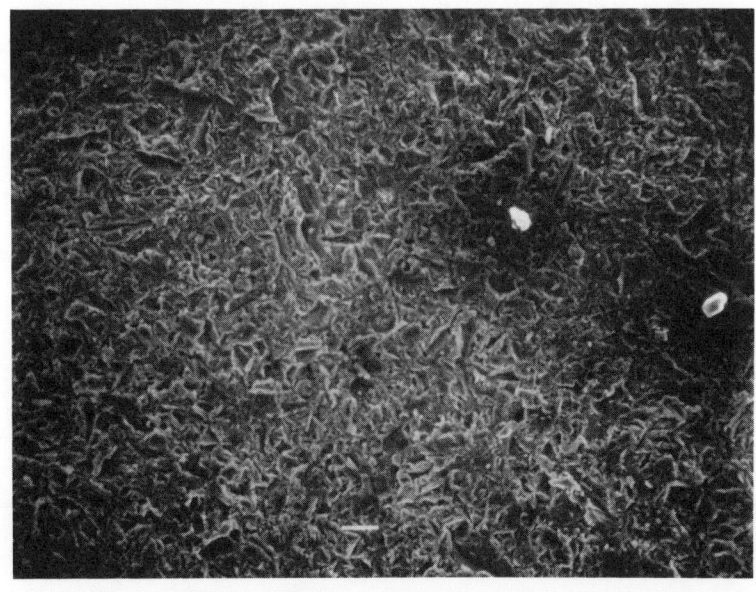

Figure 4. Surface Appearance of Specimen of 7075 Grit Blasted with Very Fine Grit, Alundum #80. (X90)

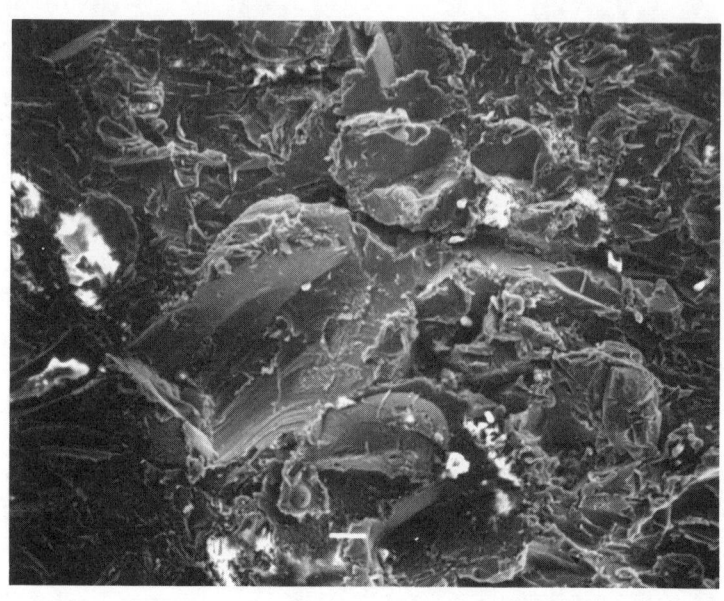

Figure 5. Surface Appearance of Specimen of 7075 Grit Blasted with Medium Industrial Crushed Slag Grit (Black Beauty). (X90)

Figure 6. Electron Micrograph (TEM) of Fatigue Fracture Surface of Alloy 7075 Tested at 20 Ksi to Failure at 5,008,000 cycles. (X5000)

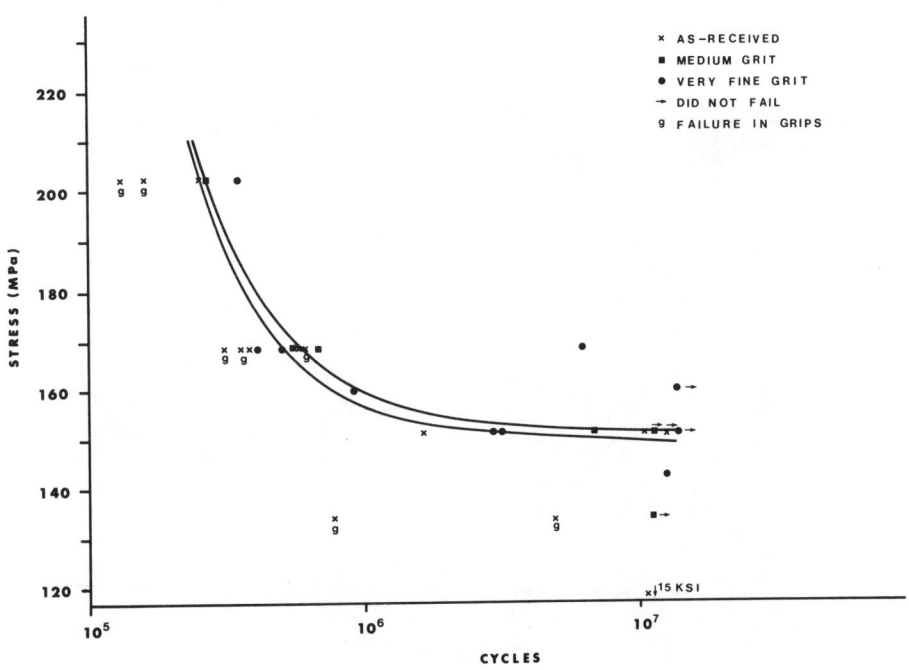

Figure 7. Effect of Grit Blasting on the Stress-Life of 7075-T6 (Lot 1) Material.

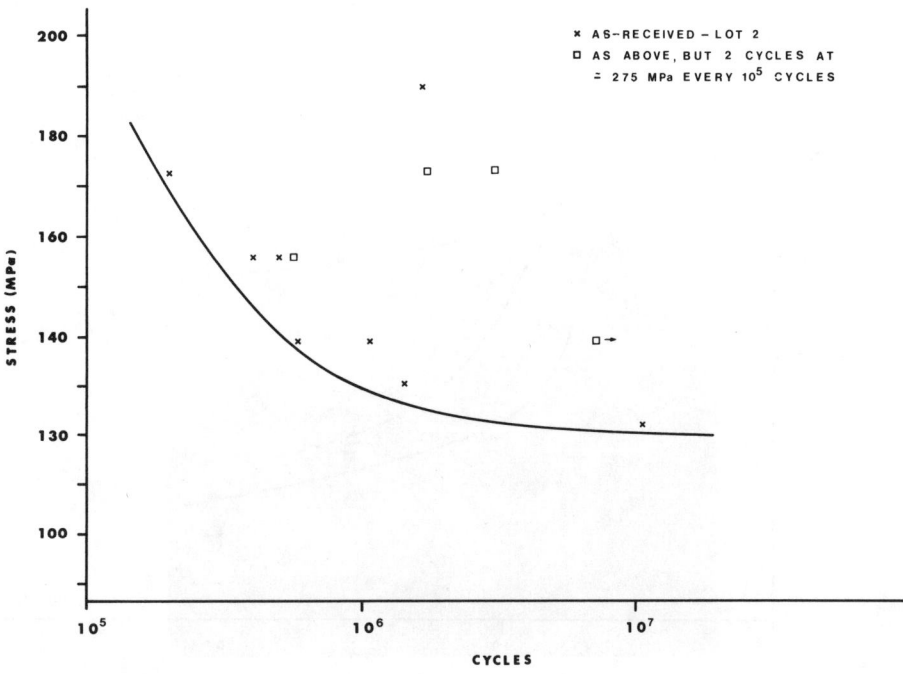

Figure 8. Effect of Overstressing on the Stress-Life of 7075-T6 (Lot 2) Material.

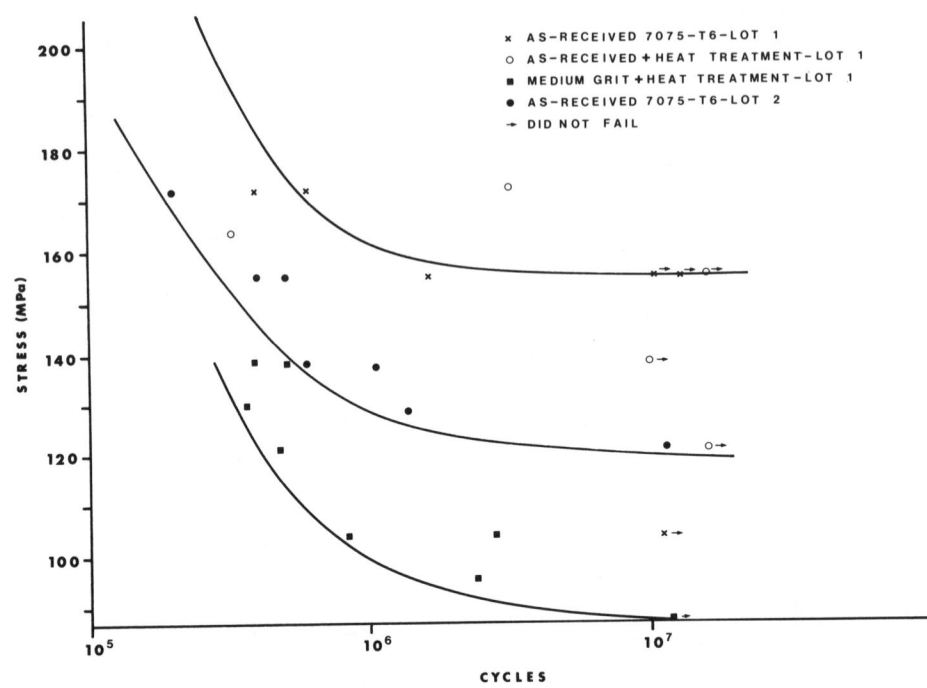

Figure 9. Comparison of Fatigue Life Results for As-Received 7075 Lot 1 and Lot 2 Materials; Heat Treated 7075 Lot 1 and Heat Treated Grit Blasted Specimens.

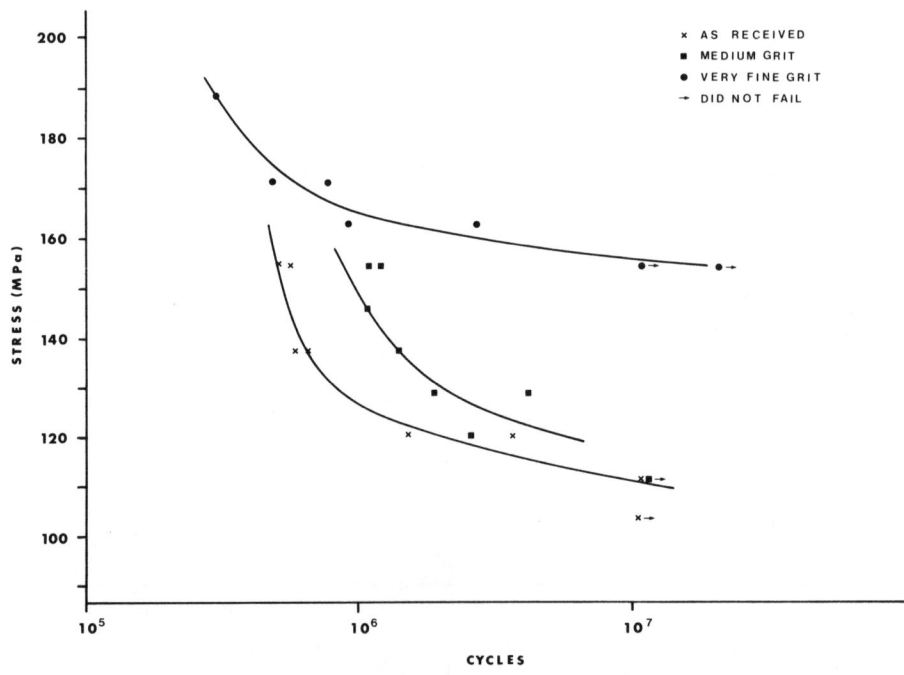

Figure 10. Fatigue Life Results for 5083 Material; As-Received, Very Fine, and Medium Grit Blasted.

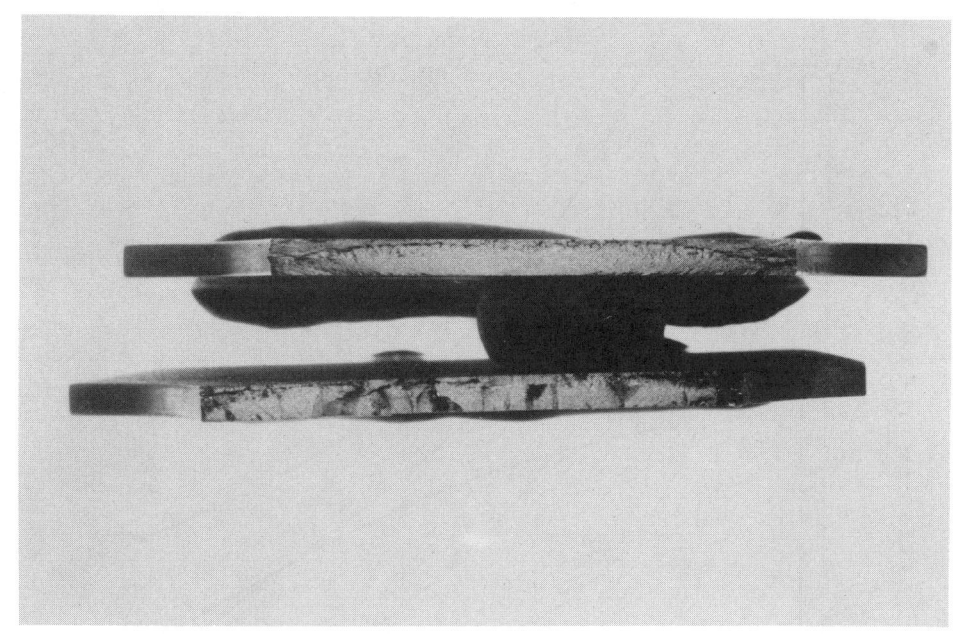

Figure 11. Fracture Surface of 7075 Material; Top, As-Received, Bottom, Grit Blasted Plus Heat Treatment. (X1.8)

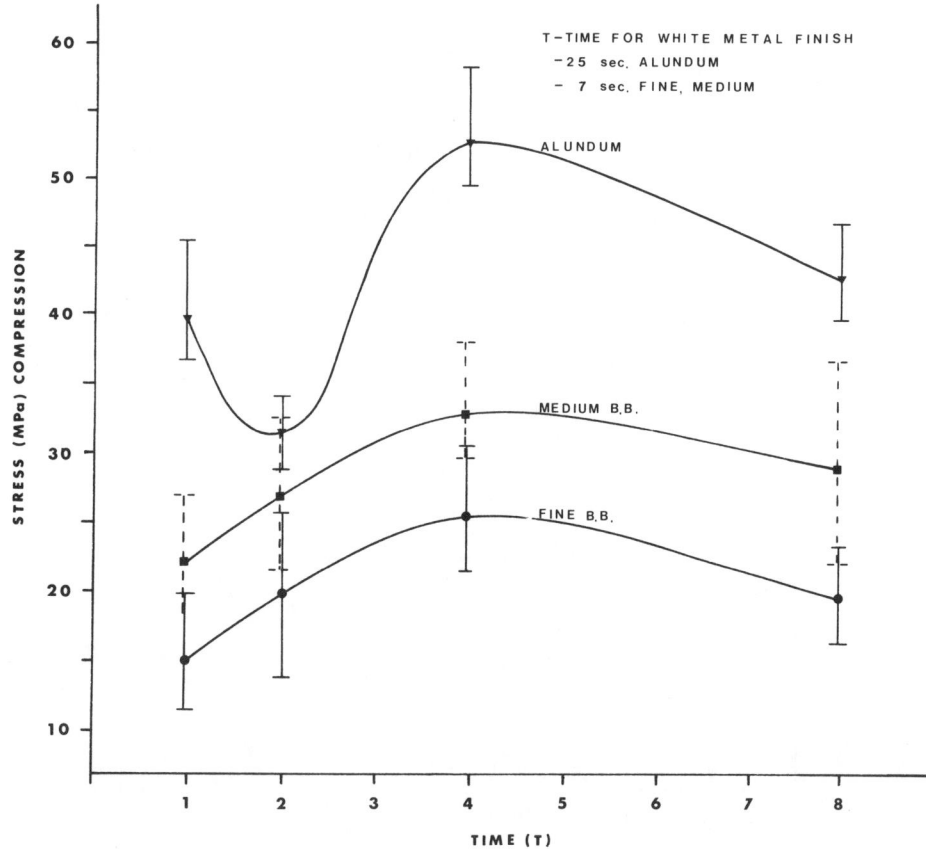

Figure 12. Residual Stress Versus Time T (Time for White Metal Finish) for Three Grit Sizes, on 5083 Material.

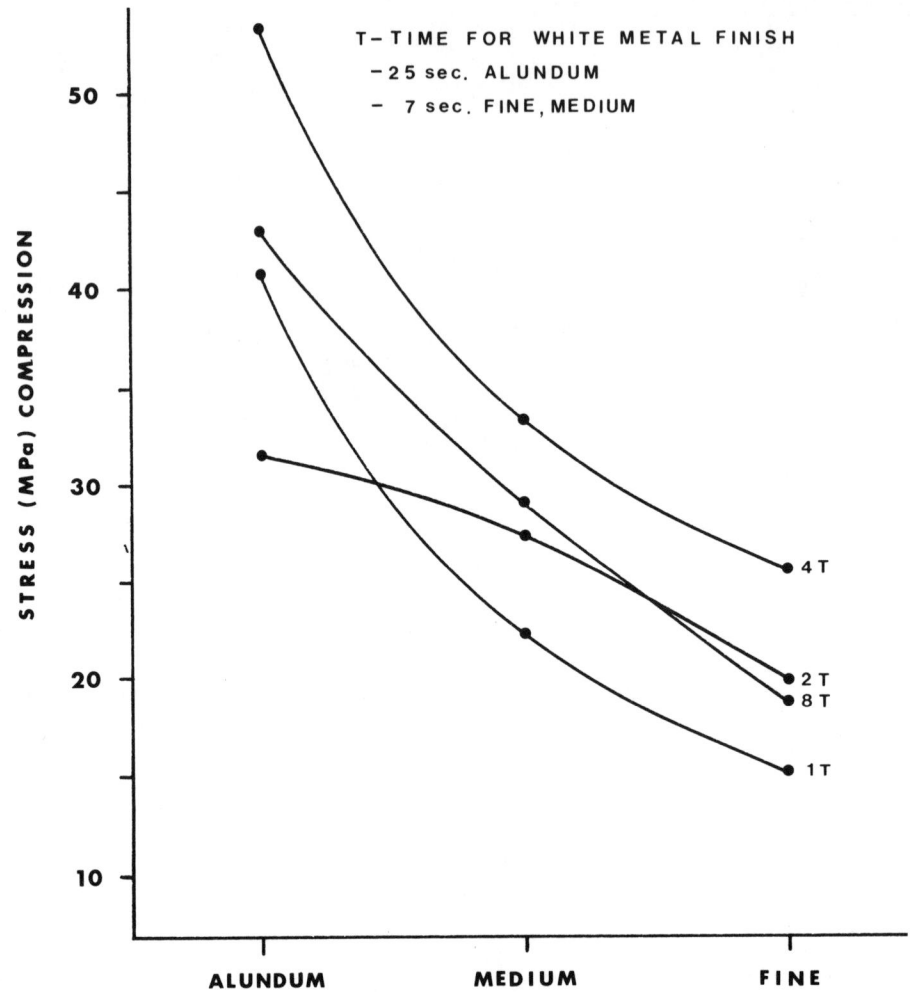

Figure 13. Residual Stress Versus Grit Size for Various Times T (Time for White Metal Finish), on 5083 Material.

RESIDUAL STRESS MAPPING OF GEARS AND BEARINGS

C.O. Ruud, D.J. Snoha
Materials Research Laboratory
The Pennsylvania State University
University Park, Pennsylvania USA

ABSTRACT

Residual stresses on raceway surfaces of bearings and contact surfaces of gears have been of keen interest to mechanical engineers and metallurgists for some time. However, there has been no convenient technique to offer practical residual stress measurement to provide extensive measurement for substantial coverage of the surface of gears and bearings. The Pennsylvania State University has recently developed a compact x-ray device that is capable of measuring residual stresses on inside diameters as small as 3 1/2 inches with unprecedented speed, precision and accuracy.

This instrument has been applied to residual stress measurements on high precision ball bearings for helicopter rotors and roller bearing surfaces for the transportation industry. It has also been used to evaluate residual stresses and cold-work induced by shot peening automotive transmission gears as well as being an aid to developing ausforming technology for high precision gears.

This paper will describe the instrumental technology and procedures and results of residual stress two-dimensional mapping on several examples of gears and bearings.

ALMOST ALL MACHINES or mechanical systems contain precision contact elements such as bearings, cams, gears, shafts, splines, and rollers. Improvements in the quality and life expectancy of such systems as automotive and truck power trains and oil well production and drilling equipment can be improved by monitoring the condition of the surfaces of the contact elements. For example, improvements in bending and pitting fatigue can contribute directly to lighter weight gears by reducing the required width and mass of the gear elements for equivalent loads. Processing of contact elements often involves carburizing, grinding, induction hardening, and/or shot peening; and it is desired that these processes are consistent with producing highly compressive residual stresses at the contact and highly stressed surfaces. However, to a large extent measurement of residual stresses has been, by nature, a destructive test, except for the application of x-ray diffraction which up until now has usually been time consuming and/or difficult to apply in confined regions of the components to be inspected, e.g. gear tooth roots.

X-Ray diffraction (XRD) remains the only reliable method for nondestructive measurement of residual stresses in spite of the considerable research that has been invested in methods such as Barkhausen Noise Analysis, ultrasonic velocity, and eddy current techniques [1]. But, the instrumentation and techniques available for the application of XRD have not provided the speed, accuracy, and ease of measurement required for general use [2, 3]. This is especially so when it is recognized that in order to adequately measure the residual stress field on any type of part ten to hundreds of measurements must be made in order to characterize the residual stress condition. However, recently developed instrumentation has provided a means to address this situation. This paper presents results from residual stress measurements on several contact elements, e.g. gears and bearings, using such an instrument [3, 4].

PROCEDURE

SAMPLES - Five types of samples were taken to provide examples of residual stress measurement on contact elements. The first two samples were steel gears, one type shot peened and the other carburized and induction hardened. The next two samples were steel bearing races, one type a double row tapered roller bearing and the other an aircraft ball bearing. The last sample was the six-inch inside diameter bearing insert region of an aluminum aircraft wheel.

X-RAY PARAMETERS - Chromium k-alpha radiation was used for all residual stress measurements reported herein. For steel the (211) crystallographic planes which diffracts x-rays at twice the Bragg angle of about 156.1 degrees was used, and for aluminum the (222) crystallographic planes which diffracts x-rays at twice the Bragg angle of about 156.7 degrees was used. The Young's modulus and Poison's ratios used were 29×10^6 psi (204×10^3 MPa) and 0.29, and 10×10^6 psi (70×10^3 MPa) and 0.33 for steel and aluminum, respectively. The instrumentation used incorporated a Position Sensitive Scintillation Detector (PSSD) system developed at Penn State University [3, 4], see figure 1. The data collection times for complete residual stress measurement were about three seconds for steel and about five for aluminum utilizing a miniature stress head, see figure 2.

RESULTS AND DISCUSSION

The first type of gear samples were a suite of four shot peened high strength low alloy steel automobile transmission gears, with about 1.5 inch (38 mm) pitch diameter, see figure 3. The four gears each had been exposed to a varying degree of shot peening as represented by the time of peening. Stresses and x-ray peak breadths were measured on the gear hub and at the tooth roots and the mean of the tooth root results are plotted versus the time of shot peening in figure 4. The amount of cold work indicated by x-ray peak breadth and the residual stress is indicated by x-ray peak shift [5]. Thus, both parameters were determined simultaneously from the XRD data. Figure 4 shows that the residual stresses in the surface of the gears become highly compressive after only a short time of peening, then remained nearly constant. On the other hand, the x-ray peak breadth gradually became broader with increasing peening times. Thus, the x-ray peak breadth would seem to offer a better means of determining the time, or amount, of shot peening for optimum effort.

The second type of gear sample was a carburized and induction hardened spur gear

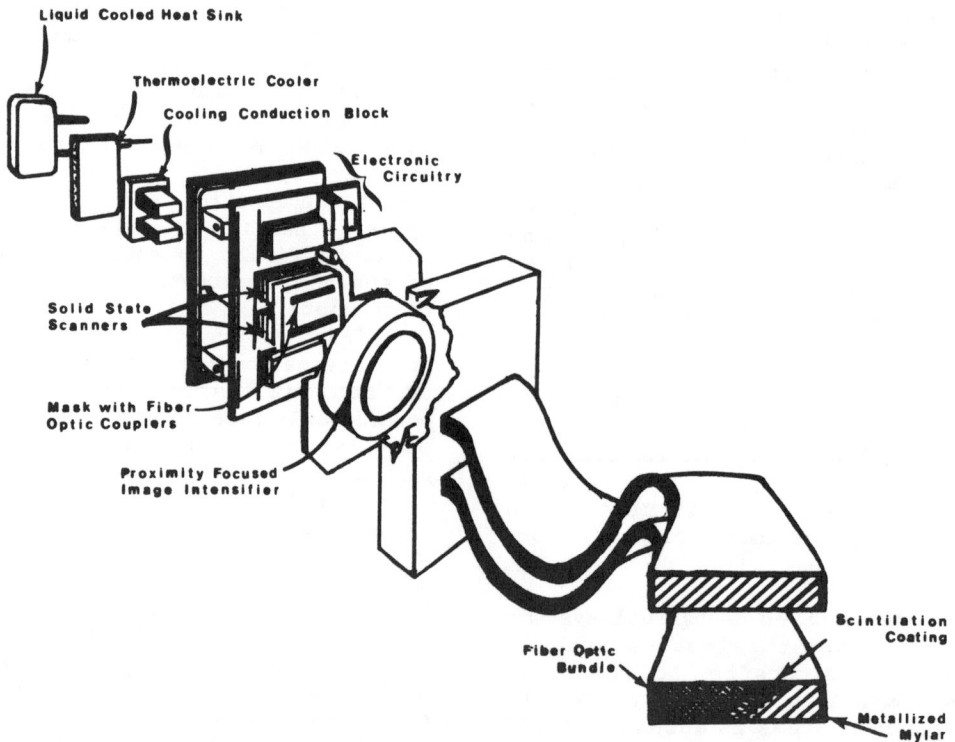

Figure 1. Exploded view of position-sensitive scintillation x-ray detector.

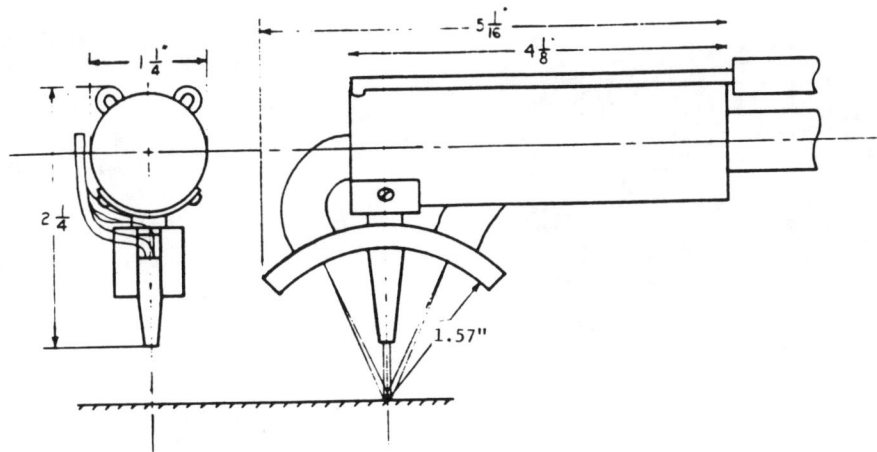

Figure 2. Engineering sketch of a miniature x-ray diffraction residual stress measurement head showing key dimensions.

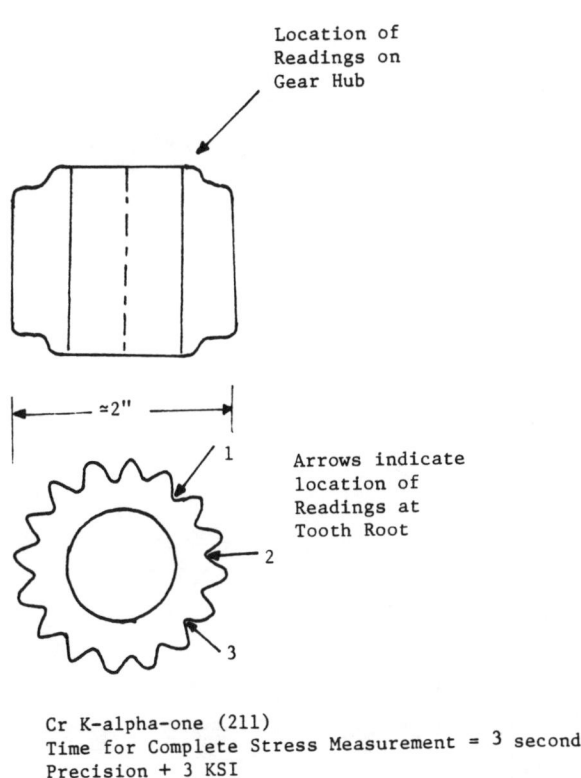

Cr K-alpha-one (211)
Time for Complete Stress Measurement = 3 seconds
Precision ± 3 KSI

SHOT PEENED GEAR

Figure 3. Shot peened automobile transmission steel gear.

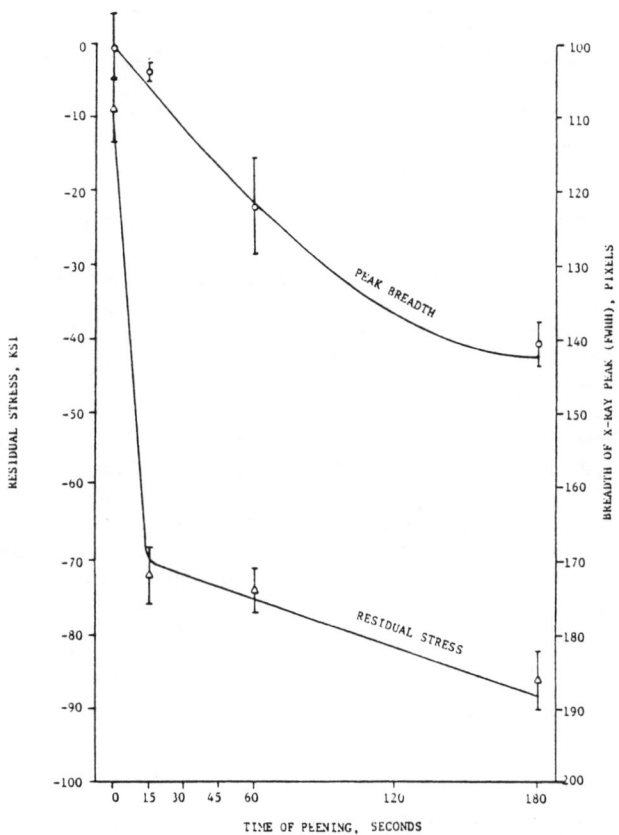

Figure 4. Plot of residual stress, or breadth of the x-ray peak, versus the time (or degree) of shot peening for high strength steel gears.

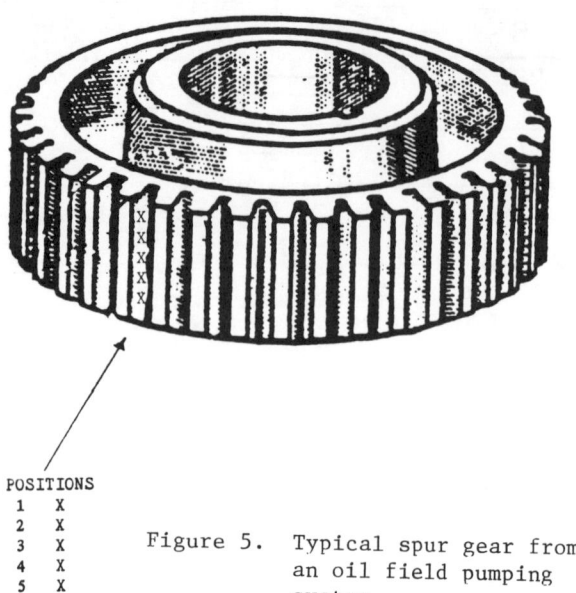

POSITIONS
1 X
2 X
3 X
4 X
5 X

Figure 5. Typical spur gear from an oil field pumping system.

from an oil field pumping system. The gear was about 12 inches (305 mm) pitch diameter and stresses were measured on several of the teeth at the pitch diameter, i.e., approximately at the half depth of the tooth, and in a radial direction to the pitch circle. The application of the single exposure technique (SET) for x-ray residual stress measurement allowed for a more compact x-ray geometry than would be available with x-ray stress measurement devices incapable of applying the SET. This compactness allowed for measurement deeper on the gear tooth than could be realized with other XRD stress measurement techniques. Figure 5 shows a typical gear and positions on a typical tooth where residual stress data was obtained at five locations along an axial traverse. Table 1 lists the residual stresses obtained on several teeth from carburized gears. Table 2 lists the residual stresses obtained on one tooth on each of four gears carburized and/or induction hardened. It should be noted that the stresses are highly compressive and that there is some variance from the edges of the teeth to their center.

TABLE 1
FOUR TEETH ON ONE CARBURIZED GEAR

POSITION	TOOTH			
	1	2	3	4
1	-56KSI	-65KSI	-76KSI	-78KSI
2	-81	-70	-82	-80
3	-83	-58	-82	-79
4	-80	-51	-79	-81
5	-82	-66	-109	-74

CR K-ALPHA ONE (211)
TIME FOR EACH COMPLETE STRESS MEASUREMENT = 3 SECONDS
PRECISION = ± 3 KSI

TABLE 2
ONE TOOTH ON EACH OF FOUR GEARS
CARBURIZED AND/OR INDUCTION HARDENED

POSITION	GEAR			
	6	7	8	9
1	-71KSI	-66KSI	-86KSI	-85KSI
2	-78	-62	-94	-80
3	-77	-66	-100	-77
4	-72	-64	-94	-76
5	-77	-66	-83	-81

The third and fourth types of contact element samples were bearings. The races of both types of bearings were measured on the inside diameter surface which required that the XRD stress measuring instrument be sufficiently compact to measure stresses in about a 5 inch (127 mm) inside diameter. The Penn State developed instrument is capable of measuring stresses in diameters smaller than 4 inches (102 mm); thus, it was suitable for the bearing inside diameter race measurement, see figure 2.

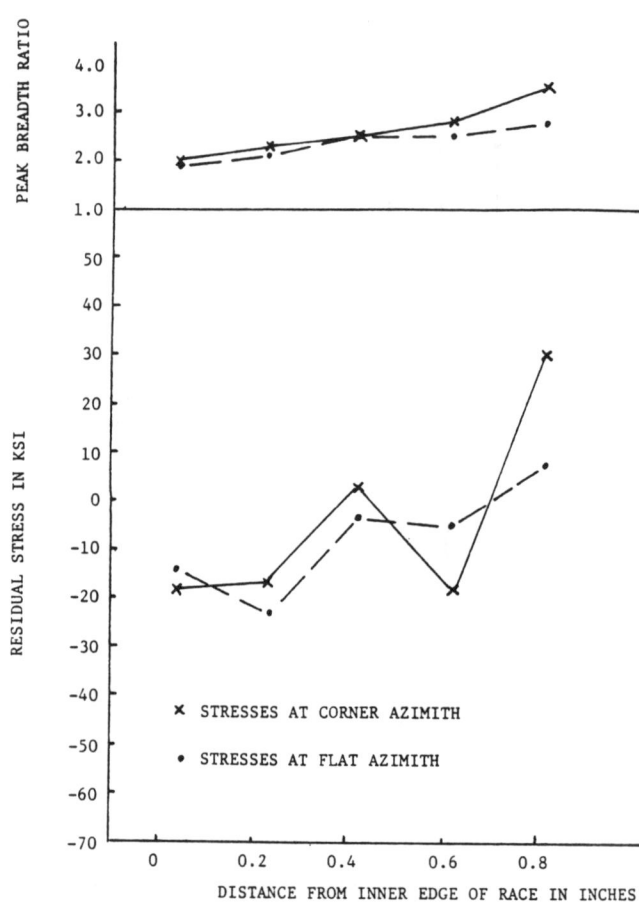

Figure 6. Hoop direction residual stresses on Bearing No. 2.

The first type of bearing sample was a steel, double race tapered roller bearing. Residual stresses were measured in the hoop and axial directions on the inside race along an approximately axial traverse from the outside edge of the race to its center. The traverses were made at two azimuths 45° apart, representing varying degrees of quenching. Figure 6 shows plots of hoop residual stresses versus the distance from the inner diameter of the bearing race at the two azimuths and figure 7 shows the axial stresses. The peak breadths are also indicated in these figures.

The second type of bearing samples was an aircraft ball bearing with an approximately 6 inch (152 mm) inside diameter outer race and a 5 inch (127 mm) outside diameter inner race. Hoop and axial residual stresses were measured on both race surfaces at threee azimuths and are shown in figure 8 from a typical bearing. As shown in the figure the residual stresses are highly compressive.

The fifth and last type of sample also required a compact XRD stress measuring instrument because it required measurements on the inside surface of a high strength aluminum aircraft wheel hub. Table 3 lists the residual stresses as measured from the hub surface by both XRD and the hole drilling technique. In order to perform the semi-destructive hole drilling residual stress measurement it was necessary to cut the wheel into three pie shaped sections. The residual stress values from both methods are seen to compare well in spite of the fact that XRD samples only about 0.0005 inches (0.012 mm) into the surface while the hole drilling technique sample a minimum of 0.006 inches (0.152 mm) deep.

TABLE 3

ALUMINUM AIRCRAFT WHEEL BEARING SURFACE X-RAY DIFFRACTION RESULTS BEFORE SECTIONING THE WHEEL AND HOLE DRILLING CORRECTED FOR SECTIONING

AXIAL STRESSES ON THE INSIDE DIAMETER

Position	20	22	24
Stress (KSI)	20±2	22±2	29±2
Hole Drilling	19	2	--

RADIAL RESIDUAL STRESSES ON THE INSIDE DIAMETER

Position	2	4
Stress (KSI)	3±2	-3±1

HOOP RESIDUAL STRESSES ON THE INSIDE DIAMETER

Position	19	21	23
Stress (KSI)	1±4	-2±4	4±4
Hole Drilling	4	2	--

Position	1	3
Stress (KSI)	25±3	20±3

Cr K-alpha-one (222)
Time for Complete Stress Measurement = 6 seconds
Precision ± 2 KSI

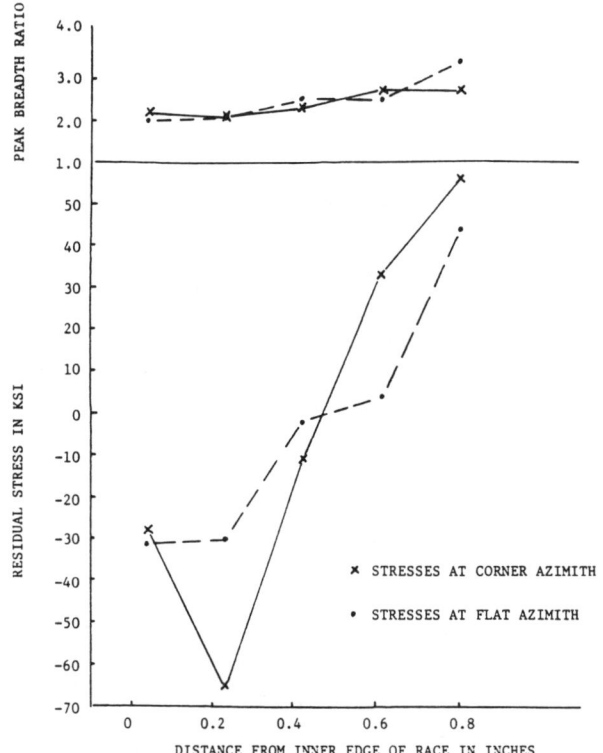

Figure 7. Axial direction residual stresses on Bearing No. 2.

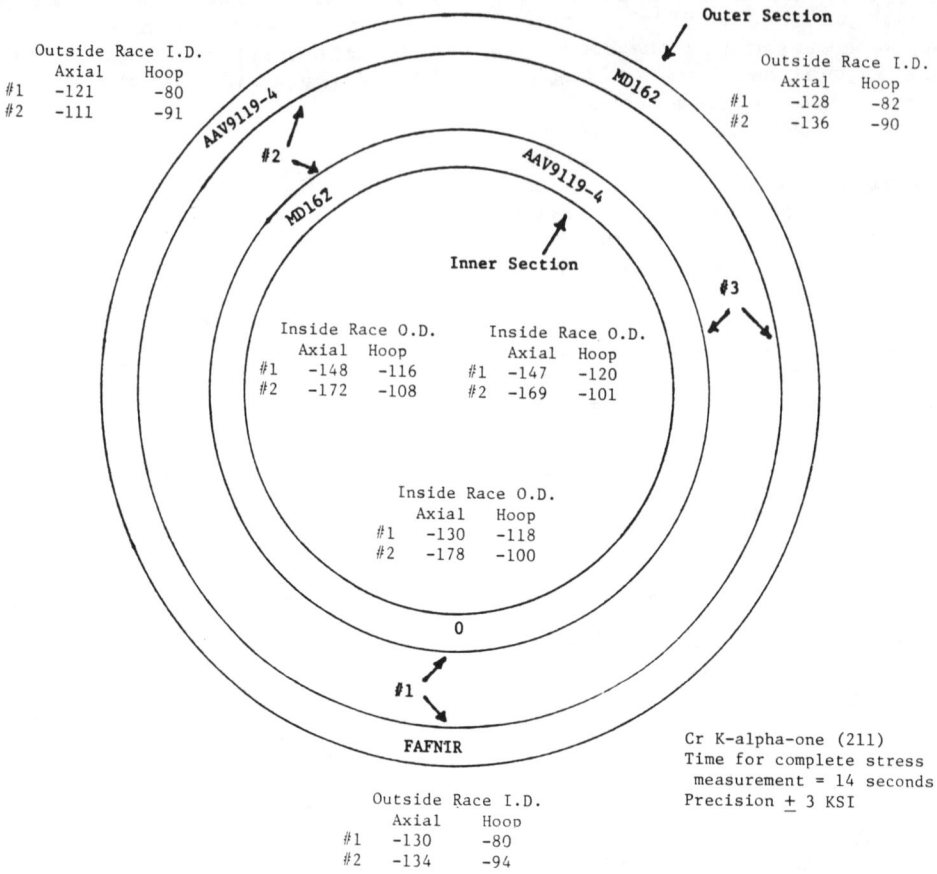

Figure 8. Sketch of an aircraft bearing showing residual stresses at center of race locations #1, #2, and #3.

CONCLUSION

The results presented herein have demonstrated the applicability of a new type of XRD residual stress measurement instrument to contact element samples, e.g. gears, bearing, and bearing surfaces.

BIBLIOGRAPHY

1. C. O. Ruud, 'A Review of Nondestructive Methods for Residual Stress Measurement,' J. of M., 33, pp. 35-40, July (1981).

2. C. O. Ruud, 'X-Ray Analysis and Advances in Portable Field Instrumentation,' J. of M., 31 (6), pp. 10-15, June 1979.

3. C. O. Ruud, P. S. DiMascio, and D. J. Snoha, 'A Miniature Instrument for Residual Stress Measurement,' Adv. in X-Ray Anal., 27 (1984).

4. C. O. Ruud, 'Position Sensitive Detector Improves X-Ray Powder Diffraction,' Ind. Res. and Dev. Jan. (1983).

5. C. O. Ruud, 'Application of an Advanced X-Ray Diffraction Instrument to Nondestructive Materials Characterization,' NDT Comm., Vol. 2, pp. 19-27, (1985).

GENERATING COMPRESSIVE RESIDUAL STRESS BY CBN GRINDING

Glenn A. Johnson
General Electric
Gotenba, Japan

ABSTRACT

Grinding with CBN has been shown to generate beneficial compressive residual stresses in the ground surface over a wide range of grinding conditions. This is in contrast to normal grinding with conventional abrasives such as aluminum oxide where tensile residual stresses are usually formed causing a reduction in the fatigue life of the ground component. In cases where surface integrity is important to the performance of a ground part CBN grinding processes have been implemented to take advantage of these benefits. The thermal properties of CBN are a major reason for the generation of compressive residual stresses in CBN grinding.

IMPROVING QUALITY AND REDUCING TOTAL GRINDING COST is a constant challenge for both manufacturing engineers and designers alike in producing mechanical components. CBN grinding has been well documented in reducing total grinding costs (wheel + labor + overhead) on hardened steels with HRc hardness greater than 50 and nickel and cobalt based superalloys with HRc hardness greater than 35.(1) It is noted, however, that in recent years there have been an increasing number of applications where CBN grinding has been cost effective on softer materials.(2)

Quality improvements achievable with CBN can be divided into two areas. 1)The dimensional accuracy of the component, i.e., roundness, flatness, length, etc. which is well documented with CBN grinding. 2)Surface integrity. There is now increasing interest in this area as the relationship between surface integrity and structural performance is better understood.

ASPECTS OF SURFACE INTEGRITY

The major surface integrity problems associated with grinding include: tensile residual stress, untempered martensite, overtempered martensite, cracks.

Excessive heat at the grinding interface is the major cause of these surface defects. Therefore, the key to achieving good surface integrity in a ground part is to prevent excessive surface temperatures from being generated during the grinding process.

TENSILE RESIDUAL STRESS - In components made of high strength materials, fatigue is often the most common cause of failure. It has been clearly shown that the peak residual stress has a key influence on the fatigue strength of such materials.(3) In figure 1, the results indicate the negative effects of increasing tensile residual stress and the benefits of increasing compressive residual stress.

Thus it is clear that a process which increases the compressive residual stress in the surface will increase the fatigue life of a component in most applications. Shot peening is a common procedure used on parts subjected to high fatigue loading. Shot peening produces compressive residual stress by the impact of the shot causing plastic compressive deformation of the top surface layers.(4) It is also known that tensile residual stress increases the susceptibility of a surface to stress corrosion. As a result, tensile surface residual stress should also be avoided in components subjected to corrosive environments.

UNTEMPERED MARTENSITE - The formation of untempered martensite (UTM) results from the overheating of the surface followed by subsequent rapid quenching. If UTM is present it will generally be found on the top layer of the surface region which is subjected to the maximum quenching rate. The untempered martensite layer will be harder and more brittle than the base material and is often the source of cracks in the surface. The UTM will show up as a white layer when nital etched.

OVERTEMPERED MARTENSITE - The formation of overtempered martensite (OTM) also results from overheating of the ground surface. Quenching at a slower rate causes an overtempering of the surface. The OTM will be softer than the base material and show up as a dark layer when nital etched. When UTM is present, a region of OTM will generally be found underneath the UTM region and the base material. In figure 2, hardness in plotted as a function of depth below the surface where both UTM and OTM regions are present.

Test results indicated that the presence of even small amounts of OTM and UTM will cause a significant reduction in the fatigue strength of the components.(5) Even with subsequent vacuum retempering, the fatigue strength cannot be significantly improved.

CRACKS - Surface cracks also result from excessive heat during grinding. The severity of the cracks will vary. In some situations the cracks will not be visible immediately after grinding but will become apparent at some later time.

It has also been observed that in some surfaces with high tensile stresses, cracks will form slightly below the surface and will not be visible until the top surface is electrically etched. Cracks always have been known to reduce fatigue life and increase the susceptibility of the surface to stress corrosion.

PAST WORK

In response to a need to develop grinding processes that would prevent surface integrity problems common in conventional aluminum oxide grinding of high strength steels, lower temperature grinding techniques have been developed. Much of this work was done in the late 50's and 60's, prior to the commercial availability of CBN grinding wheels.(6) The grinding techniques became known as Low Stress Grinding (LSG) and were designed to avoid the detrimental high temperatures generally found in normal aluminum oxide grinding.

COMPARISON OF LSG TO NORMAL ALUMINUM OXIDE GRINDING - A considerable amount of research has been done in aluminum oxide grinding comparing LSG to conventional grinding. A comparison of the impact of increasing grinding temperature by increasing the severity of the grinding condition is shown in figure 3. In this case, the surface residual stress distribution for three different grinding conditions of increasing temperature, (LSG, normal, and abusive) are shown.

The residual stress resulting from electropolishing is also included. Electropolishing is a demonstrated low temperature method of material removal with negligible alteration of the surface layer resulting in virtually no residual stress being produced. As a result, electropolishing is often used to remove successive layers from the surface for examination without causing any changes in the microstructure.

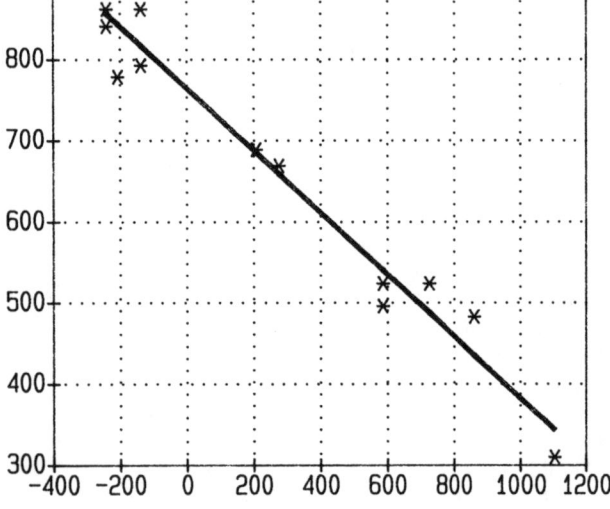

Figure 1. Correlation between peak residual stress and fatigue strength in AISI 4340 at different levels of grinding intensity with aluminum oxide.

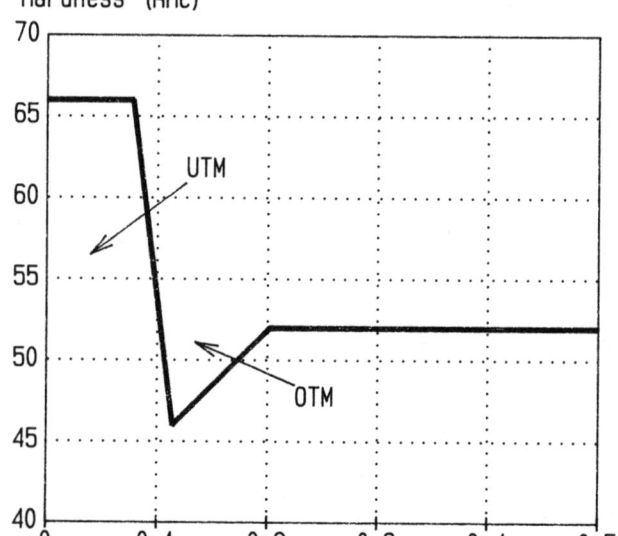

Figure 2. Example of hardness variation across UTM, OTM, and unaffected regions of cross section of 4340 Q&T that was abusively ground.

Note that the abusive and normal grinding processes produced high tensile residual stress resulting from the high temperatures involved. In contrast, the LSG surface had moderate compressive residual stress. This results from the avoidance of excessive temperatures in LSG combined with the mechanical working of the surface by the grinding process analogous to shot peening.

As expected, the fatigue life of the test pieces was directly related to the residual stress pattern, figure 4.

The characteristics of LSG with aluminum oxide wheels are sharp grains ... frequent dressing, low wheel speeds, low material removal rates, and use of oil based coolants

The disadvantages of LSG are the reduction in throughput and increased wheel cost that result from using the slower grinding conditions when compared to normal grinding. As a result the total grinding cost is significantly increased when going to a LSG process using aluminum oxide.

CBN ABRASIVES : ALTERNATIVE TO LSG

BORAZON* CBN abrasives for use in grinding wheels were introduced by General Electric in 1969. CBN's initial use was primarily in the tool room environment, grinding tool and die steels. In recent years, there has been an ever increasing level of BORAZON CBN use in production grinding, especially in Japan, figure 5.

Driven by the growing cost/quality benefits being recognized with CBN grinding, there has been a renewed interest and concentrated effort to maximize the advantages of CBN grinding in production operations. It is well established that CBN wheels wear much more slowly than aluminum oxide wheels, making control of close dimensional tolerances easier to maintain.

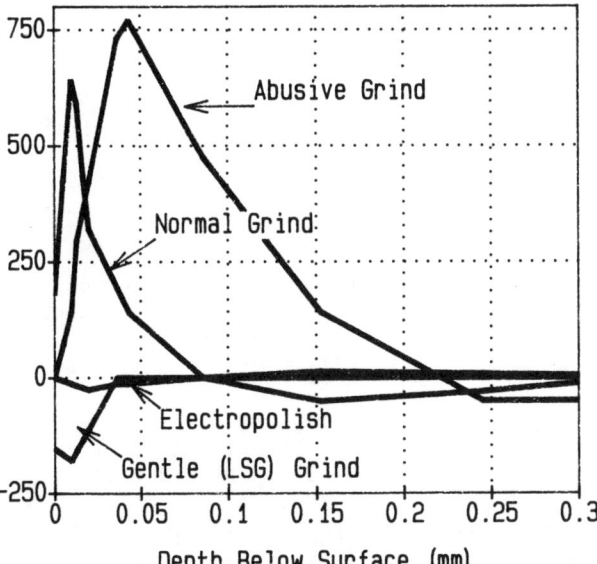

Figure 3. Comparison of residual stress patterns resulting from electropolishing and three different levels of grinding intensity with aluminum oxide.

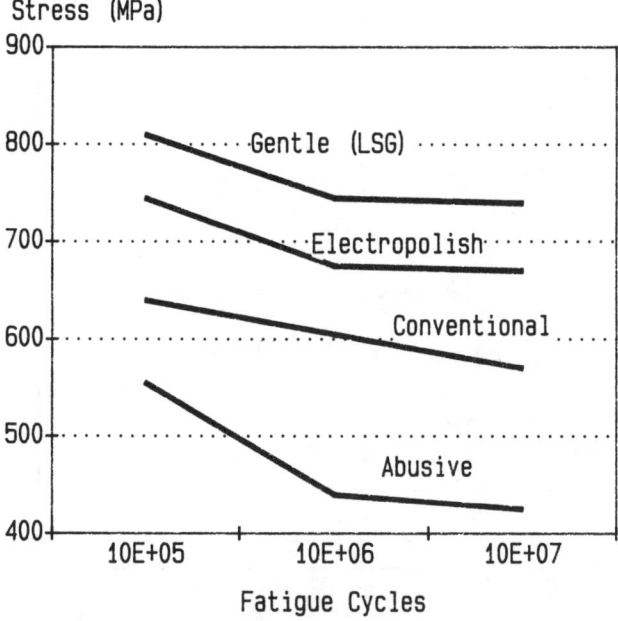

Figure 4. Comparison of fatigue strength resulting from electropolishing and three different levels of grinding intensity with aluminum oxide.

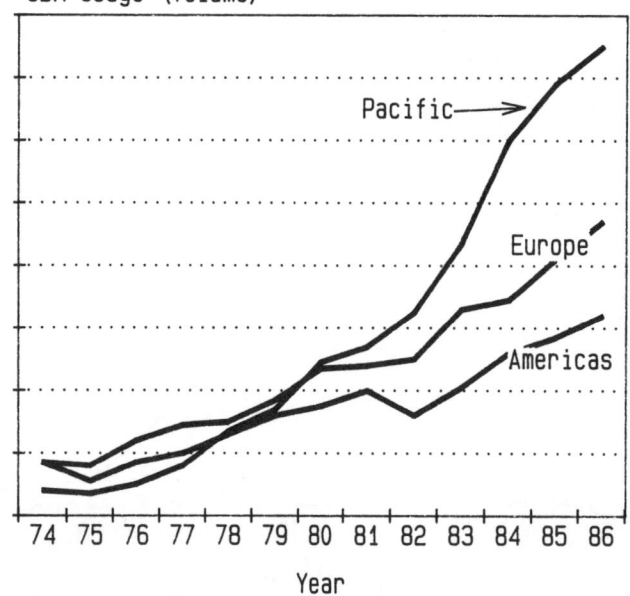

Figure 5. Growth in CBN usage by geographic region.

Another major advantage of CBN grinding is its inherently cooler grinding characteristics as compared to aluminum oxide. This aspect has proven extremely beneficial in generating excellent surface integrity properties in ground components.

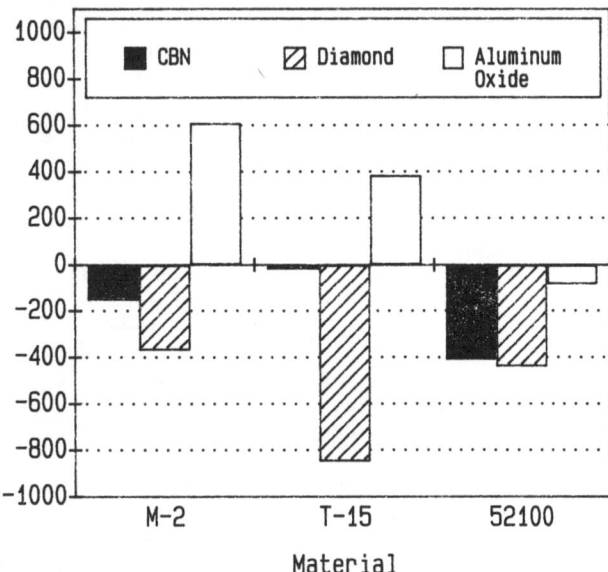

Figure 6. Effect of three different abrasives on resulting surface residual stress.

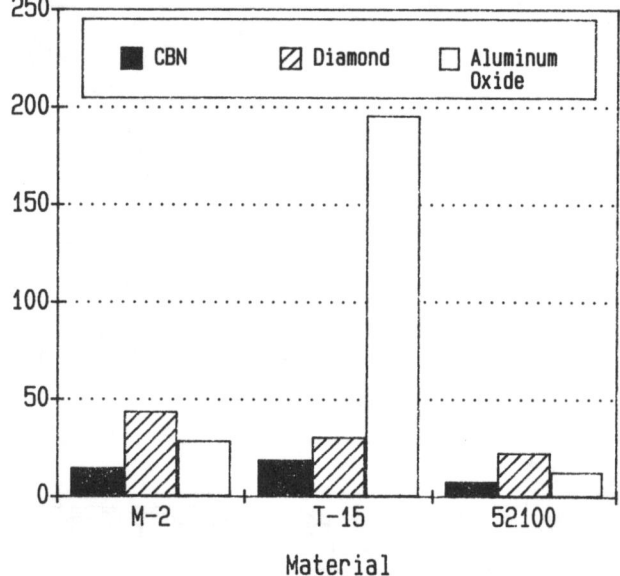

Figure 7. Effect of three different abrasives on specific energy required to remove material.

RESIDUAL STRESS STUDIES - Shortly after the introduction of BORAZON CBN, tests were carried out by General Electric Company to evaluate the surface residual stress obtained with CBN, diamond, and aluminum oxide (8)(figure 6, 7). It was observed that diamond grinding produced the highest compressive residual stress, even though as shown in the case of M2 and 52100, diamond grinding resulted in the highest specific energy.

These tests clearly showed that both BORAZON CBN and diamond must grind cooler than aluminum oxide since tensile stress results from excessively high temperatures in the grinding zone. The residual stresses from CBN and diamond grinding are more compressive in contrast to aluminum oxide being more toward tensile.

Even though diamond grinding generated the highest beneficial compressive stress CBN grinding is, with few exceptions, far more cost effective in grinding ferrous metals. Diamond wears very rapidly when grinding ferrous metals as a result of the diffusion of carbon into the workpiece.

Further fatigue testing as reported by Navarro confirmed these results.(9) Under normal grinding conditions CBN and diamond produced compressive residual stress while aluminum oxide produced tensile stresses (figure 8). In addition to the residual stress measurements, the samples were fatigue tested to confirm that fatigue strength directly related to the residual stress pattern from grinding (figure 9). The results again demonstrated the benefits of CBN grinding over aluminum oxide grinding.

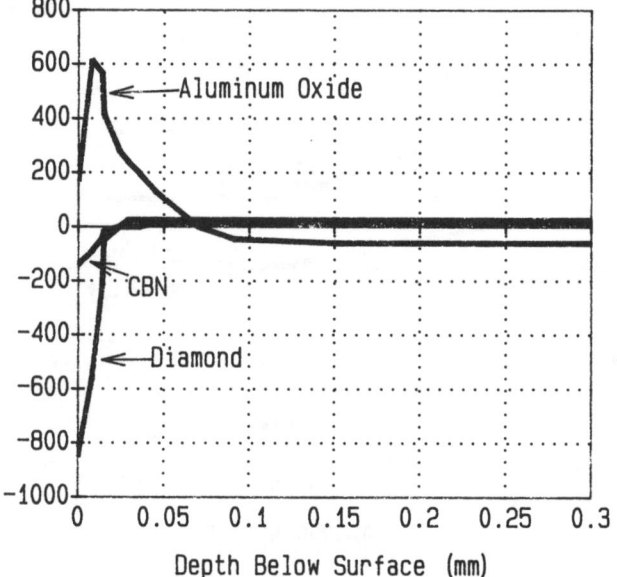

Figure 8. Resulting residual stress from identical surface grinding conditions for three different abrasives.

More recent studies in both Europe and Japan have shown that over a wide range of normal grinding conditions, CBN produces compressive residual stress. In contrast, aluminum oxide produces tensile residual stress except in the cases of using more costly LSG. (10, 11, 12, 13).

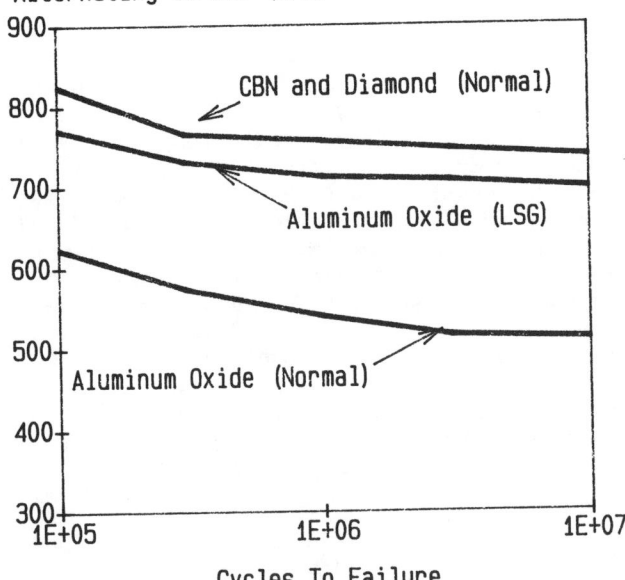

Figure 9. Resulting fatigue strength comparing aluminum oxide, normal and LSG, to CBN and diamond (same conditions as aluminum oxide normal)

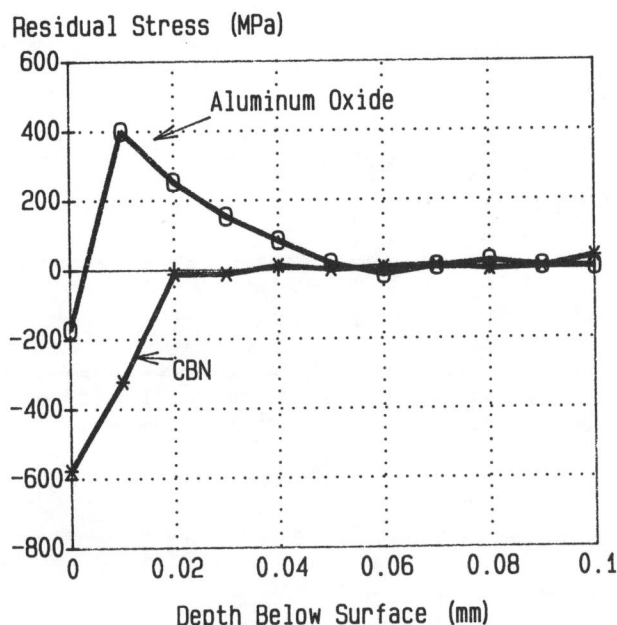

Figure 10. Effect of abrasive on residual stress generated under identical grinding conditions.

K. Yokogawa and M. Yokogawa, leading CBN grinding researchers in Japan, have been conducting many studies in this area. Their work has further clarified the phenomenon of the generation of beneficial compressive residual stresses with CBN grinding and the undesirable tensile residual stresses generally found with aluminum oxide grinding.(14) Figure 10 illustrates an example of their findings showing the compressive residual stress with CBN and tensile stress with aluminum oxide.

INFLUENCE OF WHEEL SURFACE CONDITION - An investigation of the influence of CBN wheel surface conditioning on residual stress has been conducted at General Electric Application Technology Operations.(15) A workpiece of AISI 4340, quenched and tempered HRc 50, was ground using a coarse grit (80/100), 200 concentration vitreous bonded BORAZON CBN wheel in a cylindrical plunge grinding mode. Using the edge of a special diamond rotary cup truing device, the wheel was conditioned to adjust the wheel's topography for three different grinding finishes: coarse, medium, and fine (1.0, .5, .2 um Ra). The three conditions were generated by adjusting the infeed and traverse rates of the rotary truer.

At each of the three wheel conditions, the workpiece was ground over a range of material removal rates. The residual stresses were measured using the X-ray diffraction technique. Measurements were made both parallel and perpendicular to the direction of grind. Figure 11 shows that compressive stresses in the order of 700 MPa were generated at all wheel surface conditions and material removal rates evaluated.

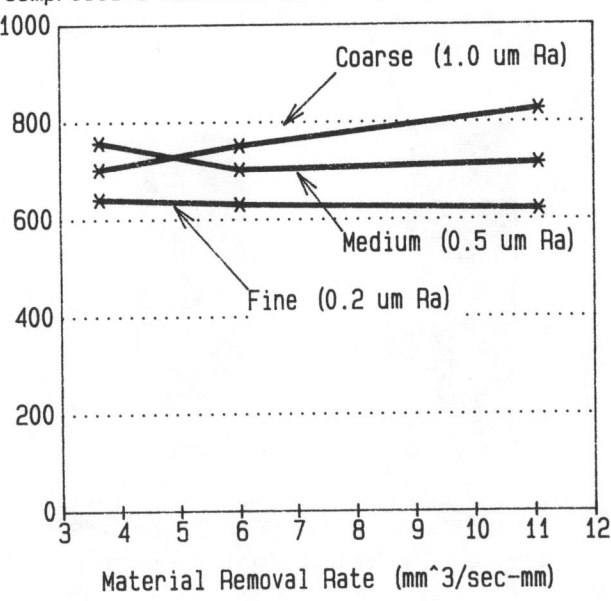

Figure 11. Effect of wheel condition and MRR on surface residual stress measured perpendicular to grind direction.

It was also observed that in all cases the compressive residual stress measured on the surface was greatest in the direction perpendicular to the grind as shown in figures 12 and 13.

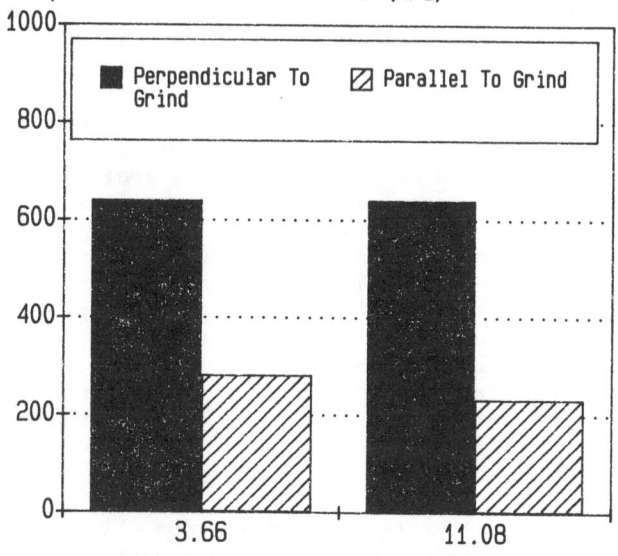

Figure 12. Residual stress measured parallel and perpendicular to the direction of grind on surfaces ground with a vitrified CBN wheel conditioned for a fine finish (0.2 um Ra)

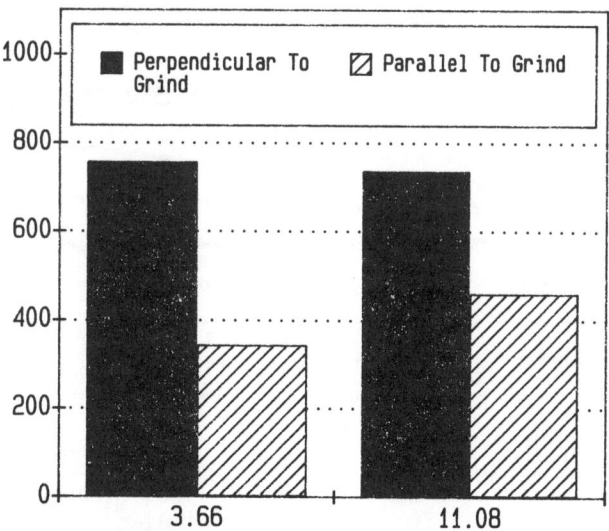

Figure 13. Residual stress measured parallel and perpendicular to the direction of grind on surface ground with vitrified CBN wheel conditioned for a coarse finish. (1.0 um Ra)

WHY CBN GRINDS COOLER

The traditional theory given for the cooler grinding characteristics of CBN wheels is that the CBN grains remain much sharper than aluminum oxide during the grinding process. An analysis of the thermal properties of CBN indicate that these properties also have a significant role in reducing the temperatures developed on the ground surface. (15, 16)

A comparison of the hardness and relevant thermal properties for CBN, aluminum oxide, diamond, tool steel, and copper are shown in figures 14 through 18. (17, 18, 19, 20)

The data clearly reveals that the thermal conductivity (the ability of a material to transfer heat) of both CBN and diamond is much greater than aluminum oxide. Diamond and CBN are the two most thermally conductive materials known to man. The thermal diffusivity property of CBN and diamond is also much greater than aluminum oxide. Thermal diffusivity of a material, k/pc_p, is the ratio of the thermal conductivity, k, to the thermal capacitance, pc_p. The larger the value, the more effective the material is in transferring energy by conduction than it is in storing it.

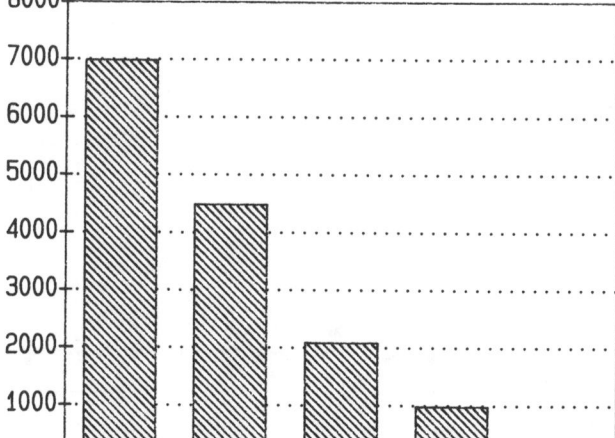

Figure 14. Comparison of hardnesses for selected materials.

THERMAL MODEL OF COOLING EFFECT OF CBN - An illustration of the dramatic effect which the thermal properties of CBN and diamond have on extracting heat from the grinding interface can be shown by a thermal model of the process.(15) A finite, elemental analysis model was derived for the cooling effect of one abrasive grain during the removal of one chip of material (figure 19). In this model, an individual grain in a 150 mm wheel is operating at 28 m/sec. At a downfeed of .025 mm, the abrasive grain will be in contact with the surface for approximately 71 micro seconds.

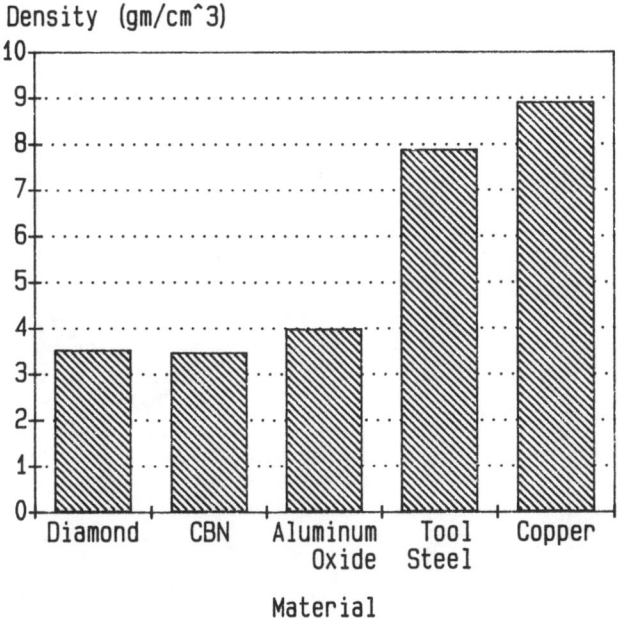

Figure 15. Density values for selected materials.

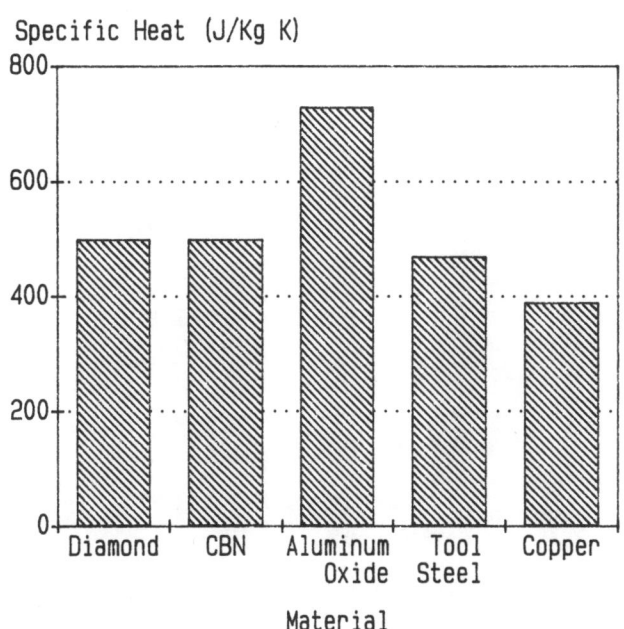

Figure 17. Specific heat values for selected materials.

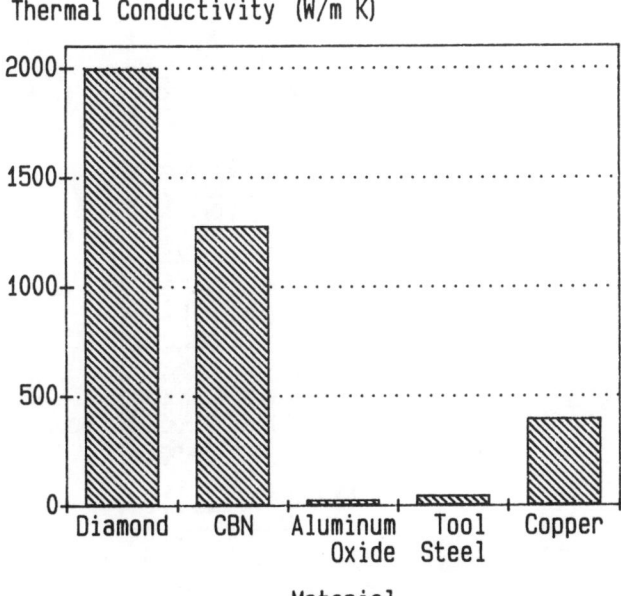

Figure 16. Thermal conductivity values for selected materials.

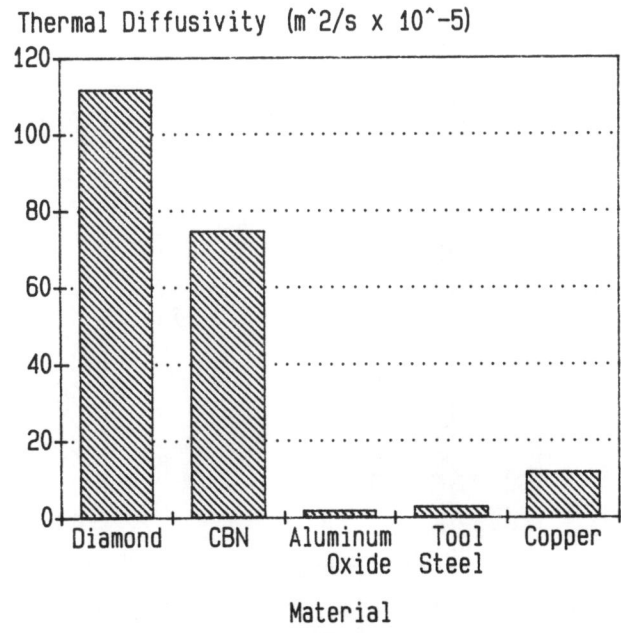

Figure 18. Thermal diffusivity (k/pc) values for selected materials.

The results of the analysis are illustrated in figure 20. The CBN and diamond abrasive are shown to have three times the effect of the aluminum oxide abrasive in reducing the surface temperature. Clearly the thermal properties of CBN and diamond allow the abrasive to behave as heat sinks in the grinding zone.

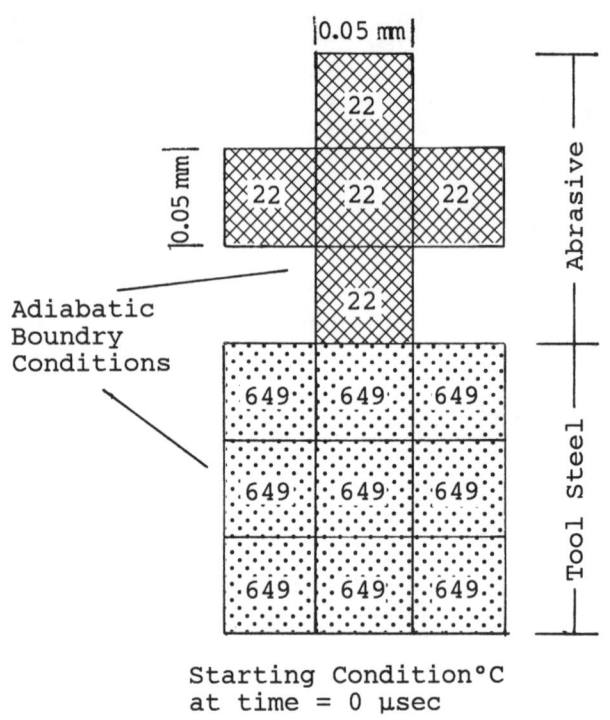

Figure 19. Finite elemental analysis model of heat transfer between workpiece and abrasive.

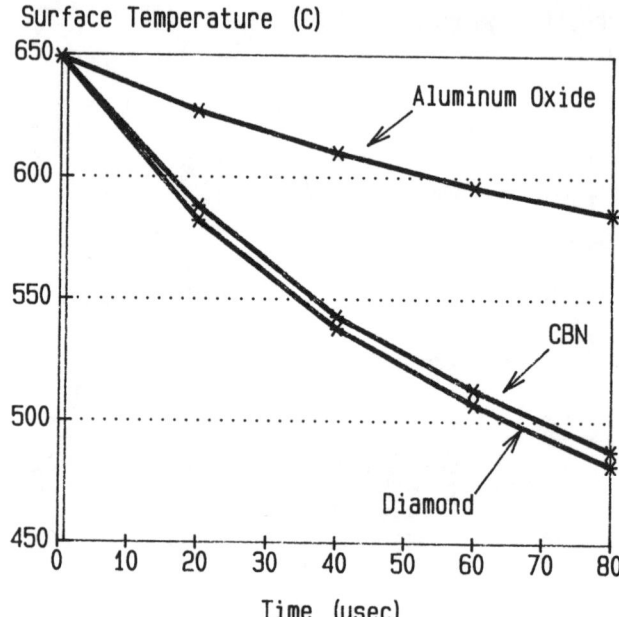

Figure 20. Effect of thermal properties of different abrasives on the surface cooling effect during the formation of a chip.

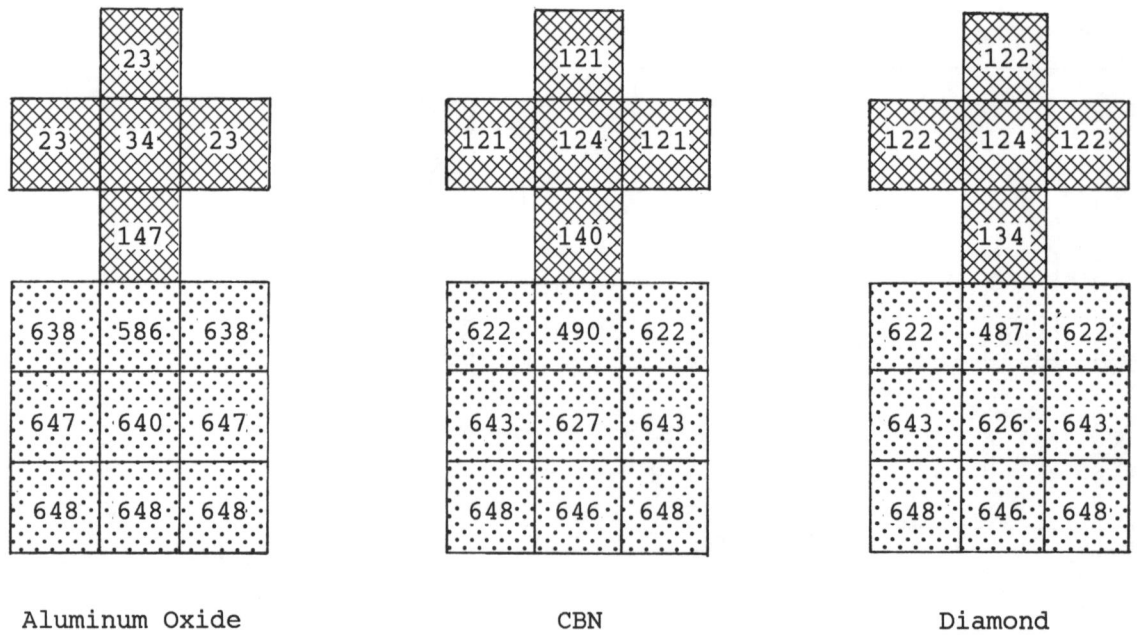

Figure 21. Comparison of temperature distribution after 80 usec of contact between workpiece and abrasive for three types of abrasive.

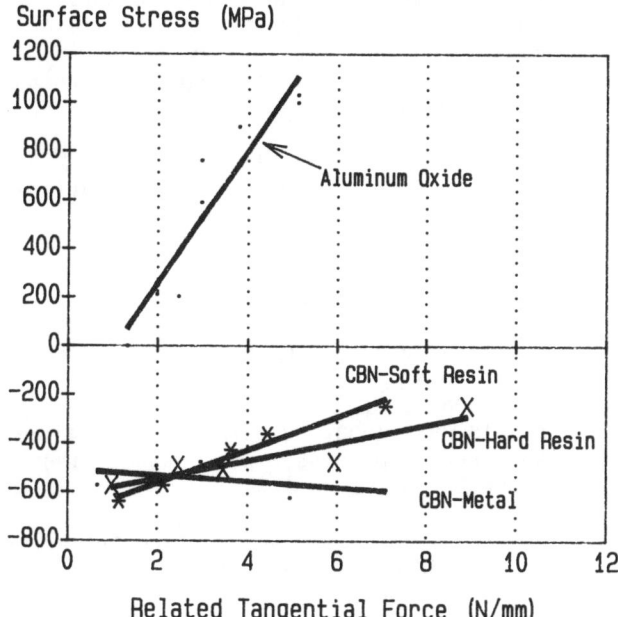

Figure 22. Surface residual stress as a function of tangential grinding force (total energy generated at grinding zone) for aluminum oxide and different CBN bonds in surface grinding bearing steel 100Cr6.

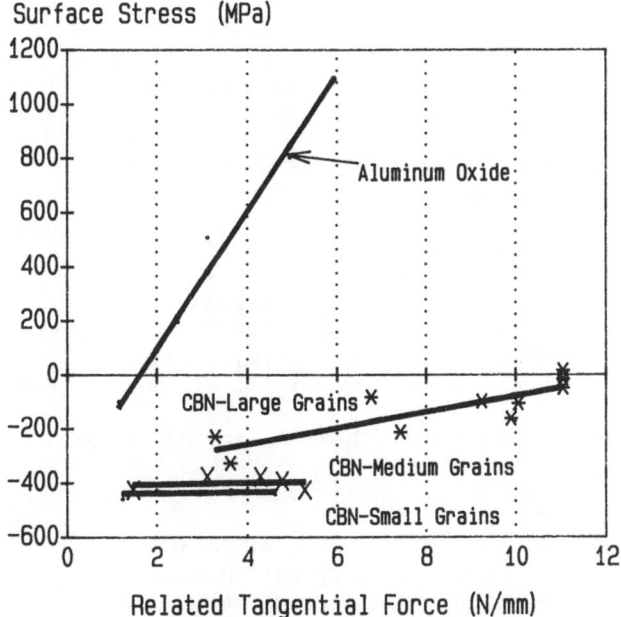

Figure 23. Surface residual stress as a function of tangential grinding force (total energy generated at grinding zone) for aluminum oxide and different CBN mesh sizes in ID grinding bearing steel 100Cr6.

An individual abrasive grain at the above condition is out of contact with the workpiece surface approximately 99.7 percent of the time. It is during this period in wet grinding that the energy absorbed by the grain during its short contact period is transferred to the coolant or bond. This returns the abrasive to the near ambient condition ready to absorb more energy during its next pass.

The remainder of the analysis indicates the temperature distribution in the grain itself. The results are shown in figure 21. Note that both the CBN and diamond grains have been heated throughout due to their extraordinary thermal diffusivity properties. In contrast, the region of the aluminum oxide grain away from the contact point has changed little in temperature over the same period.

Further evidence of the importance of the thermal properties of CBN in reducing grinding temperature and thus avoiding tensile residual stresses was demonstrated by Tonshoff.(22) Comparisons were made of Aluminum Oxide wheels to various types of CBN wheels in both ID and surface grinding modes. The grinding conditions were varied to achieve different levels of total energy being generated at the grinding zone (related tangential force). The workpieces were then evaluated for residual stress. It can be reasoned that the level of residual stress is directly related to the amount of energy going into the workpiece from the grinding process. A comparison is made between the wheels operating at the same tangential force (total amount of energy being generated) to the resulting residual stress in the work piece. At the same tangential force the more compressive the residual stress is the lower the percent to total energy going into the workpiece. As shown in figures 22 and 23 the results indicate that aluminum oxide has a much greater percent of energy going into the workpiece (higher grinding zone temperature) then CBN.

UTILIZING THE BENEFITS OF CBN

There is growing activity in utilizing the surface integrity benefits obtained with CBN grinding. Gear grinding is one area where the benefits of CBN on surface integrity are being actively developed. The Gleason Works have shown that gears ground with CBN have a dramatic increase in fatigue life especially when the critical root area of the gear is ground with CBN. (23,24) Not only is the fatigue life improved over conventionally processed gears (figure 24) but the dimensional accuracy of the gear was also improved by using the CBN grinding process.

The gear grinding industry is undergoing a revolution in the development of new machines to take advantages of the benefits derived from CBN. The emerging trend is to rough shape, heat treat, then finish grind the gear with

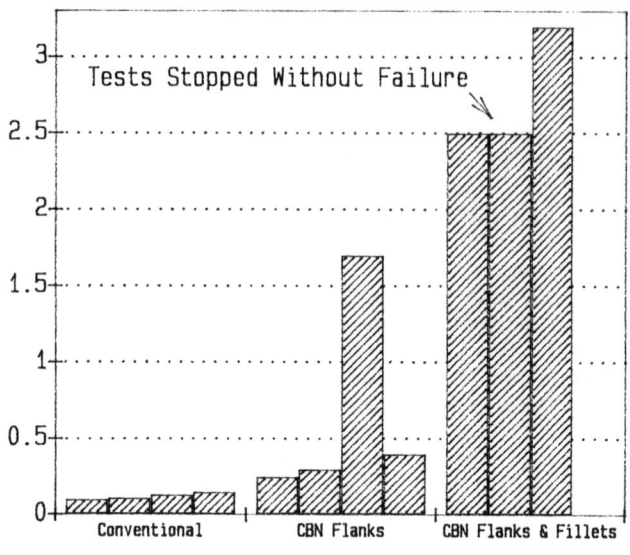

Figure 24. Fatigue life comparison of CBN ground vs conventionally processed gears.

CBN, generating both improved dimensional accuracy and surface integrity. CBN gear grinding also fits in well with the emerging technology of near net forging of gears where CBN will be used for finish grinding of the gears produced by this process. The results of these activities will be to reduce overall gear costs while improving gear performance.

The awareness on the part of designers and manufacturing engineers of the improved surface integrity with CBN grinding is still growing. With this increasing awareness more and more grinding applications will be developed to more fully utilize the benefits of CBN to improve the performance of ground components.

REFERENCES

1. R.L. Mahar, "Progress In CBN Production Grinding", Proceedings of Superabrasives' 85, April 1985, Chicago, Illinois.

2. H. Meyer, F. Klocke, J. Sauren, "CBN - New Developments In Europe", 1986 Proceedings, 24th Abrasive Engineering Society Conference, April 1986, Cincinnati, OH.

3. W.P. Koster, "Observations On Surface Residual Stress vs. Fatigue Strength", Bulletin 677-1, Metcut Research Associates Inc., Cincinnati, OH, 1977.

4. Metals Handbook, Vol. 2, 8th Edition, American Society For Metals, 1964.

5. M. Field, J.F. Kahles, "The Influence Of Untempered And Overtempered Martensite Produced During Grinding On The Surface Integrity Of ANSI 4340, HRC, Steel", Milton C. Shaw Grinding Symposium, American Society of Mechanical Engineers, Miami Beach, Florida, 1985.

6. M. Field et al, "Machining Of High Strength Steels With Emphasis On Surface Integrity", Air Force Machinability Data Center, Data Publication Number AFMDC 70-1.

7. N. Zlatin, M. Field, W. P. Koster, et al, "Surface Integrity Of Machined Structural Components", Interim Report IR-70808(III). Contract No. F33615-68-C-1003. Cincinnati, OH, Metcut Research Associates, Inc., June 1, 1968 - September 30, 1968.

8. N.P. Navarro, "Improved Surface Integrity With BORAZON CBN", Society of Manufacturing Engineers, MR71-803.

9. N.P. Navarro, "The Technical and Economic Aspects of Grinding Steel With BORAZON Type II and Diamond", Society of Manufacturing Engineers, MR70-198.

10. L.O. Chukwudebe, "Grinding Superalloys With BORAZON CBN", Proceedings DWMI, November 1975, Chicago, Illinois.

11. H.K. Tonshoff, "Influence of the Abrasive on Fatigue in Precision Grinding", Milton C. Shaw Grinding Synposium, American Society of Mechanical Engineers, Miami Beach, Florida, 1985.

12. P.G. Althaus, "Residual Stresses In Internal Grinding", IDR, 3/85.

13. T. Matsuo, H. Shibahara, "Effects of Grinding Parameters on Distortion of Surface Ground Steel Strips, Paper No. MR84-536, Society of Manufacturing Engineers, International Grinding Conference, Lake Geneva, Wisconsin, 1984.

14. K. Yokogawa, "Grindability Of CBN Grinding Wheels", Proceedings JSPE Spring Conference, March 1987.

15. G. Johnson, "Beneficial Compressive Residual Stress Resulting From CBN Grinding", SME Second International Grinding Conference, June 1986.

16. S. Ramanath, M. C. Shaw, "Role of Abrasive and Workpiece Properties Relative To Surface Temperature and Residual Stress", Proceedings 14th North American Manufacturing Research Conference Proceedings, May 1986.

17. E. Ratterman, G. Johnson, "Review of CBN Grinding Influence on Surface Integrity", Milton C. Shaw Grinding Symposium, American Society of Mechanical Engineers, Miami Beach, Florida, 1985.

18. R.M. Chernko, H.M. Strong, "Physical Properties of Diamond", General Electric Company Research And Development Technical Report 75CRD089, October 1975.

19. R.C. DeVries, "Cubic Boron Nitride: Handbook of Properties", General Electric Company Research And Development Technical Report 72CRD178, June 1972

20. Eckert, Drake, "Analysis of Heat and Mass Transfer", McGraw-Hill Inc., 1972.

21. "American Institute Of Physics Handbook", McGraw-Hill Inc., 1972.

22. H. Tonshoff, E. Brinksmeier, H. Choi, "Abrasives And Their Influence On Forces, Temperatures, And Surface", SME Second International Grinding Conference, June 1986, MR86-626.

23. H.P. Dodd, K.V. Kumar, "Technological Fundamentals of CBN Bevel Gear Finish Grinding", Society of Manufacturing Engineers, Superabrasives '85, Chicago, Illinois, April, 1985.

24. G. Kimmet, "CBN Finish Grinding Of Hardened Spiral Bevel And Hypoid Gears", Proceedings Fall Technical Meeting AGMA, October 1984, 84FTM6.

CORRELATION OF RESIDUAL STRESS TO BEARING CONDITION

Gary R. Kuhlman
NAS — North Island
San Diego, California USA

Beth S. Pardue
Technology for Energy Corporation
Knoxville, Tennessee USA

ABSTRACT

To determine whether there is a correlation between residual stress and bearing condition in reworked bearings, both properly ground and abusively ground bearings were analyzed. It was concluded that properly ground bearings could be distinguished from abusively ground bearings by their high compressive surface stresses. The abusively ground bearings had both compressive and tensile stresses, but the compressive stresses in the abusively ground bearing raceways were consistently lower than in the properly ground raceways. It was also determined as a result of the analyses that residual stresses, line broadening, and the linearity of d-spacing versus $\sin^2\psi$ plots are directly related to the grinding parameters used on the reworked bearing raceways. These factors can be used in process control as well as to aid in the selection of properly ground bearings.

BEARING RINGS from a high-performance jet engine were used to study the effects of regrinding on residual stress. Specifically, residual stresses (strains) were measured by x-ray diffraction on properly ground and abusively ground raceways made of M50 steel to determine if x-ray diffraction stress analysis could be used as an inspection tool to separate properly ground from abusively ground bearings.

HISTORY

Engines that come to the rework facility are disassembled and the components are carefully inspected. Bearings are nondestructively examined and are checked for correct dimensional tolerances. If the bearing passes inspection, small discontinuities on the surface may be removed by grinding or honing. The grinding operation may remove up to 0.005 to 0.008 cm (0.002 to 0.003 in) of material while the honing operation may remove up to 0.0008 cm (0.0003 in).

Grinding is done on a 40 to 60 grit Norton grinding wheel with about 0.00018 cm (0.00005 in) of material removed per pass. Standard grinding practice is employed; but, because many factors affect the quality of the grinding (wheel speed, abrasive media, coolant, coolant flow, coolant temperature, feed rates, etc.), variations in the reground part may occur.[1] For this study, three sets of bearing rings (Figure 1) were used. In each set, a bearing raceway that was carefully ground under controlled conditions (good) and a bearing raceway that had been intentionally abusively ground (bad) were supplied.

MEASUREMENTS

A TEC Model 1610 Portable Stress Analysis System which uses the $\sin^2\psi$ (multiple tilt) technique was used in performing these measurements. Both positive and negative psi angles were used to check for shear stresses in the raceways.[2] Measurements were first made on the surface at the point where the rolling elements contact the inner ring raceways. Residual stresses at various depths into the surface of one good inner ring raceway were then measured using standard depth profiling techniques. A chemical etchant, Tarasov's etch,* was used, not only to remove the surface layers, but also to show the areas of martensite versus over-tempered martensite. This etchant was also used on the bad bearing for the latter reason. Previous work performed by S. W. Shin and G. H. Walter[3] used this etchant on bearing rings to show clearly the areas of grinding burn. They estimated that this etchant removed less than 0.003 cm (0.001 in) of material for every fifteen-second etch cycle. The geometry of the inner ring raceways (Figure 2) made it impractical to determine the amount of material removed.

---

*4% nitric acid in water for 15 seconds, hot water rinse, 4% hydrochloric acid in acetone for 15 seconds, water rinse, air blast dried.

The abusively ground (burned) raceways had periodic axial stains that were assumed to be related to areas of martensite alternated with areas of over-tempered martensite. A very small collimated beam was used to measure the stresses associated with each area. After completing the surface measurements, one of the burned raceways was etched to reveal the over-tempered martensite. A thin, light-colored circumferential strip at the bearing contact surface was accentuated by the etchant. Axial striations similar to the stains found before etching were visible within the circumferential strip.

The good rings exhibited uniform metallic coloration. Measurements were made on the surface of each good raceway, and then one of these raceways was selected for etching and depth profiling. The Tarasov's etch uniformly blackened the raceway surface in contrast to the light and dark regions found on the burned raceway.

Additional measurements were made on the mating surface between the split inner ring sections of the burned bearing to compare this surface with the raceway. This flat surface could be measured to help determine if the complex geometry of the raceway affected the d-spacing versus $\sin^2\psi$ plots.

RESULTS AND DISCUSSION

Surface stresses in the raceways of the inner rings from the good bearings were more compressive than those from the bad bearings. Results of the x-ray diffraction stress analysis are given in Table 1 and Figure 3. For one set of bearings, the good bearing had uniform compressive stresses in both the axial and circumferential directions, -774 to -887 MPa (-112 to -129 ksi). The bad bearing inner ring raceway of this set had all tensile stresses +476 to +777 MPa (+69 to +113 ksi). In the second set of bearings, stresses in the bad bearing inner ring raceway ranged from +30 to -461 MPa (+4 to -67 ksi), while those in the good bearing of the same set were -982 to -1172 MPa (-142 to -170 ksi). Similarly, in the third set of bearings, stresses in the bad bearing were -299 to +292 MPa (-43 and +42 ksi) compared to -966 and -1044 MPa (-140 and -152 ksi) for the good raceway. The compressive stress in the bad ring was measured near the edge of the raceway.

A consistent result seen when comparing the inner ring raceways of the properly ground bearings to the abusively ground bearings was the difference in the diffraction peak full width at half the maximum intensity (FWHM). The FWHM is an indication of the amount of cold working in the surface of the material.[4] From these measurements, the FWHM was consistently lower (less cold working) in the properly ground bearings than in the abusively ground bearings. The FWHM ranged from 4.6 to 4.8° 2θ for the good bearings and from 5.1 to 5.8° 2θ for the bad bearings, indicating a higher amount of cold working in the abusively ground bearings.

Another noticeable difference between the measurements made on the good and bad inner ring raceways was the d-spacing versus $\sin^2\psi$ plots. The plots for the good bearings were generally linear (Figure 4) while those for the bad bearings were nonlinear (Figures 5 and 6). The nonlinearity may be a result of measuring microstresses (short range stresses) rather than macrostresses (long range stresses). It is speculated that the abusive grinding process contributes to the nonlinearity of these plots.

Stress measurements made after etching differed from the surface measurements. Recall that only the first set of bearings was etched with the Tarasov's etchant. A stress measurement made in a dark axial striation in the light-colored circumferential strip of the bad raceway was -41 MPa (-6 ksi). This was the only compressive stress found in the inner ring raceway of this bearing with one exception. Measurements on the mating surface between the split inner ring sections were -1096, -1158, and -1227 MPa (-159, -168, and -178 ksi), suggesting that this surface was not representative of the raceway. Stresses in the light-colored etched areas, +372 and +510 MPa (+54 and +74 ksi), were similar to the surface stresses.

The stresses on the surface of the good bearing were consistently compressive, but this was not the case for the subsurface measurements. Etching with the Tarasov etchant for an unknown time (less than one minute) quickly discolored the surface and resulted in very low stresses, 69 to +28 MPa (-10 to +4 ksi). Considering that the etchant should only remove about 0.003 cm (0.001 in) of material, it was questioned that a stress gradient of this magnitude existed. The stress was then measured on the surface adjacent to the etched area. Once again the surface stress was compressive, -786 MPa (-114 ksi).

A series of etchings followed by stress measurements were performed in this adjacent area. This time the etching procedure was carefully controlled to insure that the etching time did not exceed ten seconds per etch. This procedure resulted in a linear decrease in the compressive stress measured for the first five etching cycles (Figure 7). The etching procedure was repeated a total of seven times until the stresses were -159 to -172 MPa (-23 to -25 ksi). A one-minute etch at a separate location resulted in a low stress of +28 MPa (+4 ksi). These results led to speculation that the etchant, at dwell times greater than ten seconds, may preferentially dissolve portions of the grain boundaries around the surface grains. The areas that had been etched longer than ten seconds at a time had a rougher texture than the other etched areas and lends credibility to this speculation. If portions of the grain boundaries were dissolved by the etchant, then the surface was left artificially stress-free since the connective material necessary to support stresses was removed. Additionally, if high stress gradients really existed, then curved dspacing versus $\sin^2$psi plots of the unetched surface would be expected.[5] The d-spacing versus $\sin^2$psi plots were linear for the unetched measurements.

A section (less than one inch) of the good bearing inner ring from the first set of bearings tested was removed for scanning electron microscopy (SEM) of the etched and unetched surfaces. The intent of this investigation was to determine if the etchant had preferentially attacked the grain boundaries resulting in an artificially induced stress-free surface. The bearing raceway geometry made the examination difficult. The 700X photographs (Figure 9) are clear enough to show that the etching did not occur only at grain boundaries. It was not possible to estimate the amount of material removed by etching. Based on the SEM and the ten-second etching cycles, it is assumed that stress gradients do exist in the raceway. The gradient cannot be quantified from these data since the amount of material removed could not be measured by means employed during this work.

The ten-second etching depth profiling was also performed on the mating surface of the bad bearing of the same set. Recall that the surface measurements on the area of the bad inner ring raceway were similar to the raceway measurements on the good bearing. The depth profile in Figure 8 follows the same trend as Figure 7. Again, this indicates that the split inner ring mating surface measurements cannot be used to distinguish good from bad bearings.

Data on raceways from a previous study[6] had suggested that shear stresses (indicated by $\sin^2$psi splits) may be related to grinding burns. The current work indicates that tensile or low compressive stresses are found on the bad bearing raceways, and high compressive stresses are found on the good raceways. The d-spacing versus $\sin^2$psi plots are linear or nonlinear and do not exhibit the $\sin^2$psi split associated with shear stresses. The collimated x-ray beam was smaller for the current study. It was concluded that geometric effects from a large collimated x-ray beam and not shear stresses resulted in the $\sin^2$psi splits seen in the previous work.

CONCLUSIONS

Abusively ground bearings can be distinguished from properly ground bearings by measuring the stresses on the inner ring raceways at the point of rolling element contact. The properly ground bearing raceways have high compressive stresses, linear d-spacing versus $\sin^2$psi plots and a smaller FWHM. The abusively ground bearing raceways have tensile or low compressive stresses, nonlinear d-spacing versus $\sin^2$psi plots and a larger FWHM. It was concluded that these differences were directly related to the increased cold working imparted to the surface during abusive grinding.

The shear stresses indicated by the $\sin^2$psi splits observed during a previous study were not found on either the good or the bad bearings. Consequently, these splits were attributed to the geometric effects from the large collimator used.

High stress gradients probably exist in the surface of the inner ring raceways. The stress gradients could not be quantified; however, the stresses decreased linearly when the sample was etched under controlled conditions.

ACKNOWLEDGEMENTS

The work described was performed under Contract N00019-85-C-0419 for Naval Air Systems Command. Joe Stewart and Sue Macdougall of Oak Ridge National Laboratory operated by Martin Marietta Energy Systems, Oak Ridge, Tennessee, performed the scanning electron microscopy. Their support was invaluable to this work and the authors gratefully acknowlege their contribution.

REFERENCES

[1] James B. Pond, "Industry Focuses on Abusive Grinding," Iron Age, May 16, 1986, pp. 41-46.

[2] H. Dolle, J. Appl. Cryst. (1979), 12, pp. 489-501.

[3] S. W. Shin and G. H. Walter, "Case Histories of Residual Stress Related Component Failures," Residual Stress for Designers and Metallurgists, American Society for Metals.

[4] Paul Prevey, Diffraction Notes, Vol. 1, No. 1, Winter 1987, Lambda Research, Inc., Cincinnati, Ohio.

[5] H. Dolle, loc. cit.

[6] R. W. Hendricks, E. B. S. Pardue, and M. V. Mathis, SBIR Phase I Final Report, Contract No. N0001984-C-0026, Technology for Energy Corporation Report R-84-019, January 1985.

Table 1

Surface Stresses on Inner Ring Raceway

Bearing	Direction	Stress, MPa	Stress, ksi	FWHM, °2θ
1st Set				
Good	Circumferential	− 825.3 ∓ 150.3	−119.7 ∓ 21.8	4.6
	Circumferential	− 887.3 ∓ 161.3	−128.7 ∓ 23.4	4.6
	Circumferential	− 773.6 ∓ 78.6	−112.2 ∓ 11.4	4.8
	Axial	− 808.1 ∓ 119.3	−117.2 ∓ 17.3	4.8
	Axial	− 787.4 ∓ 69.6	−114.2 ∓ 10.1	4.7
Bad	Circumferential	+ 482.6 ∓ 193.1	+ 70.0 ∓ 28.0	5.4
	Circumferential	+ 777.0 ∓ 223.4	+112.7 ∓ 32.4	5.3
	Axial	+ 476.4 ∓ 170.3	+ 69.1 ∓ 24.7	5.3
	Axial	+ 529.5 ∓ 160.6	+ 76.8 ∓ 23.3	5.6
2nd Set				
Good	Circumferential	− 981.8 ∓ 62.1	−142.4 ∓ 9.2	4.8
	Axial	−1172.1 ∓ 84.1	−170.0 ∓ 12.2	4.8
Bad	Circumferential	− 311.0 ∓ 115.8	− 45.1 ∓ 16.8	5.2
	Circumferential	− 313.7 ∓ 115.1	− 45.5 ∓ 16.7	5.1
	Circumferential	− 268.9 ∓ 126.2	− 39.0 ∓ 18.3	5.2
	Circumferential	+ 29.6 ∓ 160.0	+ 4.3 ∓ 23.2	5.7
	Circumferential	− 33.8 ∓ 144.1	− 4.9 ∓ 20.9	5.5
	Axial	− 461.3 ∓ 89.6	− 66.9 ∓ 13.0	5.2
	Axial	− 49.0 ∓ 122.7	− 7.1 ∓ 17.8	5.8
3rd Set				
Good	Circumferential	−1044.5 ∓ 62.1	−151.5 ∓ 9.0	4.8
	Circumferential	− 965.9 ∓ 91.0	−140.1 ∓ 13.2*	4.8
Bad	Circumferential	+ 291.6 ∓ 183.4	+ 42.3 ∓ 26.6	5.7
	Circumferential	− 299.2 ∓ 115.1	− 43.4 ∓ 16.7*	5.5

*Measurement made near edge of inner ring raceway.

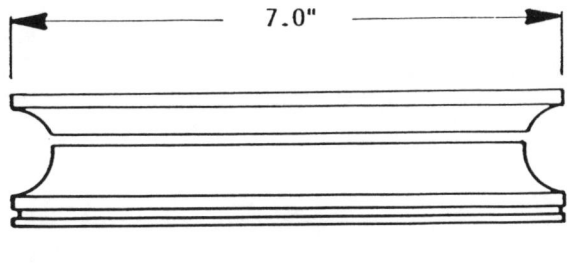

SET 1

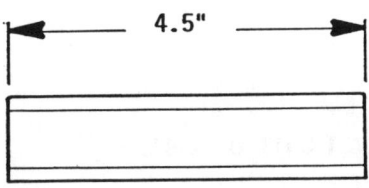

SET 2

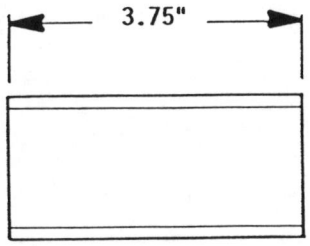

SET 3

Figure 1. Drawing of Three Bearings Used for Study.

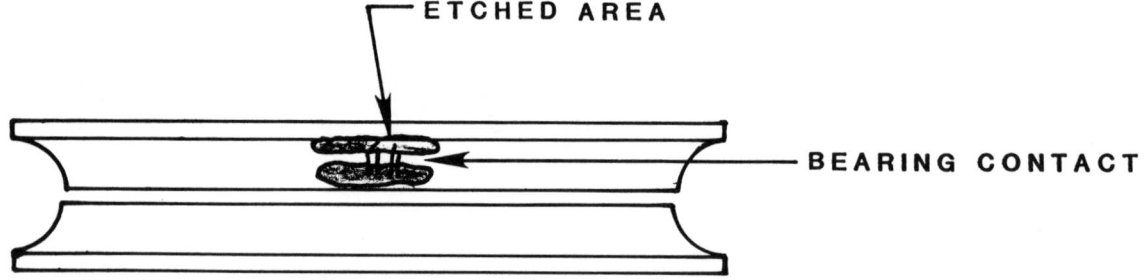

Figure 2. Geometry of the Bearing Inner Ring Raceways.

Figure 3. Results of X-Ray Diffraction Stress Analysis on Abusively and Properly Ground Bearing Raceways.

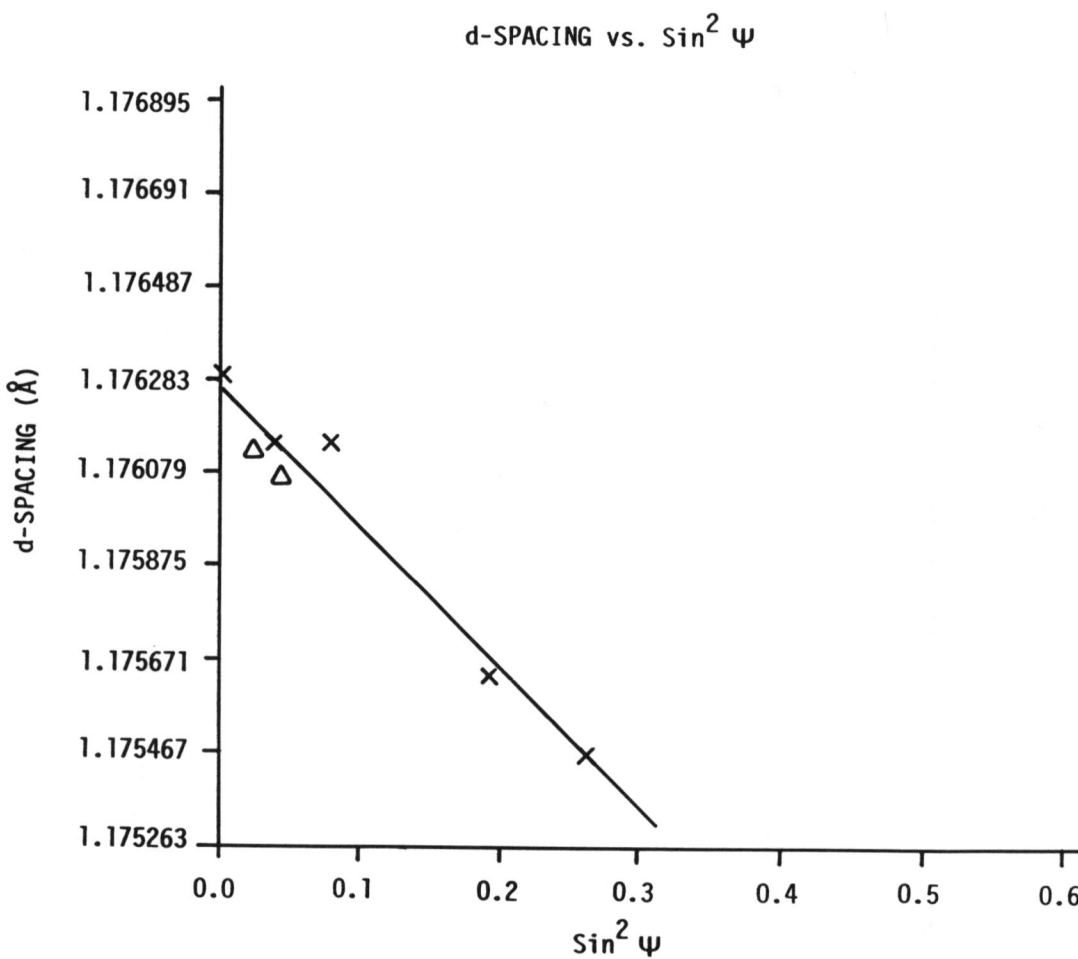

Figure 4. Linear Plot for Properly Ground Bearing.

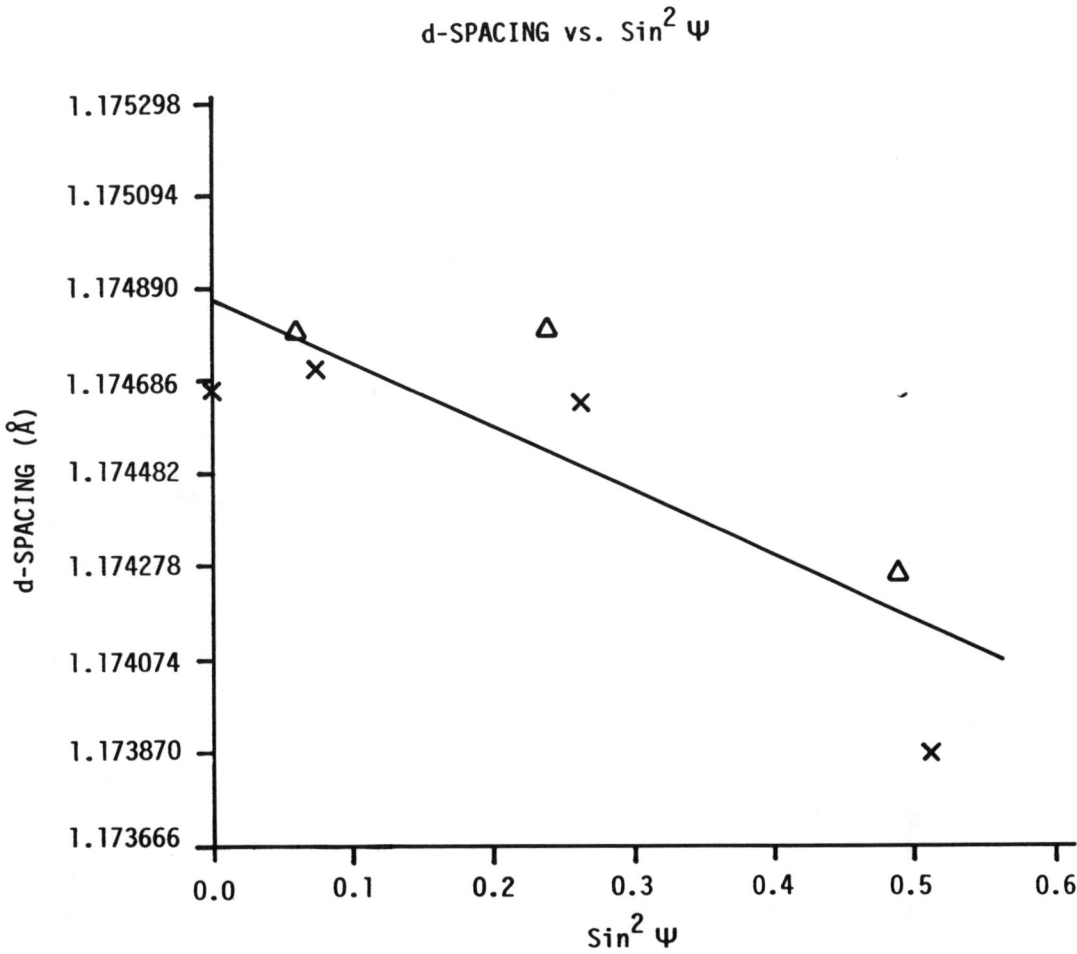

Figure 5. Nonlinear Plot for Abusively Ground Bearing (Compressive Stress).

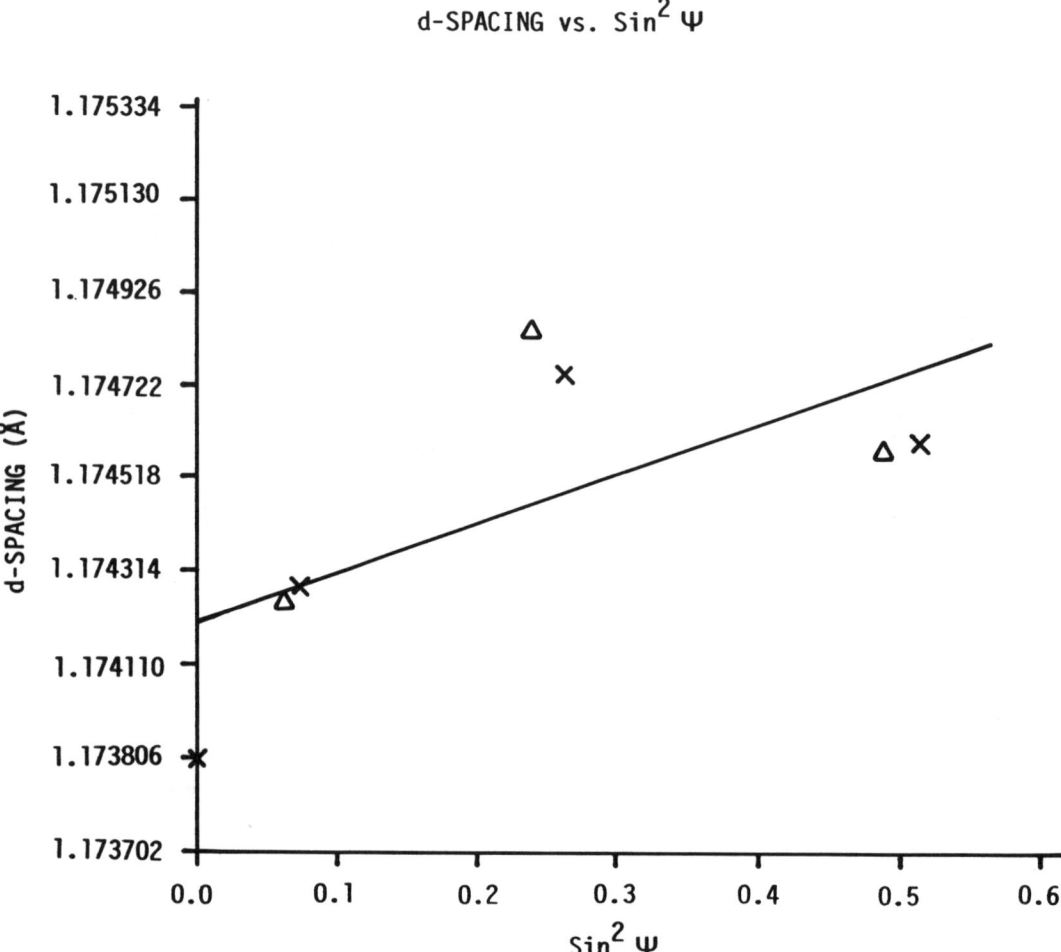

Figure 6. Nonlinear Plot for Abusively Ground Bearing (Tensile Stress).

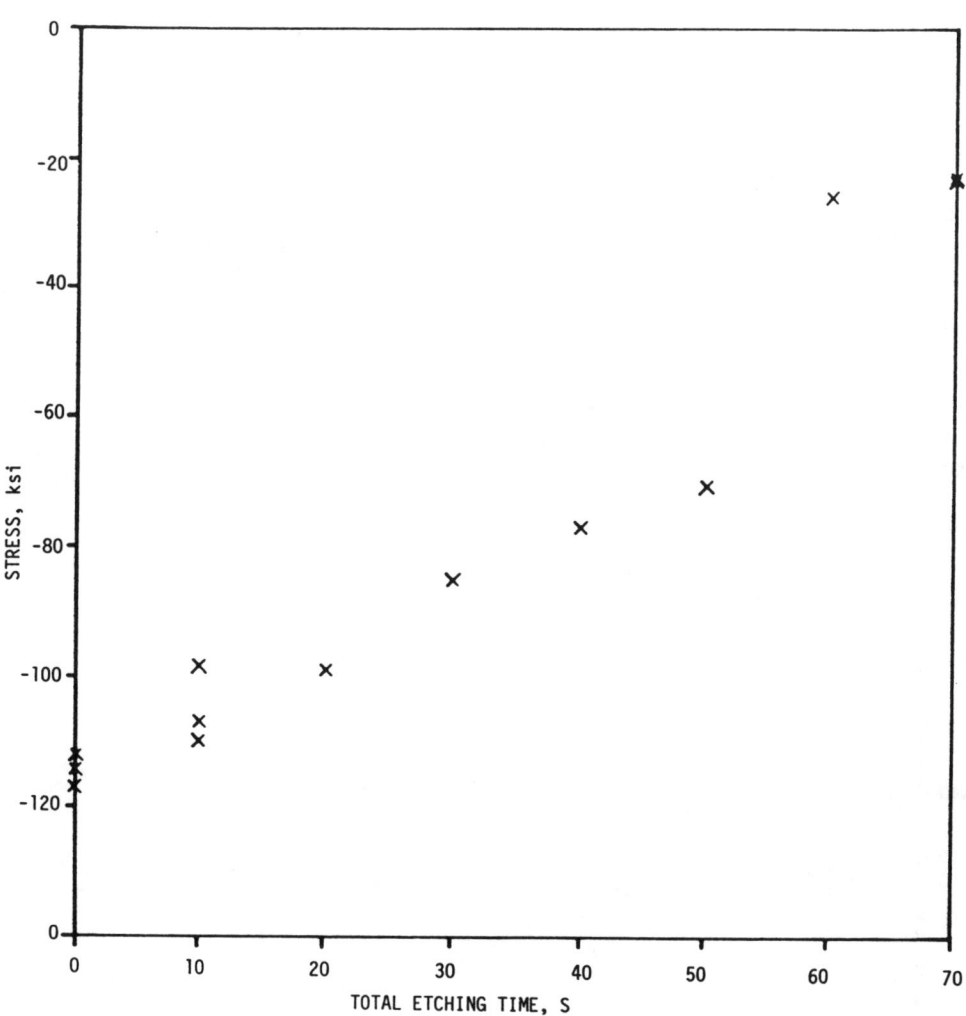

Figure 7. Residual Stress Versus Etching Time for Bearing Raceway (Good).

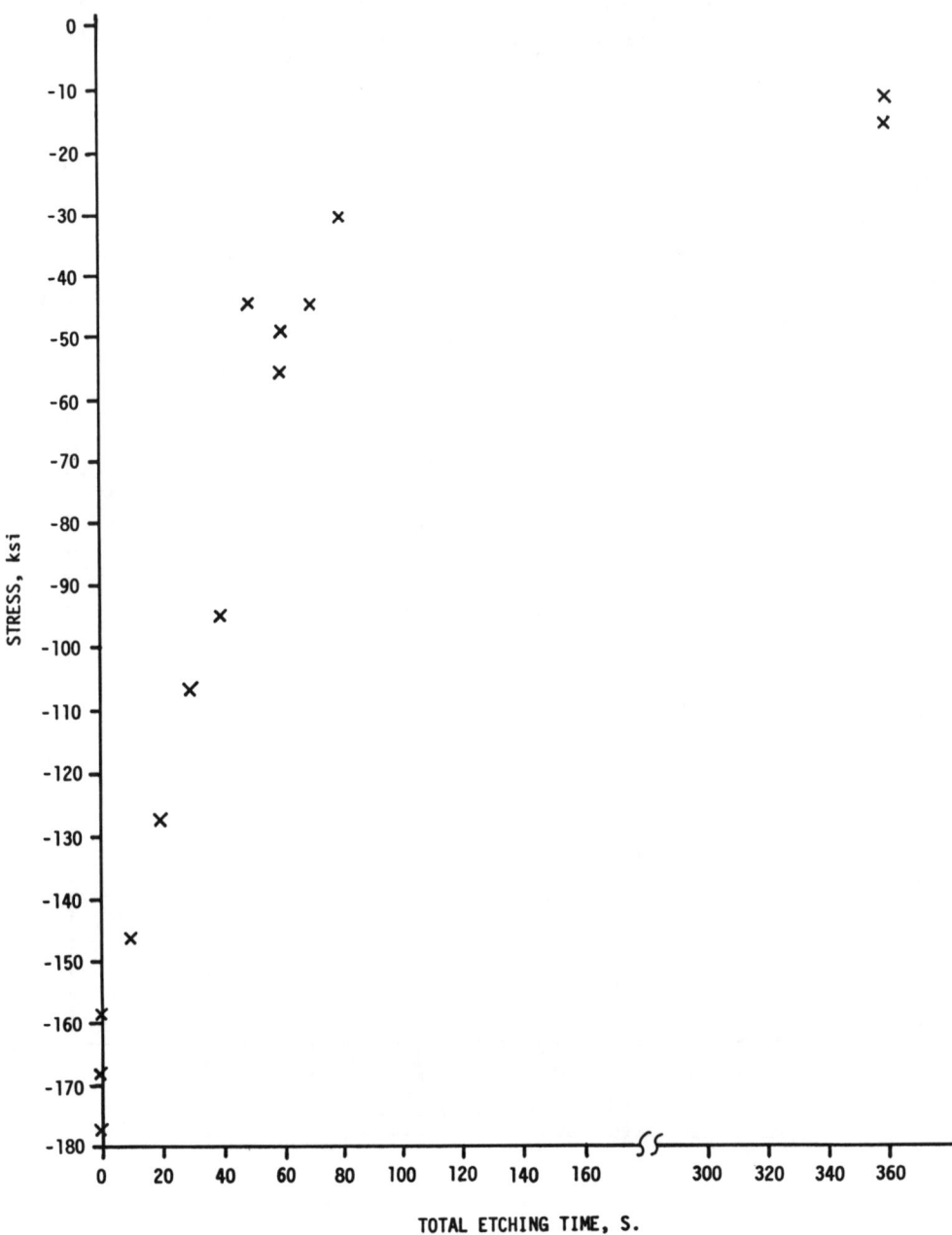

Figure 8. Residual Stresses Versus Etching Time for Inner Split Ring (Bad) Mating Surface.

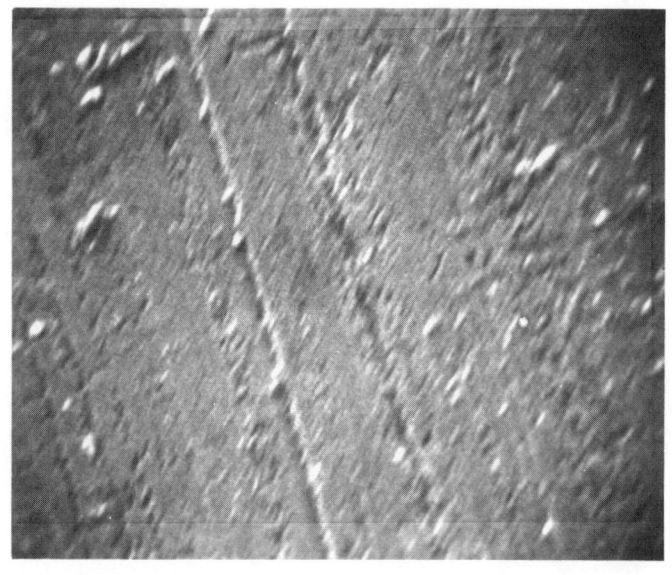

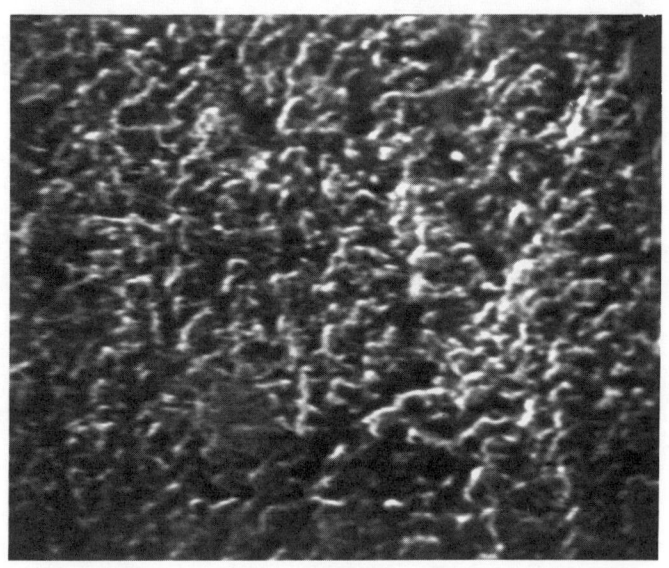

ETCHED

Figure 9. Scanning Electron Microscopy Photographs of Raceway.

RESIDUAL STRAINS IN ROLLED JOINTS

S.R. MacEwen, T.M. Holden, R.R. Hosbons, A.G. Cracknell
Atomic Energy of Canada Limited
Chalk River Nuclear Laboratories
Chalk River, Ontario, Canada

ABSTRACT

Rolled joints are used extensively by many industries to attach tubes to tube sheets or end fittings. In CANDU reactors, the zirconium alloy pressure tubes are attached to the stainless steel end fittings by such joints. The success, or failure, of a joint depends on the residual stresses produced by the rolling operation. Neutron diffraction provides the only non-destructive tool available to measure residual strains in the interior of pressure tube-end fitting components. Results for the axial and hoop residual strains in the rolled joint region of a correctly made joint, and in an incorrectly made over-rolled joint will be presented. The data will be compared to results from strain gauge measurements and will be discussed in light of the anisotropy and texture of the zirconium alloy tube.

ROLLED JOINTS ARE USED EXTENSIVELY to attach tubes to tube-sheets or endfittings, when more conventional techniques such as welding or threading are, for a variety of reasons, inappropriate. Rolled joints find a major use in heat exchangers, where they are used to attach the tubes to the tube-sheet. They are also used extensively in CANDU reactors; in a Pickering CANDU there are 780 tube to tube-sheet and 780 tube to endfitting rolled joints. All rolled joints are based on an expansion process which flares the end of the tube so that when it is constrained by the endfitting a strong, leak-tight joint results. A variety of techniques, including hydraulic forming, explosive forming and rolling have been used to make expansion joints, but rolling is by far the most commonly used technique.

The success or failure of a rolled joint depends critically on residual stresses: success if high compressive hoop stresses ensure a strong, leak-tight joint, and failure if tensile stresses resulting from the fabrication lead to stress-corrosion-cracking or, as was the case with some CANDU fuel channels, delayed hydride cracking. This presentation will describe the program carried out at CRNL to measure residual stresses in CANDU fuel channel rolled joints. Although some mention will be made to conventional techniques employing strain gauges and slitting or blind-hole drilling, the main emphasis will be on recent work using neutron diffraction to measure non-destructively the residual strains in tubes and in complete tube-endfitting assemblies.

CANDU REACTORS

Candu reactors are fuelled with natural uranium, cooled and moderated with heavy water, and use pressure tubes, rather than a single pressure vessel, to contain the primary heat transport water. Fig. 1 is a simplified schematic of the Pickering CANDU reactor, showing only three of the 390 fuel channels. Each fuel channel consists of two concentric tubes, about 6 m long. The outer is the calandria tube, whose main function is to provide a thermal barrier between the cool moderator and the hot primary coolant which flows inside the inner tube. The calandria tube has an OD of 13 cm, a wall thickness of 1.5 mm, is made from Zircaloy-2 and is attached to the tube sheets of the stainless steel end shields by rolled joints. The inner tube is the pressure tube, made from Zr-2.5% Nb with an ID of 10.3 cm and a wall thickness of 4.1 mm. The pressure tubes are attached to 403-SS endfittings, which provide for the attachment of inlet and outlet coolant feeder pipes and allow for on-power refuelling. Twelve fuel bundles lie in each of the 390 horizontal channels. During normal operation the coolant runs at a pressure of 9 MPa, giving a hoop stress of the order of 100 MPa in the pressure tube, at a temperature of about 300°C.

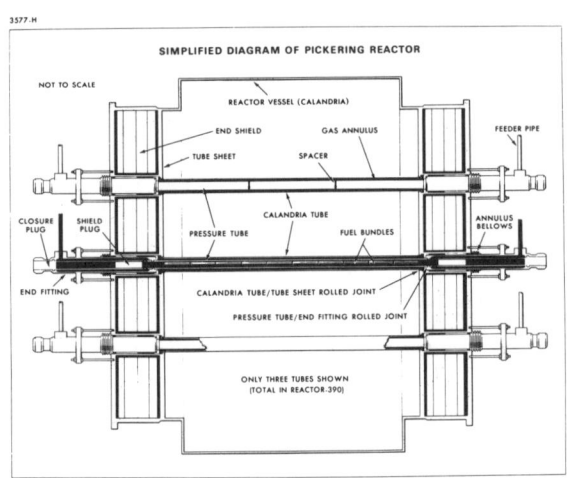

Figure 1 - Schematic of a CANDU Reactor

The pressure tubes are attached to the endfittings by rolled joints, as illustrated in fig. 2.

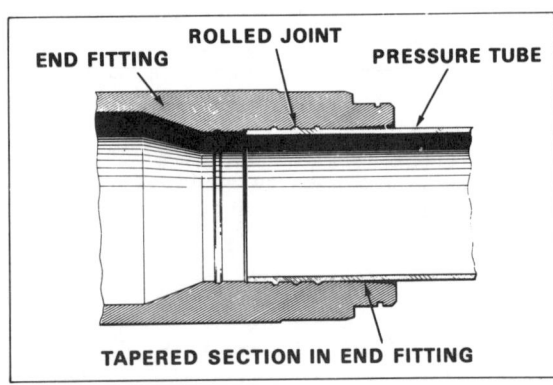

Figure 2 - Rolled Joint in CANDU Reactor

To make a joint, the tube is slid into the snout of the endfitting, which has three circumferential grooves machined into it; see fig. 3.

Figure 3 - End fitting showing grooves

To aid in the insertion of the tube into the endfitting, the inside of the inboard end is tapered by a few degrees. A rolling tool, which consists of five cylindrical rolls on a tapered mandrel, is inserted from the outboard end. Fig. 4 is a schematic illustration of the mode of operation of the rolls.

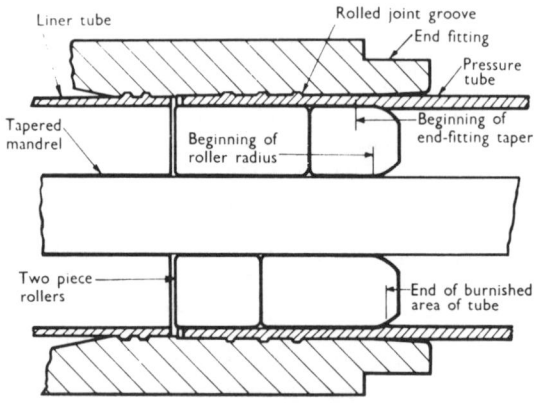

Figure 4 - Schematic of rolling procedure

The position of the roller cage, and thus the region of the tube to be rolled, is defined by a stop which comes up against the endfitting; on insertion the five rollers are compressed by the tube into the cage. The rotation of the mandrel causes an opposite rotation of the rollers. The rollers are set at a slight angle to the axis of the mandrel and as a result, during rolling, the mandrel moves inboard causing the rolls to expand and forcing the pressure tube into the three grooves of the end fitting. In a correctly rolled joint, the cage (which remains in a fixed axial position during rolling) is positioned outboard of the tapered region of the endfitting and the deformation from the rolls begins at the point where the taper in the endfitting ends. The thrust of the rollers is carried by the steel of the endfitting.

In the mid-1970's, about 70 pressure tubes had to be removed from CANDU reactors (Pickering Units 3&4) as a result of leakage near the position of the rolled joints. The pressure tubes in the leaking channels were replaced with new tubes, and the reactors were restarted with no further difficulties. The problem was traced to the residual stresses in improperly made joints. Here, due to wear of the stop that positioned the roller cage, the rolling began too far down the tube, with the result that the tube was flaired out into the tapered region of the endfitting, producing what has become known as an over-rolled joint. The nature of the residual stresses in an over-rolled joint will be discussed in a subsequent section.

EVALUATION OF ROLLED JOINTS

Because of the large number and importance of rolled joints in CANDU reactors, their strength and integrity has been the subject of ongoing investigations at Chalk River Nuclear Laboratories. Uniaxial pull tests are used to determine mechanical strength, helium leak detectors are used to test for leak-tightness, and ultra-sonics are used to check for the presence of water in the grooves of the endfitting. Residual stresses are determined destructively by a strain gauge and slitting technique, recently described by Hosbons [1]. The slitting **procedure** involves the following: The test "sample" consists of a short (50 cm) section of tube and the endfitting. Strain gauge rosettes are attached to the inside of the tube, beginning at the start of the taper of the endfitting and continuing inboard on the tube. The snout of the endfitting is then machined away, as illustrated in fig. 5, and the change in strain on the inner surface resulting from the redistribution of the residual stresses in the tube is recorded. A second string of gauges, parallel to the first but on the outside of the tube, is then installed. A small milling machine equiped with two blades separated by a spacer is then used to make two axial slits in the tube. The tongue so produced bends outwards due to the relief of residual stresses; changes in strain on both sets of gauges are measured.

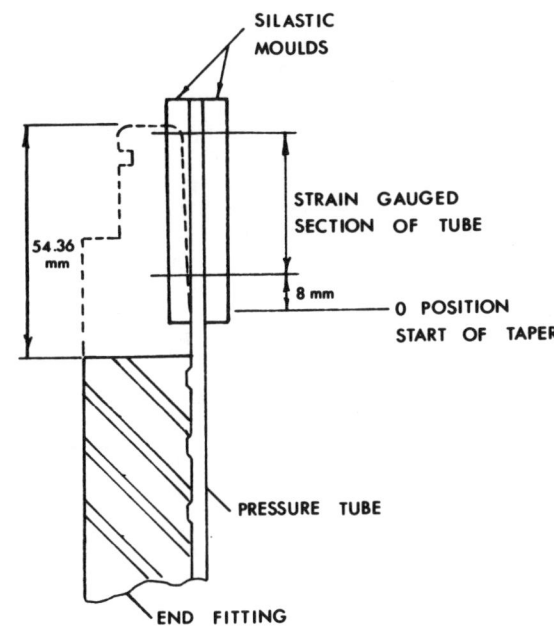

Figure 5 - Snout removal during slitting operation

The stresses are evaluated from the strains according to

$$\sigma_H = E(\varepsilon_H + \nu \varepsilon_L)/(1 - \nu^2) \qquad (1)$$

$$\sigma_L = E(\varepsilon_L + \nu \varepsilon_H)/(1 - \nu^2) \qquad (2)$$

where

σ_H = hoop stress
σ_L = longitudinal stress
ε_H = hoop strain
ε_L = longitudinal strain
E = modulus of Elasticity
ν = poisson's ratio

which assumes that the material is isotropic elastically, and that the through-wall distribution of the residual macro-stresses is linear. The first assumption is clearly incorrect, however the resultant error in the calculated stresses is likely no greater than other errors associated with the analysis. The latter assumption (linearity through-wall) has never been checked, although, as will be demonstrated in the next section, the capability now exists to do so. Figure 6 shows results, obtained using the slitting technique, for the residual stresses in an incorrectly made, over-rolled joint. Both the axial and hoop components show a peak in tensile stress located about 25 mm along the tube. From a combination of residual stress measurements such as these, metallography of the fracture surface, and an understanding of the nature of hydrogen and hydrides in Zr alloys, it was concluded that the residual tensile stresses, combined with the operating stress, led to the reorientation of hydride platelets. Some subsequently cracked, leading to a through-wall penetration and the detection of water vapour in the gas annulus between the pressure tube and calandria tube. In all cases, the reactor was shut down and the faulty tube was removed before critical crack growth was reached. It is important to note that the rolled joint itself, in all cases, was still leak-tight. The association of the cracking problem with residual stresses gave increased impetus to finding a better, more precise way to determine residual stresses, and in 1981 CRNL began what is now a most successful program: the use of neutron diffraction to measure the distributions of residual stresses in engineering components non-destructively. An over-rolled joint was the first test of our capabilities.

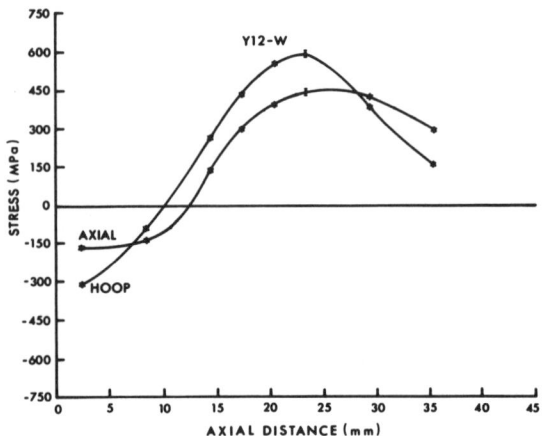

Figure 6 - Residual stresses in rolled joints measured by slitting technique

NEUTRON DIFFRACTION

The first references to the use of neutron diffraction to solve engineering problems were in 1979 by Choi et. al.[2], who used neutrons to characterize crystallographic texture. The first references on the measurement of residual strains were in 1981, by Krawitz and co-workers[3], and by Pintschovius[4]. Since then the field has expanded significantly, and now there are laboratories around the world pursuing the technique.

The scattering of thermal neutrons is analogous to X-ray scattering. When a crystalline sample is irradiated with a beam of thermal neutrons there is a general scattered background and sharp Bragg peaks. For diffraction to occur, the scattering vector Q must be parallel to the hkl-plane normal, and Bragg's law must be satisfied:

$$\lambda = 2d \sin(\theta) \qquad (3)$$

From the position of the peaks one can determine inter-planar spacings, with a precision of the order of 10^{-4} angstroms. Since residual strains cause changes in interplanar spacing, the position of Bragg peaks can be used to obtain information about macro- and grain-interaction stresses. The width of the peak gives information on particle size and micro-strains, and from the intensity of the peak one learns about crystallographic texture.

Since the interaction of a neutron is with the nucleii, and the interaction of the uncharged neutrons with the electrons is very weak, the penetration of neutrons into metals and ceramics is from 1000 to 10,000 times greater than that of X-rays. Thus while X-rays probe the first few microns of the surface of a sample, neutrons provide information that is characteristic of the sub-surface, bulk volume of the component. Instead of thinking of penetration in microns, one can think in terms of cm.

The system in use at CRNL to measure residual strains by neutron diffraction is shown schematically in fig. 7.

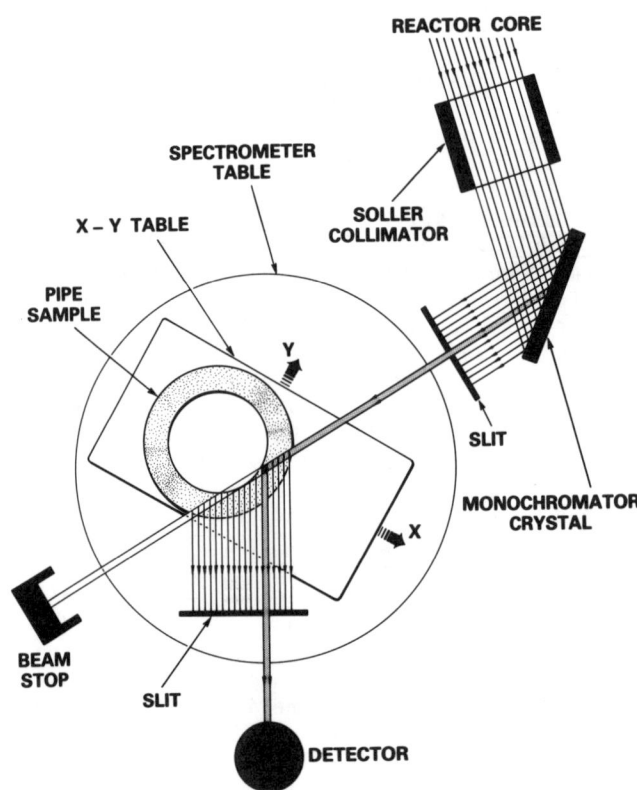

Figure 7 - Schematic of L3 spectrometer

A beam of polychromatic (white) neutrons is extracted from the reactor core. A monochromatic beam, 5 cm × 5 cm in cross-section with a wavelength of about 2Å, is obtained by the use of a Ge monochromating crystal. This is the "large-beam" that is used primarily for measurements of crystallographic texture and grain-interaction strains in uniformly deformed samples. In order to map out distributions of residual stresses, slits (typically 1 mm wide and 5 to 10 mm high) in Cd masks are used to define a narrow beam of neutrons which usually passes through the sample and reaches the beam stop. If one were to collect all of the diffracted beam, the result would give the average over the total beam path, but this is not what is required. Therefore a slit is placed in the diffracted beam, thereby defining a small, diamond-shaped prism from which all of the diffracted information originates. The diffracting volume is ideally around 10 mm³, however experiments have been performed at CRNL with volumes as small as 1 mm³. Gradients in residual strain or texture are then determined by translating the sample on computer-controlled X-Y-Z tables. The size and shape of the slits can be optimized to take advantage of the symmetry of the residual strain state. With the set-up shown in fig. 7, the through-wall distribution of the radial component of residual strain could be measured.

changing the interplanar spacing, a residual stress will alter the condition for Bragg reflection, and produce a shift in the position of the peak. If a residual strain tensor, expressed in sample coordinates, exists at a point in the sample, the strain measured in the direction of the (hkl) plane normal is

$$\varepsilon_l = l_i . l_j . \varepsilon_{ij} \qquad (4)$$

where l_i are the direction cosines of the <hkl> direction.

A particular benefit arising from the penetration power of the neutron is the ability, for most sample geometries, to measure ε_l directly by positioning the sample so that the l-direction (and therefore the hkl plane normal) bisects the incident and diffracted beams. Clearly, six measurements must be made in order to characterize the complete strain tensor. Once ε_{ij} is known, ε_{ij} can be <u>calculated</u> if certain assumptions are made concerning elastic constants.

ANISOTROPY AND TEXTURE IN ZR ALLOYS

Zirconium alloys are used extensively in the nuclear industry because of their low fast neutron capture cross-section, and in the chemical industry because of their good corrosion resistance. Zr alloys are hcp at room temperature, and are extremely anisotropic. The individual crystals are soft with respect to plastic deformation in any direction in the basal plane, and hard with respect to plastic deformation normal to the basal plane. As a result of this plastic anisotropy, Zr alloys develop extremely sharp crystallographic textures during fabrication. In general, the prism poles tend to align in the working direction, leaving the basal poles perpendicular to it. Concurrent with the development of texture is the evolution of residual grain-interaction stresses. These residual stresses are clearly manifested in the mechanical properties of Zr alloys by the Bauschinger effect and the strength differential.

In order to understand the nature of the residual stresses in over-rolled joints, one must be cognizant of plastic anisotropy (as just described) and texture. The texture of pressure tubes results primarily from the extrusion process, and is extremely sharp. Figure 8 shows the basal pole figure; the <c> directions of the individual grains are heavily concentrated in the hoop direction. There are a few basal poles in the radial direction, but none at all in the axial direction of the tube. The prism (10$\bar{1}$0) pole figure is even sharper. Every grain in the tube has an <a> direction within about 10 degrees of the axis of the tube, as seen in fig. 9.

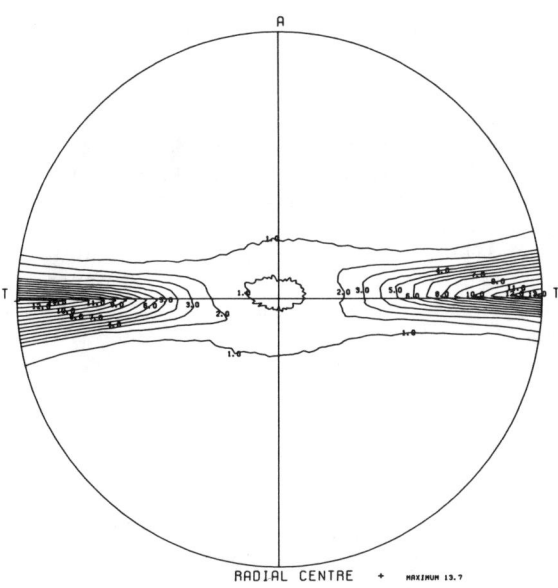

Figure 8 - Basal, (0002) pole figure in Zr-Nb pressure tube

Residual stresses are classified as macro, micro, or grain-interaction, depending on the wavelength of the stresses in the sample. Macro-stresses occur in all materials when the imposed deformation is itself nonuniform. Grain-interaction stresses occur in plastically anisotropic materials, and are present in order to maintain compatibility among grains with differing amounts of plastic deformation. However, the basis for measuring residual strains by diffraction is independent of the nature or origin of the strains, and independent of the type of diffracting wave. Basically, one uses the interplanar spacings as miniature, directional, internal strain gauges. By

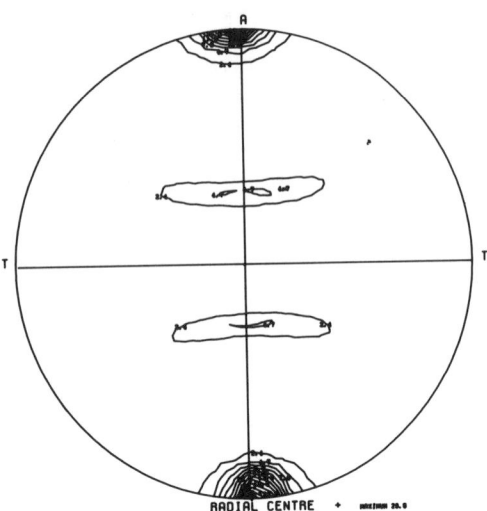

Figure 9 - Pole figure for the (10$\bar{1}$0) planes

Figure 10 illustrates how the crystallites are oriented in a tube and defines two families of grains. Grain #1 has the <c> direction of the crystal in the hoop direction of the tube. The majority of grains in a tube are oriented in this way. Grain #2 has its <c> pole oriented radially. The volume fraction of orientation #2 is about 15 times less than that of orientation #1.

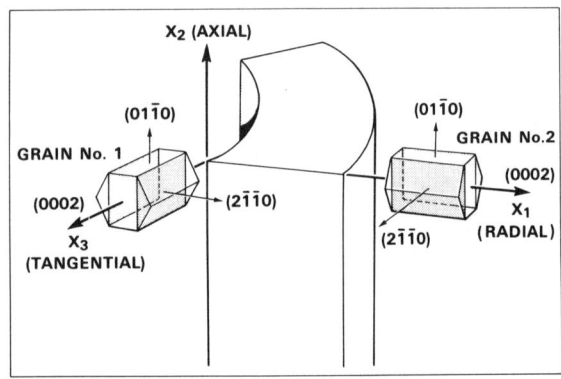

Figure 10 - Orientations of grains in pressure tubes used for diffraction measurements

DIFFRACTION MEASUREMENTS

The objective of the experiment was to use the small beam technique to measure residual hoop strains in a rolled joint that had been deliberately over-rolled in the laboratory. Figure 11 illustrates the experimental setup.

Figure 11 - Experimental setup for measuring residual stress by neutron diffraction

The tube was positioned so that the hoop direction at one position in the wall bisected the incident and (0004) diffracted beams. The incident beam was allowed to pass through the first wall, while all diffraction data originated from a position centered in the second wall. The slit size was such that the average through-wall distribution of (0004) d-spacings was obtained. The shifts in d-spacing therefore correspond to average hoop strains. By choosing the (0004) line, the diffraction data reflect strains in grains of orientation #1 (see fig. 10).

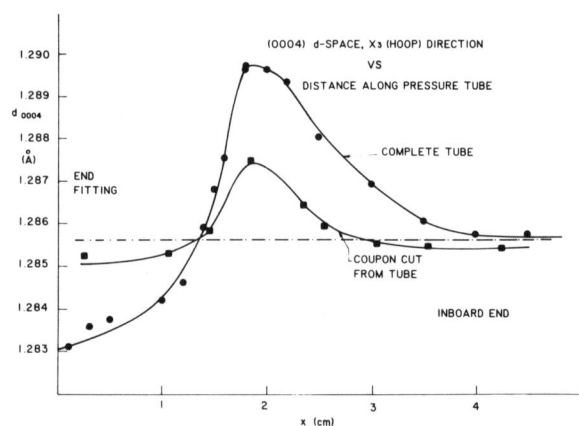

Figure 12 - d-spacing of (0004) planes as function of axial position in pressure tube

Figure 12 plots the (0004) d-spacing as a function of axial position along the tube. Near the hub the hoop strain, measured relative to the free end, is strongly compressive and the d-spacing is decreased accordingly. In the position of maximum over-rolling, about 2 cm down the tube, the residual strains are strongly tensile.

Figure 13 compares the residual hoop strains measured by neutron diffraction with the results obtained from slitting experiments, as described previously. Two sets of slitting results are shown: with the hub in place and after its removal. It is evident that the former is in reasonable agreement with the diffraction measurements.

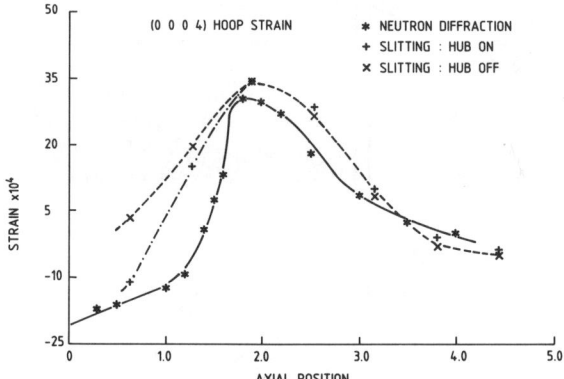

Figure 13 - Comparison of residual strains measured by neutron diffraction and the slitting technique

The tensile hoop strains shown in Figure 13 translate into residual stresses of about 380 MPa (55 ksi), roughly three times the hoop stress produced by normal operating conditions.

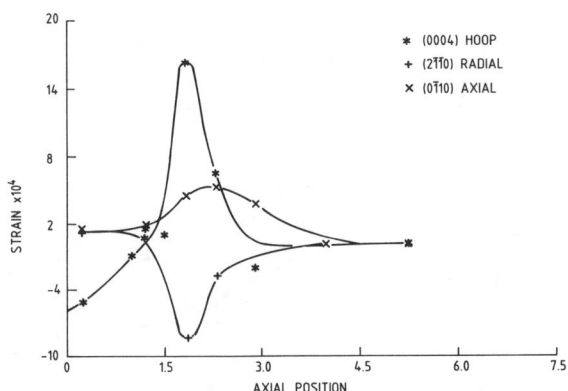

Figure 14 - Residual strains in the principal directions

The slitting procedure produced a coupon of tube about 2 cm wide, the thickness of the tube wall and about 20 cm long. An assumption of the slitting **procedure** is that all residual stresses are removed by the cutting operation;

To check this assumption the (0004) line in the X_3 direction was used again to determine the residual strains after slitting. Figure 12 shows that while much of the residual strain has been relieved, at least one-third, at the peak position, remains. It is interesting to compare this result with those of other techniques. Hole-drilling and trepanning techniques which investigate the first 25 to 250 µm from the surface showed residual strains that were close to zero. X-ray diffraction, which sampled only the near-surface region of the tube, gave the most unexpected result that the surface was 175 MPa (25 ksi) in compression. It was concluded that a compressive surface layer was most probably due to an oxide effect, and that while the X-ray results were valid, they were not relevant to the cracking problem.

The remainder of the experiment was designed to investigate the nature of the residual strains remaining after slitting in more detail. Specifically, we wished to determine if grain-interaction strains were present, as would be expected from a knowledge of the anisotropic behaviour of Zr alloys. By measuring strains in particular directions in the body, we can - as described previously - determine the triaxial, residual strain tensor. The diffraction lines, chosen to investigate the strain in each direction, were selected so that each originated from a crystallographic plane in grain #1. Figure 14 shows the residual strains in the three principal directions of the tube measured from diffraction lines that come from planes in grain #1. In the hoop direction, X_3, (0004) is measured and, as was shown previously, the residual strain is tensile. The radial direction X_1 is sampled using a $(2\bar{1}\bar{1}0)$ line, and indicates strong compression. The axial, X_2, direction is measured using the $(0\bar{1}10)$ line and here one sees residual, tensile strains that are about 1/3 of those in the X_3 direction. Because of the very sharp texture, the (0004) and $(2\bar{1}\bar{1}0)$ lines can unambiguously be associated with grain #1. The $(0\bar{1}10)$ line, however, contains intensity originating from all grains having rotational symmetry about $<0\bar{1}10>$, and therefore includes grain #2, and all others with their $<c>$ directions in the X_1X_3 plane. However, the fact that the distribution of $<c>$ poles in X_1X_3 is strongly weighted to the X_3 direction suggests that most of the intensity in $(0\bar{1}10)$ comes from grains near the grain #1 orientation. The data of Figure 14 gives no information about the shear components of the strain tensor. To investigate shear terms, one must measure d-spacings for directions other than the principal directions of the coordinate system. To do so, the specimen was considered as if it were a single crystal, in the present case with the orientation of grain #1. One can then determine the rotations that are required to bring a given plane into diffraction position.

Table 1 lists the lines, direction cosines in tube coordinates, and residual strains used to characterize the complete, residual grain-interaction strain tensor for grain #1. A six-parameter linear regression fitting routine was used to calculate the residual strain tensor.

$$\varepsilon_{ij} = \begin{pmatrix} -8.6 & -3.1 & 1.5 \\ -3.1 & 4.9 & 1.2 \\ 1.5 & 1.2 & 17.0 \end{pmatrix} \times 10^{-4} \quad (11)$$

TABLE I

LINE	l_1	l_2	l_3	$\varepsilon_R \times 10^4$
$(02\bar{2}0)$	0	1	0	4.4
$(01\bar{1}0)$	0	1	0	4.7
$(02\bar{2}1)$	0	.965	.263	5.6
$(02\bar{2}\bar{1})$	0	.965	-.263	5.0
$(01\bar{1}1)$	0	.878	.478	10.1
$(01\bar{1}\bar{1})$	0	.878	-.478	7.1
(0004)	0	0	1	16.2
$(2\bar{1}\bar{1}0)$	1	0	0	-8.3
$(20\bar{2}0)$	.886	.5	0	-8.4
$(11\bar{2}0)$	.5	.866	0	0.1
$(\bar{1}2\bar{1}0)$	-.5	.866	0	4.9
$(01\bar{1}0)$	0	1	0	3.1
$(02\bar{2}0)$	0	1	0	5.4
$(2\bar{1}\bar{1}0)$	1	0	0	-8.3
$(2\bar{1}\bar{1}2)$	.846	0	.523	-0.6
(0002)	0	0	1	17.4
(0004)	0	0	1	17.3
$(\bar{2}112)$	-.846	0	.532	-2.5

Since the measurements were done on a small coupon it is reasonable to assume, and hole drilling indeed confirmed, that the macro-stresses have been relieved, thus the strain tensor shown above is entirely due to grain interaction caused by anisotropic plasticity. Since the residual strain tensor is expressed in crystal coordinates, the elastic stiffness coeficients, C_{ijkl}, can be used to calculate residual the residual stress tensor using Hooke's law:

$$\sigma_{ij} = \begin{pmatrix} 22.8 & -11.2 & 4.7 \\ -11.2 & 120.2 & 3.9 \\ 4.7 & 3.9 & 256.7 \end{pmatrix} \text{MPa} \quad (12)$$

PROTOTYPE ROLLED JOINT

An obvious advantage of neutron diffraction over other techniques for measuring residual strains is that the sample is completely unaltered and can, therefore be used for further experimentation. Recently, preliminary measurements have been made to characterize residual stresses in correctly made, prototype joints, with the ultimate aim being to be able to monitor possible reduction of residual stresses due to in-service relaxation. In these experiments, the incident beam passed through the steel of the end-fitting, through the first wall of the pressure tube, and across the inside of the tube to the second wall. The slits were positioned so that the intersection diamond was centered in the second wall. The diffracted beam passed through the steel of the end-fitting to the detector, as shown schematically in fig. 15.

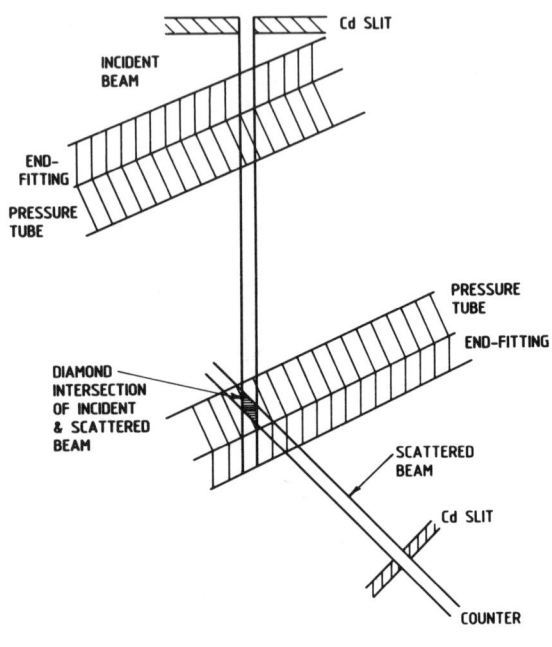

Figure 15 - Schematic of the experimental arrangement for the measurement of axial strain

Figures 16 and 17 show the axial distribution of the through-wall average of the axial and hoop strains respectively.

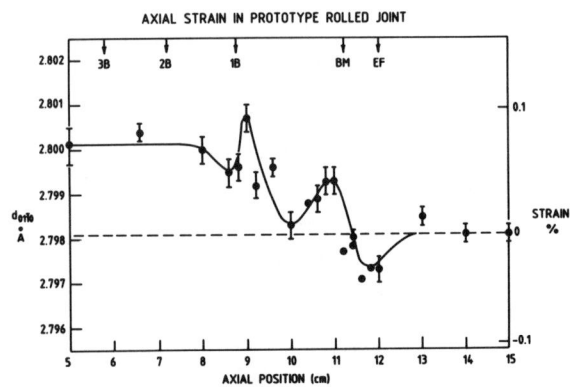

Figure 16 - Axial distribution of axial strain

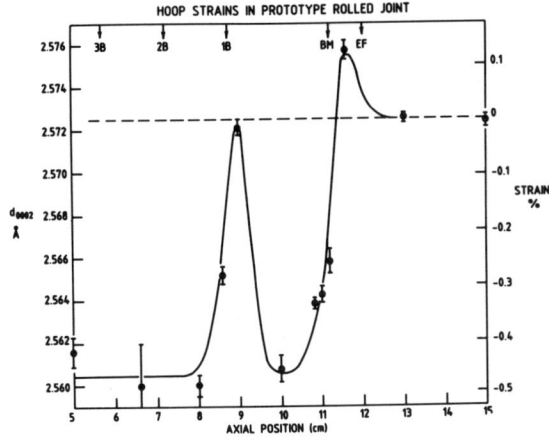

Figure 17 - Axial distribution of hoop strain

The axial component is tensile for positions within the steel endfitting; variations in magnitude are seen, associated with the positions of the grooves in the endfitting. Beyond the constraint of the endfitting, one observes slightly compressive residual strains. The hoop residual strains are highly compressive where the tube is constrained by the endfitting and again variations associated with the grooves in the steel are observed. Inboard of the endfitting one observes a small tensile peak. The non-destructive nature of the measurement, and the resulting ability to use the same sample for an entire series of experiments, is a great advantage of neutron diffraction over other techniques used to determine residual stresses.

CONCLUSIONS

1) Neutron diffraction has been used to measure residual strains in correctly and incorrectly made rolled joints. The results are consistent with strain gauge determinations of macro-stresses.

2) The complete, triaxial residual strain and stress tensors have been determined at the position of maximum over-rolling in an coupon cut from an incorrectly fabricated joint. These result from grain-interaction and can not be determined by hole-drilling or slitting techniques.

3) It has been demonstrated that the penetration power of the neutron beam is sufficient to make measurement of residual strains in the Zr alloy tube, even when the steel endfitting is left in place.

REFERENCES

1. Hosbons, R.R., Proceedings of SEM 1987 Spring Conference, to be published

2. Choi, C.S., Prask, H.J., and Trevino, S.F. J. App. Crys. 12, p. 327-331, 1979

3. Krawitz, A.D., Brune, J.E., and Schmank, M.J. 'Residual Stress and Stress Relaxation', Proceedings of 28th Sagamore Army Materials Research Conference, p. 139-156, (1982)

4. Pintschovius, L., Jung, V., Maucherauch, L., Schefer, R., and Vohringer, O., Ibid. p. 467-482

COLDWORKING FASTENER HOLES — THEORETICAL ANALYSIS, METHODS OF COLDWORKING, EXPERIMENTAL RESULTS

William A. McNeill, Alex W. Heston
SPS Technologies
Jenkintown, Pennsylvania, USA

RATIONALE FOR COLDWORKING FASTENER HOLES

Coldworking a fastener hole refers to a process in which the hole is expanded to such a degree that a plastic, or permanently deformed, region exists near the hole. This expansion may be induced by an interference-fit fastener, by forcing a tapered mandrel through the hole or by stress coining[11]. The mandrel method is the primary focus of this paper.

It has been shown theoretically and experimentally that radially expanding the fastener hole into the plastic zone results in residual compressive hoop stress near the hole after the mandrel is removed. When the plate through which a hole passes is subjected to in-plane tension, a stress concentration factor having a maximum value of three occurs at the hole boundary.(Fig. 1). The residual compressive hoop stress resulting from coldworking act to offset the high tensile stress near the hole boundary. Fatigue cracks are known to initiate at the hole boundary and grow in the presence of tensile stress in fatigue loading. These fatigue cracks can lead to failure of the joint. Thus it is clear that the compressive stresses caused by the coldworking process, in offsetting the tensile stresses near the hole caused by external loading, can substantially improve the fatigue strength of the joint.

Application of a coldworked hole at the tip of an existing crack has been shown to stop the growth of the crack under certain circumstances. Experimental and theoretical research has been done on this important topic[12].

Certain limitations must be recognized in coldworking fastener holes. It is possible to cause buckling of thin plates in the vicinity of the hole. In some cases the tapered mandrel causes an undesirable flow of material axially along the hole surface, which may extrude from one side. In some cases it has been shown that the interaction of the coldworked-induced stress fields of closely spaced holes may be undesirable. A crucial question in any coldworking application is: How much coldworking should be done, so as to have neither too great nor too small residual stress. Certainly the expense associated with the coldworking process must be taken into account in deciding whether or not to use the process.

In this paper the topics outlined above and other related theoretical and experimental results will be presented. Fundamental theoretical methods of treating the coldworking problem will be described, and more elaborate theoretical and experimental methods, presented by a number of different investigators, will be discussed. The purpose of this paper is thus twofold: (1) to provide an introduction to the coldworking process together with some classical theory, and (2) to present an up-to-date literature survey of coldworking technology.

METHODS OF COLDWORKING

Residual compressive stresses may be induced in fastener holes in a number of ways. Interference fit fasteners, stress coining, shot peening the hole surface, and driving a mandrel through the hole are methods which have been used.

A recent description of a method of

mandrel coldworking known as the split sleeve technique is provided by Ozelton and Coyle[13]. In the split sleeve method, the sleeve is placed in the hole to be expanded, and a tapered mandrel is pulled through the sleeve. (Fig. 2) The expanding sleeve coldworks the hole. The undeformed region of the hole at the sleeve split line could be a site for fatigue crack initiation, and so it is customary to ream the hole after expanding, to remove the site. Ozelton and Coyle present detailed geometric considerations regarding the relationships among mandrel diameter, sleeve thickness, ream diameter, etc.. A specific application presented by them is instructive:

"For a .25-mm hole expansion in 7050-T7451 Aluminum, a typical retained expansion is Ø.125 mm, that is, 50% of the initial hole expansion. Thus for a total tool diameter of 6.24 mm and a .25 mm expansion, the coldworked hole diameter is 6.24 minus Ø.125 mm, that is, 6.115 mm. To attain a final hole size of 6.35 mm, a further Ø.235 mm was reamed from the hole diameter following coldworking. The hole was deburred by means of a hand-held tool with a 90° cutter."[13]

A coldworking method utilizing a continuous (non-split) sleeve has been developed by J.O. King, Inc., Atlanta, Georgia. In this method a tapered mandrel is pulled through a sleeve inserted in the hole that is to be expanded. the sleeve is flared on the pulling end and a mild steel washer is fitted into the flare. (Fig. 3) The sleeve may be left in place to receive the fastener after the mandrel is removed.

THEORETICAL ANALYSIS

MANDREL IN - In this section solutions are presented for the initial stage of coldworking, i.e., with the mandrel in, causing pressure to be applied at the hole. In subsequent sections solutions will be presented for the second stage of the operation, in which the plate "recovers" to some extent as the mandrel is removed. Figure (4) shows the geometry for the coldworking process. Expansion is assumed to be symmetric with respect to θ. The radii a and b denote inner and outer radii of the plate, with r denoting the radial position. The condition of plane stress is frequently assumed in order to facilitate a solution. The question of the accuracy of the assumption of plane stress has been addressed by several investigators, including Poolsuk[1]. Boundary conditions are as follows:

$$u|_{r=a} = u_a$$

where u denotes displacement in the radial (r) direction.

This condition may be replaced by

$$\sigma_r|_{r=a} = -P_i$$

where σ_r is radial stress and P_i is radial pressure at r=a.

The assumption that the radial stress is zero at the outer radius is expressed as:

$$\sigma_r|_{r=b} = 0 \ .$$

The limiting situation b >> a proves to be a useful special case, as will be seen later.

Assuming axial symmetry and plane stress, the equation of equilibrium in the radial direction is:

$$\frac{\partial \sigma_r}{\partial r} + \frac{\sigma_r - \sigma_\theta}{r} = 0 \qquad (1)$$

where σ_r and σ_θ denote radial and hoop stress, respectively.

The definition of radial and tangential strain leads to:

$$\epsilon_r = \frac{du}{dr} \ , \quad \epsilon_\theta = \frac{u}{r} \qquad (2)$$

(a) Elastic Solutions.

If the internal pressure P_i is small enough that the plate remains elastic throughout, the solution utilizing the pressure boundary condition at a, is[2]:

$$\sigma_r = \frac{a^2 P_i}{b^2 - a^2} \left[1 - \frac{b^2}{r^2} \right] \qquad (3a)$$

$$\sigma_\theta = \frac{a^2 P_i}{b^2-a^2}\left[1 + \frac{b^2}{r^2}\right]. \quad (3b)$$

The hoop stress σ_θ assumes its maximum value at $r=a$, given by:

$$\sigma_\theta\big|_{max} = \frac{P_i[a^2 + b^2]}{b^2 - a^2}. \quad (4)$$

The maximum value of the hoop stress σ_θ is always numerically greater than P_i, and it approaches P_i as b increases.

Utilizing the displacement boundary condition at $r=a$ leads to the following elastic solution:

$$\sigma_r = \frac{E\, u_a}{[1-\nu]a + [1+\nu]\frac{b^2}{a}}\left[1 - \frac{b^2}{r^2}\right] \quad (5a)$$

$$\sigma_\theta = \frac{E\, u_a}{[1-\nu]a + [1+\nu]\frac{b^2}{a}}\left[1 + \frac{b^2}{r^2}\right] \quad (5b)$$

$$\epsilon_r = \frac{u_a}{[1-\nu]a + [1+\nu]\frac{b^2}{a}}\left[(1-\nu)-(1+\nu)\frac{b^2}{r^2}\right] \quad (5c)$$

$$\epsilon_\theta = \frac{u_a}{[1-\nu]a + [1+\nu]\frac{b^2}{a}}\left[(1-\nu)+(1+\nu)\frac{b^2}{r^2}\right] \quad (5d)$$

For $b/a \gg 1$ and $b/r \gg 1$, the above expressions for σ_r and σ_θ reduce to

$$\sigma_r = \frac{-a\, E\, u_a}{[1+\nu]\, r^2} = -\sigma_\theta. \quad (6)$$

The quantities E and ν are the modulus of elasticity and Poisson's ratio, respectively. From the above expressions for σ_r and σ_θ, it is noted that for $b \gg r$, $\sigma_r \to -\sigma_\theta$. When this is used in conjunction with the expression for ϵ_z in plane stress

$$\epsilon_z = -\frac{\nu}{E}(\sigma_r + \sigma_\theta), \quad (7)$$

the result is $\epsilon_z = 0$.

When u_a (or P_i) is large enough, a plastic zone will be created. In order to investigate the onset of plasticity, the von Mises-Hencky, or distortion-energy, yield criterion may be used[3]. In the case of plane stress, the Von Mises criterion is:

$$\sigma_y^2 = \sigma_r^2 + \sigma_\theta^2 - \sigma_r\sigma_\theta. \quad (8)$$

Yielding is assumed to occur when this equation is satisfied, where σ_y is the yield stress of the plate material as determined in an axial tension test, and the yield stress in compression is assumed equal to that in tension.

Evaluating equations (5a) and (5b) for σ_r and σ_θ at $r=a$, the location of the largest stress, and substituting into the Von Mises equation, yields a u_a value

$$u_a\big|^{Max}_{Elast}$$

which will produce incipient yield. This value is

$$u_a\big|^{Max}_{Elast.} = \frac{\sigma_y\, a}{E\left[1+3\frac{b^4}{a^4}\right]^{\frac{1}{2}}}\left[1-\nu+[1+\nu]\frac{b^2}{a^2}\right]. \quad (9)$$

The corresponding internal pressure is obtained by substituting equation (9) into equation (5a) evaluated at $r=a$. The result is

$$P_i\big|^{Max}_{Elast.} = \frac{\left[\frac{b^2}{a^2} - 1\right]\sigma_y}{\left[1 + 3\frac{b^4}{a^4}\right]^{\frac{1}{2}}}. \quad (10)$$

The limiting value of this expression as $b/a \to \infty$ is

$$P\big|^{Max}_{Elast.} = \frac{\sigma_y}{\sqrt{3}}. \quad (11)$$

For purposes of comparison, the Tresca, or Maximum Shear Yield theory, expressed as $\sigma_y = \sigma_\theta - \sigma_r$, yields

$$u_a\big|^{Max}_{Elast} = \frac{\sigma_y\, a\left[1 + \nu + (1-\nu)\frac{a^2}{b^2}\right]}{2E}. \quad (12)$$

Once the internal pressure exceeds the elastic limit, a region bounded by a radius r_p will exist, within which stresses will be plastic for $r < r_p$. The magnitude of the plastic radius r_p is determined in accordance with an assumed constitutive relation, as shown in the next section. To conclude the elastic solution, however, it is assumed that r_p is known. The solution for the elastic region $r > r_p$ will be obtained. The general solution to the equations of equilibrium for this problem is

$$\sigma_r = \frac{A}{r^2} + 2C$$

$$\sigma_\theta = -\frac{A}{r^2} + 2C \, ,$$

where A and C are constants to be determined by the boundary conditions. The boundary conditions for the elastic region with $r \geq r_p$ are

$$\sigma_r\big|_{r=r_p} = -\frac{\sigma_y}{\sqrt{3}} \quad (13a)$$

(i.e., the onset of yield occurs at $r=r_p$)

$$\sigma_r\big|_{r=b} = 0 \, . \quad (13b)$$

With A and C determined by these conditions, the solution to the elastic problem for $r \geq r_p$ and large b is[1]

$$\sigma_r = -\frac{\sigma_y}{\sqrt{3}} \left[\frac{r_p}{r}\right]^2 \quad (14a)$$

$$\sigma_\theta = \frac{\sigma_y}{\sqrt{3}} \left[\frac{r_p}{r}\right]^2 \quad (14b)$$

$$\epsilon_r = -\frac{\sigma_y}{E\sqrt{3}} \left[1+\nu\right]\left[\frac{r_p}{r}\right]^2 \quad (14c)$$

$$\epsilon_\theta = \frac{\sigma_y}{E\sqrt{3}} \left[1+\nu\right]\left[\frac{r_p}{r}\right]^2 \quad (14d)$$

$$u = \frac{\sigma_y}{E\sqrt{3}} [1+\nu] \frac{r_p^2}{r} \, . \quad (14e)$$

(b) Plastic Region Solutions.

Nadai[4] first applied the equations of coldworking to the problem of plasticly expanding boiler and condenser tubes into existing holes in head plates. It should be noted that ϵ_z is not zero in the plastic region because $\sigma_r \neq -\sigma_\theta$ in that region. Nadai assumed a perfectly-elastic/ perfectly-plastic material behavior, and obtained solutions in the plastic region $r \leq r_p$.

Nadai's solutions for deflection and radial and tangential stresses in the plastic region, under load at the inner radius are

$$\sigma_r = \frac{\sigma_y}{\sqrt{3}} \left[-1+2\,\ln\!\left(\frac{r}{r_p}\right)\right] \quad (15a)$$

$$\sigma_\theta = \frac{\sigma_y}{\sqrt{3}} \left[1+2\,\ln\!\left(\frac{r}{r_p}\right)\right] \quad (15b)$$

$$u = \left[\frac{3}{2}\right]^3 \frac{u_a\big|_{\substack{Max\\Elast.}} \frac{r}{r_p}}{\left[\frac{3}{2} + \ln\!\left(\frac{r}{r_p}\right)\right]^3} \, . \quad (15c)$$

These stresses are shown in Figure (5). In developing this solution, Nadai employed a linear approximation to the Mises/Hencky yield criterion.

Other investigators have obtained plastic region solutions under other assumptions regarding material stress/strain behavior and yield criteria. In this regard the work of Hsu and Forman[5], Swainger[6], Taylor[7], Alder and Dupree[9] and others is available. Poolsuk[1], Emery[8] and Mann and Jost[10] have presented comprehensive reviews of the work of investigators of this problem. Finite-Elements computer techniques have been brought to bear on this problem by Adler and Dupree[9] and others.

The value of the plastic radius r_p may be expressed in terms of the amount of coldworking at the hole radius $r=a$ by substituting the expression previously developed for:

$$u_a\big|_{\substack{Max\\Elast}}$$

into Nadai's plastic solution, Equation (15c), and setting $r=a$. The result is expressed nondimensionally in terms of a variable:

$$\frac{\sqrt{3}\,u\big|_{r=a}\,E}{a\,\sigma_y} \, .$$

Evaluating Equation (15c) at $r=a$ and using Equation (9) for large b, gives

$$\frac{\sqrt{3}\,u\big|_{r=a}\,E}{a\,\sigma_y} = \frac{\left[\frac{3}{2}\right]^3 [1+\nu]}{\frac{r_p}{a}\left[\frac{3}{2} - \ln\!\left(\frac{r_p}{a}\right)\right]^3} \, .$$

Figure 6 shows the relationship between these variables. It is interesting that the curve approaches a maximum. Grandt and Tupper[15] point out that this maximum point is a candidate for an optimum degree of coldworking.

MANDREL OUT - When the mandrel is removed after plastically expanding the hole, a relaxation of the hole takes place to allow for the fact that radial stresses at the inner radius must return to zero. Nadai and the investigators listed above have considered the final "mandrel out" state of the expanded hole under a number of different assumptions and techniques of solution.

One straight-forward method of obtaining a mandrel-out solution is to superimpose onto the previous solutions an assumed distribution of stress which will guarantee that the radial stress is zero at r=a. On such distribution is[1,8]

$$\sigma_r = \sigma_m \left[\frac{a}{r}\right]^2$$

$$\sigma_\theta = -\sigma_m \left[\frac{a}{r}\right]^2,$$

where σ_m is the magnitude of radial stress at r=a due to loading.

When these stresses are subtracted from Equations (15a) and (15b), the stress distribution shown in Figure 7 results. The significance of the final, mandrel-out, stress distribution is clear from Figure 7. The residual hoop stress has a very large negative, or compressive value.

Sharpe[16] has pointed out that this assumption of elastic unloading is not correct, that the residual stresses violate the yield criterion near the hole, if the elastic assumption is used. He refers to the Hsu/Forman[5] theory for a treatment which takes into account yielding in unloading.

EXPERIMENTAL RESEARCH ON COLDWORKING

Poolsuk[1] has performed detailed measurements aimed at locating the elastic/plastic interface radius r_p. His experimental methods included the use of strain gauges and careful measurement of the change in plate thickness with radius. It is easily shown that the transverse strain

$$\epsilon_z = \frac{1}{E}\left[\sigma_z - \nu(\sigma_r + \sigma_\theta)\right]$$

will be zero in the elastic region, since in that region $\sigma_r = -\sigma_\theta$, and $\sigma_z = 0$. Thus by experimentally determining the radius at which ϵ_z becomes non-zero, he was able to determine r_p.

Poolsuk used the J.O. King coldworking process, with 7075-T6 aluminum plates. He found that the Nadai equations for prediction of r_p were adequate for the thinner (1/8") plate tested. However, he determined that other theories are preferable for thicker plates. He reported that the coldworking process used did not produce radially symmetric deformation, and he observed shearing deformation of the hole due to the pulling action of the mandrel.

Emery[8] conducted experiments to more nearly duplicate radially symmetric expansion. She found that the assumption of plane stress did not appear to be totaly justified. She also found that the theoretical assumption of small strain is invalid in typical cold-working processes, in which strains of up to eighteen percent occur. She concluded that existing theories are inadequate to completely describe the complex coldworking process.

Sulaimana[14] performed experiments to study the effect of in-plane compressive loads and geometric interaction effects on coldworking. He measured strain distribution by means of an optical Moiré technique. He found that in-plane compressive loads decreased residual hoop strains parallel to the load and increased residual hoop strain in the transverse direction, but that the in-plane loads have little effect on radial strains. His experiments showed that coldworking a hole or row of holes near a plate edge produces an undesirable shift of the distribution of the residual strains. He recommends a minimum separation of 2 diameters between coldworked holes.

Sharpe[16] compared theory and experiment for the distribution of strains around a coldworked hole in the presence of in-plane tension.

The above investigators and others stress the approximate nature of theoretical treatments, the lack of symmetric coldworking strains, undesirable interaction effects, and the possibility of buckling of thin sheets under coldworking.

EXPERIMENTAL APPLICATIONS TO IMPROVE FATIGUE LIFE. OPTIMUM LEVEL OF COLDWORKING

In spite of the fact pointed out by numerous investigation that the coldworking process is not fully understood theoretically, and is complex and non-symmetrical, it has been shown that the process does improve the fatigue life of fastener holes. Grandt and Tupper[15] have presented recommendations for the optimum amount of coldworking, based on the work of Nadai and others.

Sharpe[16] conducted fatigue tests in a configuration comparable with the theoretical predictions of Adler and Dupree[9]. He used a microscope to observe crack growth as a function of the number of fatigue cycles. He reported cases in which the fatigue performance of non-coldworked holes was superior to that with coldworking. In this regard he emphasises the fact that the geometry and size of the joint, versus the degree of coldworking, is crucial to the performance of the joint in fatigue. He observed buckling of the sheet around the hole in the case of thin (1/16") sheet, and he reported that strains produced around the hole were inhomogeneous and varied at the hole surface in the axial direction. He concluded that, for the thin sheet, coldworking reduced fatigue life rather than increasing it, because the coldworking process served to initiate fatigue cracks at the hole surface.

Hsu, McGee and Aberson[18] performed a theoretical and experimental study of crack growth rate in coldworked versus non-coldworked holes. Their experiments showed that coldworking reduces the rate of crack growth, and that the reduction in rate is greater with more coldworking. They found that the beneficial effect of coldworking is most pronounced for small crack lengths, and decreases with increasing crack length. The beneficial effect of coldworking is associated with a decrease in stress intensity factor for a crack of a given length. They found that the presence of a small amount of fastener load transfer increases the corresponding stress intensity factor significantly. Their theoretical work indicated that if the initial crack length is smaller than the length of the compressive zone the crack should not propogate under constant amplitude cyclic loading. Their experimental work indicated the opposite conclusion, however. They postulated that the residual compressive strains, due to coldworking, relax after the cyclic load is applied, and that the net hoop stress reverts to tension.

Van der Kinder[19], in a study of fastener fatigue under aeronautical loadings modeled by the FALSTAFF and MINITWIST programs, found that a clearance fit in a coldworked hole gave a 20-50% improvement in fatigue life compared with a medium interference fit in a non-coldworked hole.

Moore[20], in an extensive experimental study, reported that coldworking always enhanced fatigue life.

Chang[21] has studied the rate of crack growth from coldworked fastener holes using fracture mechanics. He developed finite-element computer techniques to predict the rate of crack growth as a function of the degree of coldworking. In addition, he performed experiments using the split-sleeve method to test the results predicted by analysis. His analytical predictions of crack growth rate were accurate to within 15% compared with experiment. He does not present comparisons of fatigue strength between coldworked versus non-coldworked holes.

Petrak[22], performed a series of tests to study the retardation of crack growth afforded by coldworking and by the use of interference-fit fasteners. Using a 1/4" thick 7075-T6 aluminum plate, he predrilled 1/8" holes, which were precracked with a saw cut and then were grown through fatigue loading. Final drilling to 1/4" diameter was then done.

Fatigue tests were then conducted with, zero-load transfer, double dog-bone and double-tab shear specimens, with an R ratio of 0.5. Petrak reported definite improvement in crack growth rate when coldworking is used. He reports that a crack length of about 0.1 inches is the maximum length which will be improved by coldworking. He found that the improvement in crack growth retardation afforded by interference-fit fasteners is, for the most part, related to that of coldworking in proportion to the amount of final coldworking in the two procedures. Some exceptions to this rule were noted by Petrak, however. He found that edge distance to hole diameter ratios as small as 1-1/2 do not diminish crack retarding capability of the systems studied, provided that a fastener is in the hole and is properly torqued. He also concluded that the amount of load transfer by a fastener does not affect its crack-retarding capacity.

Cathey and Grandt[23] performed analytical and experimental work aimed at predicting crack growth rate in coldworked holes. They employed the Hsu-Forman[5] finite-element solution for the prediction of the elastic/plastic stress field, and then investigated the crack from the standpoint of fracture mechanics. They reported that the procedure used in their investigation is a simple and effective method for estimating the fatigue life of cracked coldworked holes.

Armen, Levy and Eidinoff[24] have also performed a detailed analysis, utilizing finite-elements and fracture mechanics techniques. They focused on the effect of remote compressive in-plane loads, which act to reduce the residual compressive stress due to coldworking. In one of their calculations, a fatigue life increase of five-fold was predicted analytically, and the corresponding experimental increase was found to be ten-fold.

Ozelton and Coyle[13] performed a detailed experimental study aimed at quantifying the following factors on the fatigue life of 7050-T7451 aluminum:

  i) Level of coldworking
 ii) Level of post-coldwork reaming
iii) Effect of precracks prior to coldworking
 iv) Flaws introduced through split-sleeve coldworking
  v) Interference-fit fasteners

They presented optimum expansion and reaming levels for a 6.35 mm diameter hole, and report that flaws introduced during split sleeve coldworking do not significantly influence fatigue life. They found that for existing cracks of less than 3 mm, the fatigue life can be increased by coldworking. They also reported that interference fit fasteners provide a greater fatigue life improvement than split sleeve coldworking, and that using interference fit fasteners together with coldworking is better than using either separately.

Landy, Armen and Eidinoff[12] have performed an experimental and analytical investigation, using elastic/plastic finite elements methods and fracture mechanics techniques, to study the effect on arresting existing cracks in 7075-T651 aluminum by drilling and coldworking a hole at the crack tip. They report reasonable correlation of analysis with test data. They determined that stop-drill plus expansion by coldworking increases fatigue life by a factor of 10. The further addition of an interference-fit fastener to the stop-drilled hole increases the fatigue life by an additional order of magnitude.

Rich and Impellizzeri[25] have performed a detailed analytical and experimental study of the improvement in fatigue due to coldworked and interference-fit fasteners. Their experiments were performed on 2024-T851 aluminum and 6A1-4V titanium. They reported favorable agreement between analysis and experiment. Their analysis utilizes a crack initiation model which strongly distinguishes two stages in fatigue life, namely crack initiation and crack growth. Closed-form approximate solutions for residual stress are compared with finite-element solutions.

Philips[26] has performed an extensive experimental study, in which a number of parameters were varied, including mandrel taper angle, mandrel pulling force, hole finish, etc.

Fatigue tests were reported using aluminum, Titanium and steel.

Jongesbreur and Koning[17] have performed an analytical and experimental investigation on the effect of coldworking on lug holes in aluminum, using the split sleeve method. They reported that the fatigue life was significantly improved, even for edge distances as low as one hole diameter. They found that the position of the slit in the sleeve is critical in the unsymmetrical lug, with the proper position being on the back side of the lug hole.

SUMMARY AND CONCLUSIONS

From the classical solutions presented in this paper, to more complex finite-element and fracture mechanics considerations, it is clear that the coldworking problem is challenging and demanding from the analytical viewpoint. It is fair to state, based on recent work, that analysis agrees reasonably well with experiment. The fact that commercially available coldworking tools introduce nonuniformity, plastic shear, etc., suggests that there are limits to what can be accomplished analytically, and theses limits may have already been reached.

Experimental results show very clearly that coldworking improves fatigue life of the structure dramatically. Engineering considerations of optimum coldworking and reaming levels have been addressed by some investigators, but much remains to be done in presenting optimum coldworking recommendations for a given joint, of a given material under given in-plane loads.

The question of whether to use coldworking, interference-fit fastneres, or both, has been addressed by some investigators, but general guidelines have not been presented. Likewise, questions of the interaction effects on coldworking closely-spaced holes, edge effects, and special geometries such as lugs have been addressed at a research but not a design level.

An important application of coldworking to stop existing cracks, by stop-drilling and expanding, has been studied. General engineering guidelines for this process have not been presented, however.

A question only alluded to by many investigators is the following: What is the cost of coldworking in terms of labor, time, etc.? How do we guage that cost against the benefit in terms of improved fatigue life? Admittedly, this question could be answered differently in different specific applications. However, it would appear that some general work could be done in this area to provide general guidelines.

Most of the experimental work reported has dealt with aluminum, and the reason for this is clear, given the great importance of fatigue life in aluminum aircraft structures. However, considerable insight and practical benefit could be gained by comparing the results of experiments performed using a variety of materials.

In summary, coldworking has been proven effective in increasing fatigue life. it is reasonably well-understood from an analytical standpoint. The next step is called for, namely, one in which the coldworking process is described in the literature in terms of general engineering design recommendations, with consideration of the ultimate trade-off of cost versus gain included as one of the design factors.

As an aid in a systematic review of this subject, the following tabulated information lists references, by number, which emphasize specific topics in coldworking:

Theoretical Analysis - 1,8,14,24, 23,4,15,21,10,18,16,2,3,12,13,6

Experimental Stress/Strain Measurements - 1,8,18,17,16

Fatigue Testing - 20,24,23,26,21,10, 22,18,17,13

Edge Effects or Other Complex Geometry - 14,20,24,26,22,17,12

ACKNOWLEDGEMENTS

The helpful discussions between the

authors and Dr. Harry Antes of SPS Technologies are acknowledged. The authors also gratefully acknowledge the help provided in this project by Robert Port of Drexel University.

REFERENCES

1. Poolsuk, Saravut, <u>Measurement of The Elastic - Plastic Boundary Around Coldworked Fastener Holes</u>, Dissertation, Dept. of Metallurgy, Mechanics and Materials Science, Michigan State University, 1977.

2. Timoshenko, S.P., Goodier, J.N., <u>Theory of Elasticity</u>, 3rd edition, McGraw - Hill.

3. Shigley, J.E., <u>Mechanical Engineering Design</u>, 4th ed., McGraw - Hill.

4. Nadai, A., "Theory of Expanding of Boiler and Condenser Tube Joints Through Rolling", ASME Journal, Vol. 65, Nov., 1943, pp. 865-880.

5. Hsu, Y.C., and Forman, R.G., "Elastic - Plastic Analysis of an Infinite Sheet Having a Circular Hole Under Pressure", J. of Applied Mechanics, 42.2, 1975, pp. 347-352.

6. Swainger, K.H., "Compatibility of Stress and Strain in Yield Metals", Phil. Mag., Series 7, 36, 1945, pp. 443-473.

7. Taylor, G.I., "The formation and Enlargement of a Circular Hole in a Thin Plastic Sheet", Quart. Journal of Mechanics and Applied Mathematics, Series 7, 1, 1947, pp. 103-124.

8. Emery, S.A., <u>The Residual Strain Distribution Around a Fastener Hole Coldworked with a Tube Expander</u>, Master's Thesis, Applied Mechanics, Michigan State University, 1978.

9. Adler, W.F., and Dupree, D.M., "Stress Analysis of Coldworked Fastener Holes", AFML-TR-74-44, Air Force Materials Laboratory, Wright-Patterson Air Force Base, Ohio, 1974.

10. Mann, J.Y., and Jost, G.S., "Stress Fields Associated with Interference Fitted and Cold-Expanded Holes", Metals Forum, Vol. 6, No. 1, 1983, pp. 43-53.

11. Speakman, E.R., "Fatigue Life Improvement Through Stress Coining Methods", ASTM STP 467, 1970, pp. 209-227.

12. Landy, M.A., Armen, H. Jr., and Eidinoff, H.L., "Enhanced Stop-Drill Repair Procedure for Cracked Structures, ASTM STP 927, 1986, pp. 190-220.

13. Ozelton, M.W. and Coyle, T.G., "Fatigue Life Improvement by Cold Working Fastener Holes in 7050 Aluminum", ASTM STP 927, 1986, pp. 53-71.

14. Sulaimana, R.A., <u>An Experimental Study of Large Compressive Loads Upon Residual Strain Fields and the Interaction Between Surface Strain Fields Created by Coldworking Fastener Holes</u>, Dissertation, Michigan State University, Department of Mechanical Engineering, 1980.

15. Grandt, A.F., Jr., and Tupper, N.G., "An Analysis of Residual Stresses and Displacements Due to Radial Expansion of Fastener Holes", Air Force Materials Laboratory Report, No. AFML-TR-79-4048, 1978.

16. Sharpe, W.N., "Measurement of Residual Strains Around Coldworked Fastener Holes", Air Force Office of Scientific Research, Report no. AFOSR-TR-77-0020, 1976.

17. Jongebreur, A.A., and de Koning, A.U., "Results of a Study of Residual Stresses and Fatigue Crack Growth in Lugs with Expanded Holes", National Aerospace Laboratories NLR,

Amsterdam, The Netherlands, Report No. NLR-MP-83024-U, 1983

18. Hsu, T.M., McGee, W.M. and Aberson, J.A., "Extended Study of Flaw Growth at Fastener Holes", Air Force Flight Dynamics Laboratory Report No. AFFDL-TR-77-83, vol. 1, 1977.

19. Van der Linden, H.H., "Fatigue Rated Fastener Systems in Aluminum Alloy Structural Joints", National Aerospace Laboratory, Amsterdam, The Netherlands, NLR-MP-83045U, 12th ICAF Symposium, Toulouse, May 1983.

20. Moore, T.K., "The Influence of Hole Processing and Joint Variables on the Fatigue Life of Shear Joints", Air Force Materials Laboratory Report No. AFML-TR-77-167, Vol. 1, 1977.

21. Chang, J.B., "Prediction of Fatigue Crack Growth at Coldworked Fastener Holes", Journal of Aircraft, Vol. 14, No. 9, Sept. 1977, pp. 903-908.

22. Petrak, G.J., and Stewart, R.P., "Retardation of Cracks Emanating from Fastener Holes", Engineering Fracture Mechanics, Vol. 6, No. 2, 1974, pp. 275-282.

23. Cathey, W.H., and Grandt, A.F., Jr., "Fracture Mechanics Considerations of Residual Stresses Introduced by Coldworking Fastener Holes", Journal of Engineering Materials and Technology, Vol. 102, Jan. 1980, pp. 85-91.

24. Armen, H., Levy, A., and Eidinoff, H.L., "Elastic-Plastic Behavior of Coldworked Holes", Journal of Aircraft, Vol. 21, No. 3, March 1984, pp. 193-201.

25. Rich, D.L., and Impellizzeri, L.F., "Fatigue Analysis of Coldworked and Interference Fit Fastener Holes", ASTM STP 637, 1977, pp. 153-175.

26. Phillips, J.L., "Sleeve Coldworking Fastener Holes", Air Force Materials Laboratory Report AFML-TR-74-10, 1974.

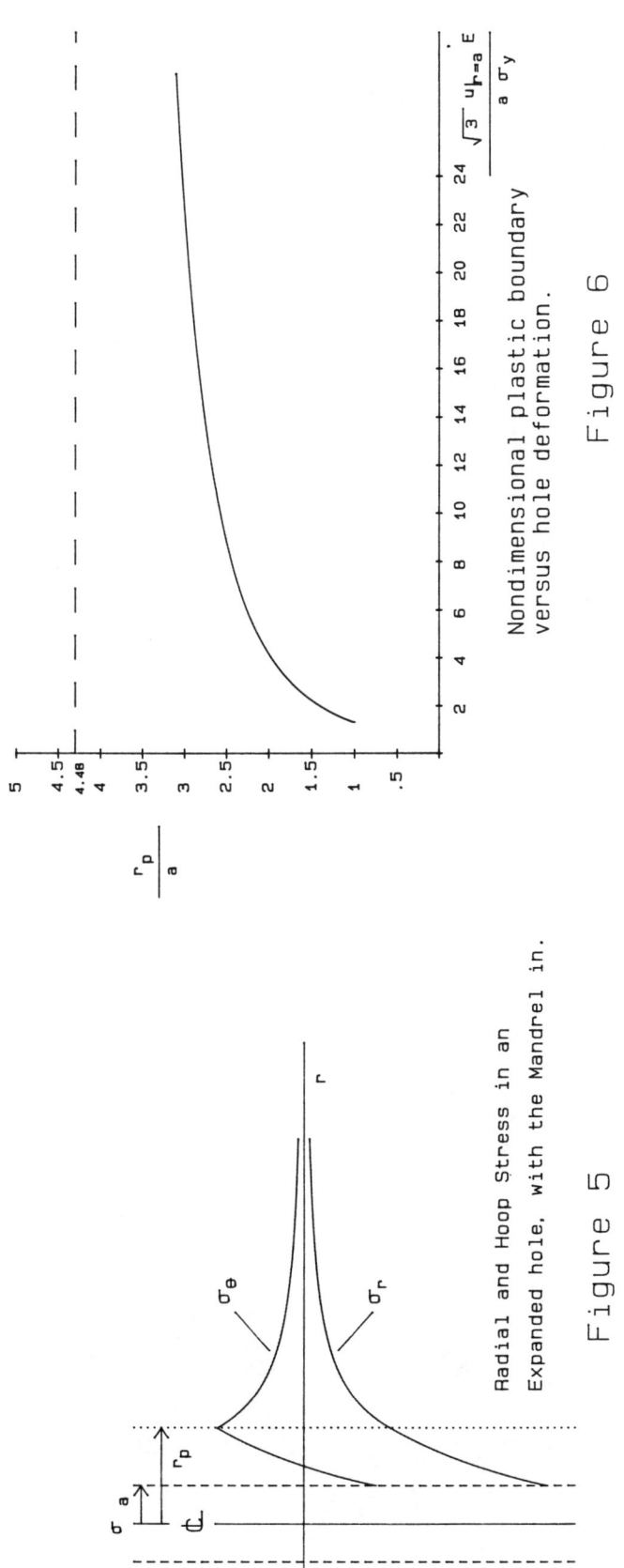

Figure 6

Nondimensional plastic boundary versus hole deformation.

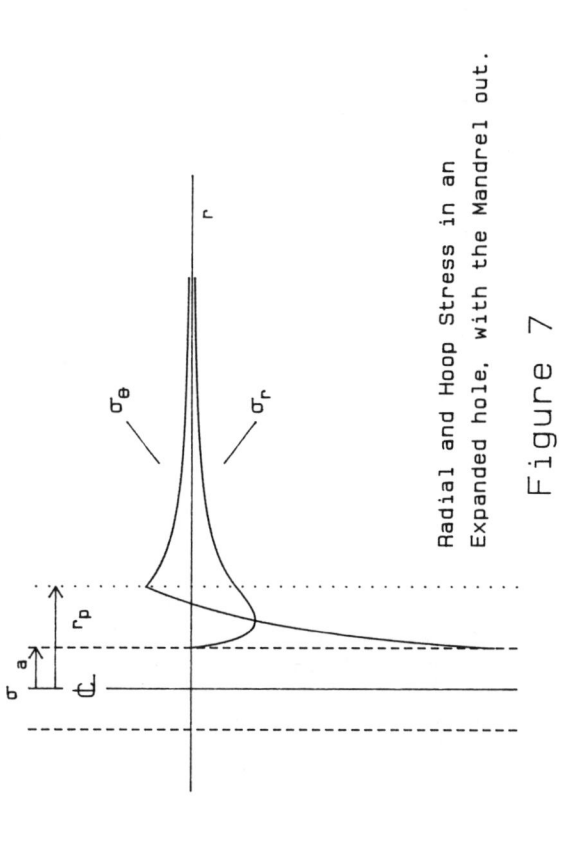

Figure 5

Radial and Hoop Stress in an Expanded hole, with the Mandrel in.

Figure 7

Radial and Hoop Stress in an Expanded hole, with the Mandrel out.

Stress Distribution Around
a Hole in a Plate in tension

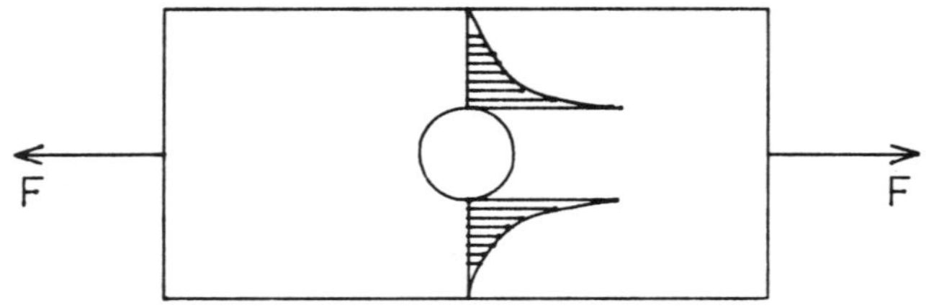

Figure 1

Split Sleeve Coldworking Method

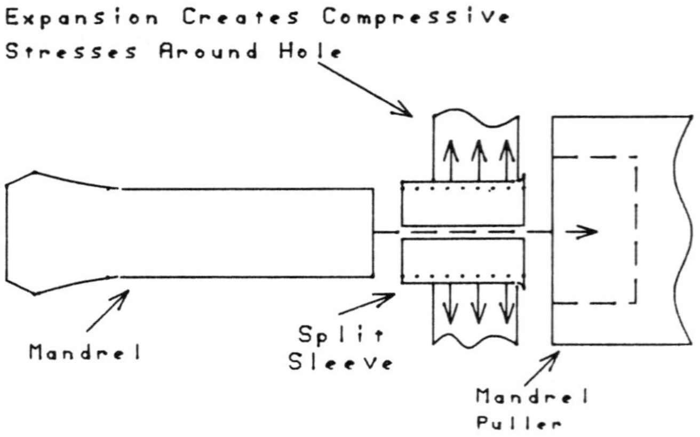

Expansion Creates Compressive
Stresses Around Hole

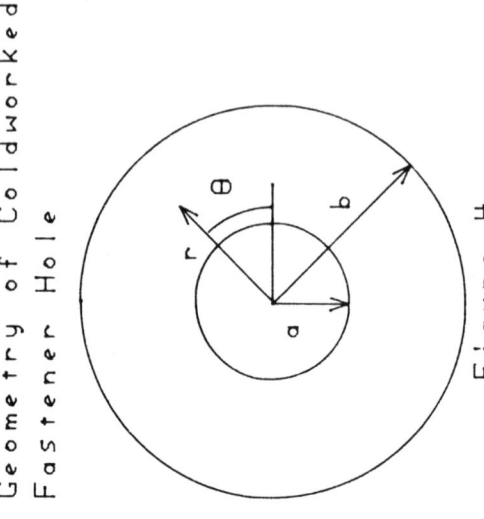

Figure 2

King Method for Coldworking
Non-Split Sleeve

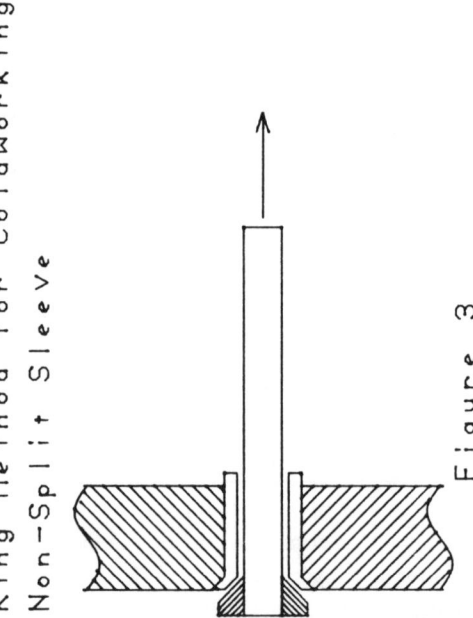

Figure 3

Geometry of Coldworked
Fastener Hole

Figure 4

CALCULATION OF RESIDUAL STRESSES IN RAILROAD RAILS AND WHEELS FROM SAWCUT DISPLACEMENT

M.W. Joerms
Association of American Railroads
Chicago, Illinois USA

ABSTRACT

When a railroad rail is roller straightened, tensile residual stresses may be created in the base and the head. If a saw cut is made along the vertical centroid, these stresses will cause the head and base pieces to separate. If the curvatures of the head and base parts are used in a finite element model, it is possible to obtain an estimate of the residual stress present in the rail before it was cut.

In like manner the opening or closing of a radial saw cut in a freight car wheel has been used to estimate the state of residual stress in the wheel. If the cut opens, it is assumed that a tensile circumferential residual stress was present in the wheel rim. If the cut closes, compressive circumferential stress is assumed. A good estimate of the magnitude and distribution of the residual stresses in a wheel can be obtained from a three-dimensional finite element calculation using displacements measured along the length of the cut.

RAILROAD COMPONENTS are simple in design and construction, but often have service lives measured in decades. Wheels and rails are constantly subject to mechanically and thermally induced plastic deformation, particularly due to contact forces. Fatigue failures can arise due to the combination of residual stresses generated either during manufacture or in service and stresses arising from vertical and lateral loads. This paper describes the development of a simple method for computing residual stresses in rails and freight car wheels, using measurements taken from partially saw cut components.

Since there is no inexpensive method to nondestructively determine the residual stresses in thick sections, the practice of saw cutting components suspected of having high residual stresses has developed. The displacements resulting from the saw cut can then be measured and a qualitative estimate made of the stress state. However, up until this point no practical method had been available to analyze displacement data in these components quantitatively. Since this method is based on gross deformations caused by the cutting, the expense of instrumentation is minimized.

RAIL

Several years ago, a problem was encountered with catastrophic web crack induced shattering of roller-straightened CrV rails[1]. A method was required for determining the stress field near the tip of the crack in sample rails. This would give some idea about the forces driving the crack. Some of these samples had been cut partially through along the neutral axis (Fig. 1) and the two sections had spread apart by up to 1/8 inch. This behavior indicated that the head and base of the rail were originally in longitudinal tension and the web in compression.

The procedure developed to perform this analysis was to determine the radius of curvature of a section, using the length of the cut and the opening displacement, construct a finite element model with this distortion, and then force the model back to the undeformed state. This method relies on the following assumptions:

- The radius of curvature is computed from the opening. It is assumed to be a constant, which implies little variation of stress with position along the longitudinal axis.
- The stress is fully relieved in the cut sections.
- The deflections of the head and base are symmetrical. (This could be avoided by modeling both sections simultaneously.)

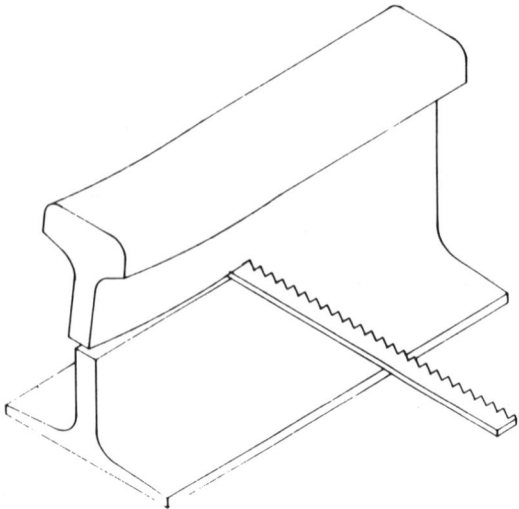

Fig. 1. Rail Sawcutting Process

These assumptions imply the following limitations:
- The stress distributions will be an average value for a particular area.
- Distributions across the width of the section will not be accurately modeled. This is more of a problem in the base and the head than in the web of the rail.

The finite element model used for this is shown in Fig. 2. Since the rail possesses a plane of symmetry, a half model can be used. The low mesh density is considered acceptable in light of the very simple distributed load. The analysis was performed using the ANSYS finite element program.

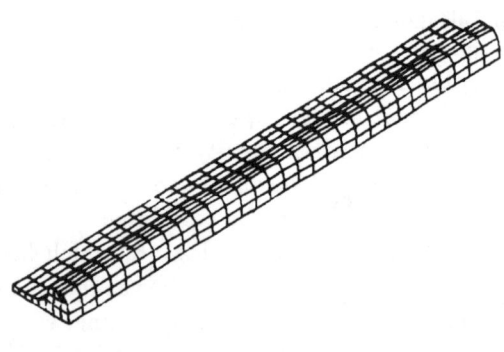

Fig. 2. Finite Element Mesh for one quarter model of rail

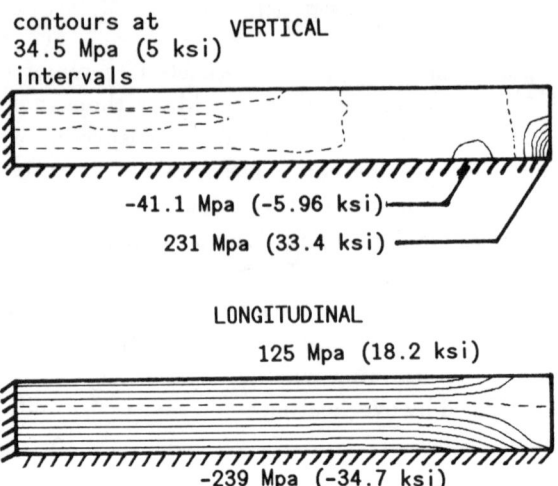

Fig. 3. Calculated Vertical and Longitudinal Stresses at End of Rail

RESULTS - Longitudinal and vertical stresses calculated by this method are shown in Fig. 3. Note the distributions near the end of the rail. A substantial vertical stress is needed for equilibrium in the web at the end of the rail. This stress field follows the crack tip (Fig. 4) and does not diminish as the crack progresses. This effect was verified experimentally by placing strain gages near the saw cut and observing the changes as the saw blade passed. These end effects would probably not be present to this extent as the rail came from the manufacturer, because the straightening process does not extend all the way to the ends of the rail. They would appear if the rail were cut to length during installation. These stresses could help explain the rail cracking phenomenon.

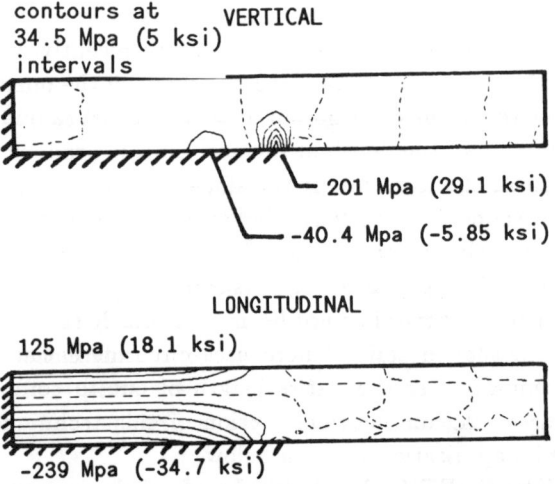

Fig. 4. Calculated Vertical and Longitudinal Stresses at End of Cut in Partially Saw Cut Rail

WHEELS

Replacement of overheated wheels is a major expense in the railroad industry[2]. Many thousands of wheels are removed every year due to suspected overheating. Overheating can cause high tensile residual stresses in the rim of the wheel, which may lead to failure and possible derailment. The current removal criterion is the "four inch rule", a simple check of whether the plate of the wheel has become discolored more than four inches from the inner edge of the rim. A better method of determining whether a wheel has actually been damaged by overheating could save many millions of dollars each year.

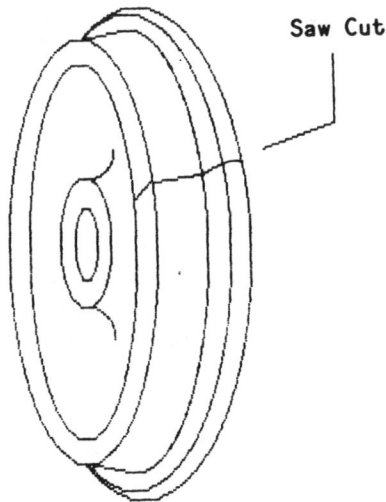

Fig. 5. Wheel Sawcutting Process

This economic incentive has led to the recent examination of a number of wheels removed for overheating. Many of these wheels have been cut inward radially from the rim and the distortion observed, a relatively inexpensive procedure (Fig. 5). More recently, when wheels were saw cut, a record was made of the opening at the top of the cut (measured at the largest radius) as a function of the depth of the cut. This is produced in the form of a x-y plot. If the slope is positive (i.e, the cut was opening), the saw blade was passing through a region of circumferential tensile stress. Similarly, if the cut was closing, the region was compressive[3]. Analysis of these curves has not been particularly successful. The method described here uses circumferential displacements over the entire length of cut to compute stresses. A complete description of this technique[4] and its application is given below.

THEORETICAL BASIS - It is hypothesized that a valid stress distribution can be obtained by modeling the final state of the saw cut and forcing it back to the uncut state. This distribution may not be the one that actually exists in the wheel, but it is one that could exist and satisfy all of the known boundary conditions.

Some of the advantages of this method over the older "work backward from the x-y plot" method include:
- an efficient non-iterative solution; and
- more complete information about the stress distribution inside the rim. Other methods do not allow the estimation of internal stresses.

Assumptions
- The stresses in the vicinity of the saw cut are completely relaxed.
- Only the circumferential displacements need be modeled.
- The plane of the cut below the bottom of the saw cut can be considered fixed.

Note: Most of these assumptions require that the relaxation owing to the saw cut be contained close to the cut.

Comments
- Stresses will be reliable only near the cut because stresses far from the cut will not be fully relieved.
- Since only exterior displacements are currently available, the following procedure is used to obtain displacements on the surface of the cut. The exterior displacements are applied to the surface of the model and a potential (thermal) solution method is used to obtain the interior displacements. While these are probably not the actual interior displacements, they are the most probable values based on the available information.
- If the cut closes, it is possible that the cut width can be increased by the saw blade rubbing against the edges of the cut. This would then cause the analysis to report erroneous stresses. Obviously, if the cut closes completely at any point, then the analysis can't be performed at all.

APPLICATION - A "cookbook" approach to performing this analysis is given below. The preceding assumptions and limitations should be considered before using this method.

Finite Element Model - Quarter wheel (90 degree) models of fairly low mesh density have typically been used for this analysis. Tests on 90 and 180 degree models show little difference in the calculated values near the cut. A sample is shown in Fig. 6, which has 95 nodes and 69 elements per plane. It was found to be more economical to use uneven slice spacing, with the higher slice density near the cut. Since relaxation owing to the saw cut decreases with distance from the cut, nothing is actually lost by doing this. How far this approach can be carried is dependent on how well the

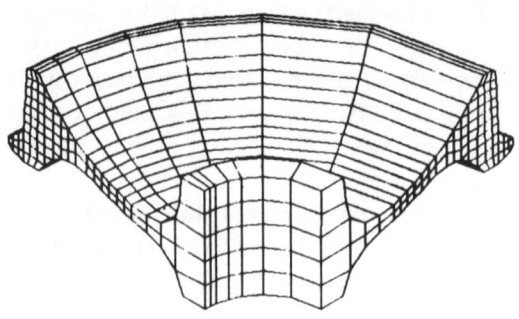

Fig. 6. Finite Element mesh for one quarter model of wheel

finite element system used handles high aspect ratio elements. More detailed models have also been tried, but the quality of the input data has not been sufficient to realize any advantage.

Boundary Conditions

- The hub is left free to move. (The wheel is not mounted on an axle at this time.)
- The fixed end (90 degrees from cut) is constrained in the circumferential direction only.
- The boundary conditions on the free (saw cut) end are fixed in the circumferential direction below the saw cut. Above the cut, the measured displacements are applied.
- One node on the fixed end is constrained in the axial direction to prevent any free body motions.

Experimental Measurement - Measurement of circumferential displacements would be easy if the cutting device removed a constant specified amount of material while leaving a smooth, easy-to-measure surface. Unfortunately the cutting methods presently employed leave a narrow, rough surfaced cut which is difficult to measure. Furthermore, the narrow cut width makes measurement inside the cut impossible. The following procedure has proved to be useful for obtaining the displacements needed for this analysis.

Measure front, back and tread opening displacements at least twice. (Typically measurements are made about every 1/4 inch along the cut.) Average the measurements for each point. Now subtract the saw cut width and plot the displacements as a function of distance along the length of the cut. The cut width is usually assumed to be the width at the bottom of the cut. Draw a reasonably smooth curve through the points (there is usually some scatter).

Interpolation - From the surface displacements along the saw cut, the computer is used to approximate the interior displacements. Since only half of the cut is modeled, the displacements should be divided by two. To perform the interpolation:

1. Generate a two-dimensional model from the cross-section of the three-dimensional model. A potential (thermal) solution method will be used to compute the internal displacements.
2. Apply the surface displacements to the surface nodes as fixed temperature boundary conditions, leaving the interior nodes unconstrained.
3. Solve for the interior displacements as though they were temperatures. Since a steady state solution is desired, any constant values for conductivity and specific heat will work.

Now there is a displacement for each node on the cut surface, which can be applied to the three-dimensional model and stresses obtained.

Results - Results for two wheels are shown below. The displacements and circumferential residual stresses for wheels 235 and 420 are shown in Figs. 7 - 10. The large tensile stresses found in the back side of the rim are consistent with those predicted by theoretical analysis[5] and coincide with one of the primary crack initiation sites in failed wheels.

CONCLUSIONS AND RECOMMENDATIONS

1. The calculations have provided useful approximations of the residual stresses in railroad wheels and rails. They are also inexpensive enough to apply to a large number of samples.
2. The major limitation to this calculation is measurement accuracy, which is closely related to the cutting process. A better cutting process must be found to further advance this method.

REFERENCES

[1] "Derailment of AMTRAK Train No 21 (the Eagle) on the Missouri Pacific Railroad, Woodlawn, Texas, November 12, 1983", NTSB Railroad Accident Report NTSB/RAR-85/01

[2] D. H. Stone, "Wheel Failures Due to Brake Heating: Experience and Solutions", 8th International Wheelset Congress (Madrid), 1985

[3] H. N. Jones, III, "The Characterization of the Residual Stress State of Railroad Wheels by the Saw Cut Method", ASME RAIL TRANSPORTATION SPRING CONFERENCE PROCEEDINGS, April 1985, pp. 15-19

[4] M. R. Johnson, R. R. Robinson, A. J. Opinsky, M. W. Joerms, and D. H. Stone, "Calculation of Residual Stresses in Wheels From Saw Cut Displacement Data", ASME Paper 85-WA/RT-17, November 1985

[5] A. J. Opinsky, M. W. Joerms, D. H. Stone, and M. R. Johnson, *"Effect of Brake Shoe Position on the Development of Residual Stresses in Freight Car Wheels as a Result of Simulated Drag Braking"*, ASME Paper 86-WA/RT-3, December 1986

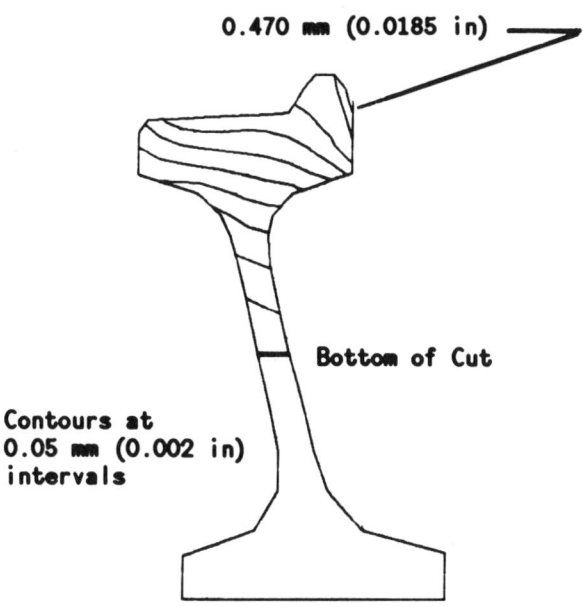

Fig. 7. Wheel 235 interpolated displacements

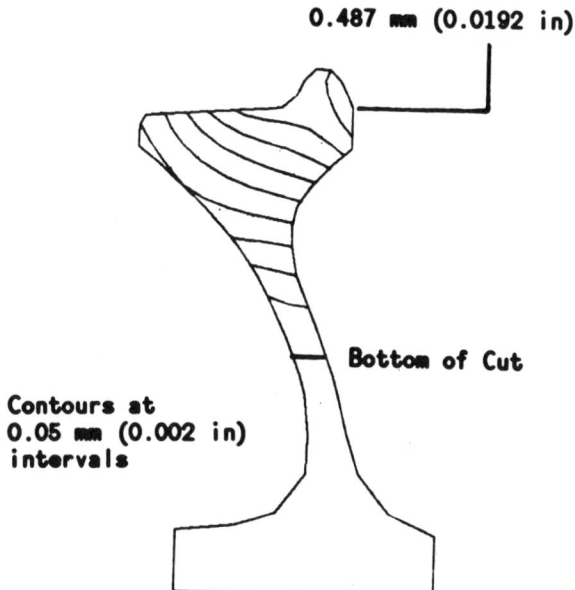

Fig. 9. Wheel 420 interpolated displacements

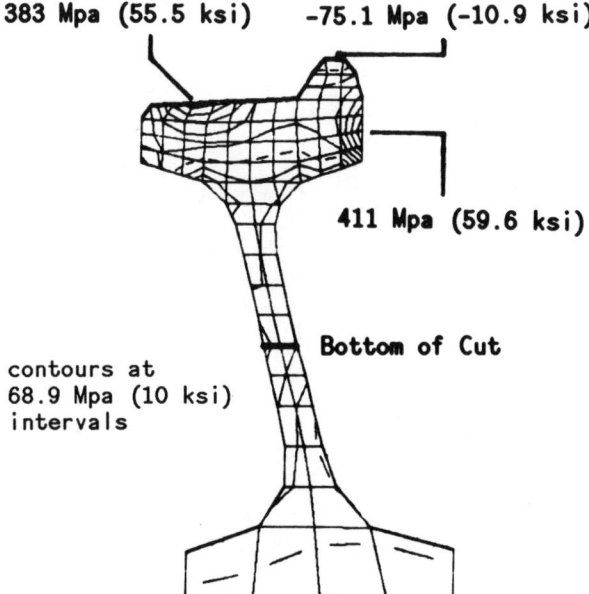

Fig. 8. Wheel 235 Circumferential Stresses

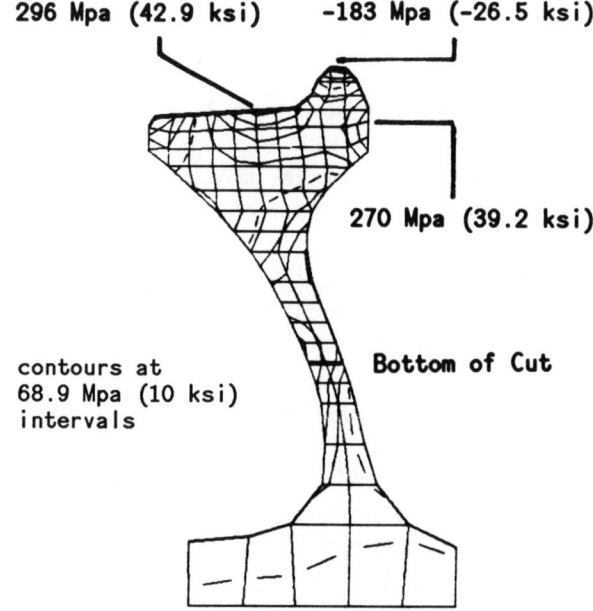

Fig. 10. Wheel 420 Circumferential Stresses